note about the author...

ROGER J. M. DE WIEST brings to this book experience in all phases of civil engineering, including structures, soil mechanics and hydraulics. He received his Engineering Diploma from the University of Ghent, Belgium, and served as an Engineer for the Belgian Government and as a Hydraulic Engineer for SOFINA, Brussels. He took an M.Sc. degree in Civil Engineering from the California Institute of Technology and received his Ph.D. degree in Civil Engineering from Stanford University. Since 1959, Dr. De Wiest has served on the faculty of Princeton University where he is presently Professor of Hydraulics and Hydrology.

Professor De Wiest, a Freeman Fellow of the ASCE, has served as a Lecturer in N.S.F. Summer Institutes in Hydrology, is the author of numerous publications in American and foreign scientific journals, and is the translator of the Russian classic on ground water, *Theory of Ground Water Movement,* by P. Ya. Polubarinova-Kochina.

GEOHYDROLOGY

GEOHYDROLOGY

ROGER J. M. De WIEST

Freeman Fellow ASCE
Professor of Hydraulics and Hydrology
Princeton University

John Wiley & Sons, Inc., New York · London · Sydney

SECOND PRINTING, APRIL, 1967

Library of Congress Catalog Card Number: 65-16427
Printed in the United States of America

To

ASCE

and

SOFINA

Preface

My experience in teaching flow of ground water to civil and geological engineering students during the past several years has indicated the need for a book which presents the basic principles of this subject matter in a rigorous, mathematical manner. Although several excellent books exist, they have been found too broad in scope or too specifically directed toward advanced topics to be used as a text in these courses. The material in this book has been used in my lectures at several recent Summer Institutes sponsored by the National Science Foundation, and the encouragement of my fellow teachers at these Institutes was an important factor in its completion. The range of topics selected has been influenced and in part restricted owing to the author's work with Professor Stanley N. Davis of Stanford University on a second book entitled Hydrogeology. In Hydrogeology the reader will find chapters on the physical and chemical properties of water, water quality, radionuclides in water, methods of ground-water exploration, water in igneous and metamorphic rocks, water in unusual climatic environments, and so forth. The two books compliment each other, and should provide a coverage of these topics useful to readers representing a wide variety of professional interests and backgrounds. *Geohydrology* is written by a civil engineer primarily for the use of civil engineers who teach and practice ground-water hydrology; however it will be useful to teachers and scientists in related disciplines such as geology, agronomy, geophysics, petroleum engineering, and so forth, and to professional people working in watershed management and water resources in general.

It has been my experience in the classroom that efficiency is greatest and participation of students keenest when eight to ten lectures are spent on the elements of surface hydrology. Therefore this material has been presented in Chapter 2. Chapter 3 deals with fundamental concepts and definitions and with some descriptive and legal aspects of ground water. In Chapter 4 an attempt is made to explain the theory of ground-water flow in a way that will be understandable to students whose background in calculus and fluid mechanics is rather limited. This means that concepts of the so-called hydraulic theory are predominant over rigorously developed hydrodynamic derivations, although hydrodynamic derivations are not entirely excluded. Indeed, hydrodynamic principles and formulations, interwoven in a more elementary text, prepare the student for the reading of literature on flow in porous media, a topic which has become increasingly theoretical. The mathematical techniques, used in these hydrodynamic formulations, are introduced from basic principles whenever necessary. Thus, in the derivation of the equation for the conservation of mass I make use of partial derivatives, but expertness in partial differentiation is neither required nor expected from the reader. Complex variables are not introduced, although this puts a severe restriction on the subject matter and precludes the use of powerful methods of analysis. Paragraphs containing more advanced material that may be skipped in a beginner's course are preceded by an ornamental line and set in reduced type. Chapter 5 illustrates practical applications chosen largely from the discipline of civil engineering, and Chapter 6 deals with the mechanics of well flow, including the theory of the leaky aquifier. In Chapter 7 I applied the concepts of Hubbert's force potential and those of the late Norbert Lusczynski on various kinds of water heads to the problem of salt water intrusion. These theories I singled out as being most suitable for undergraduate instruction and as being essential to the scientific background of every ground-water hydrologist. Finally, Chapter 8 deals with model analysis and more specifically with the Hele-Shaw apparatus and the R-C analog model because they are among the most versatile tools of investigation and also because they were particularly useful in my own research.

I am indebted to my former professors, Norman Brooks, Jack McKee, and Vito Vanoni at the California Institute of Technology, and to Ray K. Linsley, Jack Vennard, Byrne Perry, and Joseph B. Franzini of Stanford University who, through their excellent instruction, contributed much to my background in hydrology; to C. V. Theis, M. King Hubbert, and D. K. Todd for their helpful correspondence; to C. E. Jacob, Daniel Dicker, and Morris Muskat for their encouragement and advice, and to Wilfried Brutsaert and Chester Kisiel for reviewing part of the manuscript; to my secretary, Mrs. Judith Rodgers, for a flawless typing of the text, and to my colleagues. In particular, I am grateful to my friend and colleague Stanley N. Davis of Stanford University, whom I consulted throughout the writing of the book and who made many valuable suggestions. Chapter 1 was written jointly by Professor Davis and me, and we agreed on the general outline of the two textbooks.

Finally, I would like to dedicate this book to the Société Financière de

Transports et d'Entreprises Industrielles (SOFINA), my former employer in Brussels, Belgium, and to the American Society of Civil Engineers (ASCE), who, through their awards, contributed much to my achievement as a scholar and as a teacher.

ROGER J. M. DE WIEST

Princeton, N. J.
January, 1965

Contents

Chapter 1 Introduction 1

1.1. Scope of Topic 1
1.2. History of Hydrogeolgy and Geohydrology 2
Ground-Water Utilization 2
The Origin of Subsurface Water—Early Theories 4
The Founders of Hydrogeology and Geohydrology 7
Modern Hydrogeology and Geohydrology 8
1.3. Hydrology and Human Affairs 10
References 12

Chapter 2 Elements of Surface Hydrology: The Hydrologic Cycle—Relationship between Surface Water and Ground Water 14

♦ The Drainage Basin 17
2.1. Physiography of the Basin—Quantitative Study 18
2.2. Physiography of the Basin—Descriptive Features 22
♦ Precipitation 24
2.3. Types 24
2.4. Measurement of Precipitation 25
2.5. Average Precipitation over Area 27
2.6. Depth-Area-Duration Analysis 29
2.7. Variation in Precipitation 37
♦ Evapo-Transpiration 37
2.8. Factors Affecting Evaporation 39
2.9. Reservoir Evaporation 41
Determination from Water Budget 41
Measurement by Means of Evaporation Pans 42
Influence on Runoff 42
2.10. Monthly Evaporation 43
2.11. Transpiration 47
♦ Runoff 49
2.12. Runoff Cycle 49
Infiltration 50
2.13. Streamflow 52
Rating Curves 54
Extension of Rating Curves 59
Sources and Interpretation of Runoff Data 60
2.14. Hydrographs 63
Hydrograph Composition 63

Separation of Hydrograph Components 65
Hydrograph Shape 70
2.15. RAINFALL-RUNOFF RELATIONS 74
Initial Moisture Conditions 75
Coaxial Relations for Total Storm Runoff 77
2.16. UNIT HYDROGRAPHS (UNIT GRAPHS) 79
Unit Hydrographs of Different Durations 81
Synthetic Unit Graphs 83
2.17. STREAMFLOW ROUTING 87
Routing through a Large Reservoir 89
Routing in River Channels 95
2.18. FREQUENCY ANALYSIS OF RUNOFF DATA 99
The N-Year Event 101
Theoretical Distribution of the Return Period 105
Flow-Duration Curves 109
Rainfall Intensity—Duration—Frequency Curves 109
2.19. RESERVOIRS 112
Design of a Storage Reservoir 114
Capacity Loss Due to Sedimentation 116
♦ APPLICATIONS TO HYDROLOGIC PROBLEMS IN NEW JERSEY 118
2.20. DEVELOPMENT OF THE PENSAUKEN AQUIFER NEAR PRINCETON, N. J. 118
2.21. PRELIMINARY STUDY TO DETERMINE THE 1,000-YEAR DESIGN FLOOD FOR THE SPILLWAY OF SPRUCE LAKE RESERVOIR (N. J.) 122
REFERENCES 128

Chapter 3 Ground-Water Flow 129

♦ ELEMENTARY CONCEPTS AND DEFINITIONS 129
3.1. INTRODUCTION 129
3.2. MAJOR SUBDIVISIONS OF GROUND-WATER FLOW 129
Dimensional Character 129
Time Dependency 131
Boundaries of Flow Region or Domain 132
Properties of the Medium and of the Fluid 133
3.3. GEOLOGIC FORMATIONS IN GROUND-WATER FLOW—NOMENCLATURE 133
3.4. ROCK AND SOIL COMPOSITION 134
Porosity, Void Ratio 134
3.5. SIMPLE MEDIUM PROPERTIES 135
Moisture Content 135
Degree of Saturation 135
Specific Weight 137
Specific Gravity 137
Density 137

Specific Surface 137
3.6. Pore Pressure and Intergranular Stress; Soil Moisture Tension; Piezometric Head; Water Table 138
3.7. Saturated and Unsaturated Zones of Flow—Capillary Fringe 140
3.8. Various Kinds of Subsurface Water 142
♦ Interrelationship between Ground Water and Surface Water 143
3.9. Introduction 143
3.10. Sustained Yield 145
3.11. Artificial Recharge 147
♦ Ground-Water Laws in the United States 154
References 158

Chapter 4 Elementary Theory of Ground-Water Movement 161

4.1. Introduction 161
4.2. Poiseuille Flow 162
4.3. Porosity 167
4.4. Darcy's Law 168
4.5. Hydraulic Conductivity K 169
4.6. Measurement of K 172
4.7. Velocity Potential Φ 175
4.8. Derivation of Darcy's Law 176
4.9. Range of Validity of Darcy's Law 178
4.10. Main Equation for Conservation of Mass 179
4.11. Ground-Water Flow in Homogeneous Isotropic Medium 183
4.12. Extension of Main Equation of Conservation of Mass 187
Barometric Efficiency 189
Tidal Fluctuations 191
4.13. Streamlines and Equipotential Surfaces—Hubbert's Force Potential 192
4.14. Other Forces Causing Water Movement 196
4.15. Capillary Forces 199
References 201

Chapter 5 Steady State Flow 204

5.1. Introduction 204
5.2. Flow Nets—Boundary Conditions 206
Boundaries of Constant Head or Constant Potential 206
Impervious Boundaries 206
Free Surfaces 207
Surfaces of Seepage 207
5.3. Properties of Flow Nets 208
5.4. Construction of Flow Net 210
5.5. Seepage Rate 213

5.6. Dam on Infinitely Thick Stratum 215
5.7. Sheet Pile in Infinitely Deep Stratum 218
5.8. Singular Points in Flow Nets 219
5.9. Flow Nets with a Free Surface 222
5.10. Dam with Sloping Discharge Face—Depuit's Assumptions 223
5.11. Soils of Different Hydraulic Conductivity 226
5.12. Anisotropic Soil 228
5.13. Multilayered Soil 231
5.14. Dupuit's Assumptions and $V^2h^2 = 0$ for Unconfined Flow 233
5.15. Flow Nets for Wells 237
References 238

Chapter 6 Mechanics of Well Flow 239

♦ Steady Flow 239
6.1. Steady Radial Flow to a Well 239
6.2. Several Wells 244
6.3. Flow Between a Well and a Recharge Well 247
6.4. Cylindrical Sink and Source Flows—Uniform Flow 249
6.5. A Well in a Uniform Flow 252
6.6. Method of Images 255
♦ Unsteady Flow 260
6.7. Radial Flow to a Well in an Extensive Confined Aquifer 260
Theis' Method 263
Jacob's Method 266
Theis' Recovery Method 269
6.8. Boundary Effects on Unsteady Well Flow 270
Method of Images 270
♦ Leaky Aquifers 271
6.9. Modification of Laplace's Equation, Taking Leakage into Account 272
6.10. Determination of the Formation Constants of Leaky Aquifer's 275
(A) Type Curve Method for Nonsteady-State Time Drawdown 275
(B) Jacob's Method for Steady State Drawdown 278
(C) Hantush's Method 280
6.11. Nonidealized Boundary Conditions 282
♦ Partial Penetration of Wells 283
References 284

Chapter 7 Multiple-Phase Flow. Dispersion 287

7.1. Movement of Oil and Gas under Hydrodynamical Conditions 287

(A) Primary Migration 287
(B) Movement in Reservoir Rocks 291
(C) Slope of the Hydrocarbon-Water Interface 293
7.2. Salt Water Encroachment 295
(A) Ghyben-Herzberg Hydrostatic Conditions 295
(B) Hubbert's Hydrodynamical Approach 297
(C) Glover's Model 298
7.3. Dispersion in Salt Water Encroachment 299
7.4. Flow of a Nonhomogeneous Fluid in a Porous Medium 304
(A) Point-Water Head 304
(B) Fresh-Water Head 307
(C) True Environmental-Water Head 307
7.5. Correlation between Lusczynski's and Hubbert's Work 310
7.6. Recent Developments 313
References 315

Chapter 8 Numerical and Experimental Methods in Ground-Water Flow 318

8.1. Introduction 318
8.2. Grapho-Numerical Analysis—Finite Differences 318
8.3. The Parallel-Plate Model or Hele-Shaw Apparatus 322
(A) Introduction 322
(B) Horizontal Models—Scales 327
(C) Vertical Model—Scales 330
8.4. Resistance—Capacity Network Analogs 331
Conversion Factors 333
Design of Electric Circuit Elements 333
Extension to Three-Dimensional Flow. Anisotropic and Leaky Artesian Conditions 334
Boundary Conditions 335
Excitation-Response Apparatus 336
Cost and Accuracy of R-C Network Analog 337
8.5. Electrical Resistance Networks 339
Boundary Conditions 339
Unsteady State Flow 341
8.6. Conductive-Liquid and Conductive-Solid Analogs 342
8.7. Other Analogs 346
References 346

Appendix A. Proof of Hantush's Method 349

Appendix B. Functions Occurring in the Theory of Leaky Aquifiers 351

Index 361

CHAPTER ONE

Introduction

1.1 *Scope of Topic*

The study of subsurface water generally includes a consideration of its chemical and physical properties, geologic environment, natural movement, recovery, and utilization. For many years this subject has been called hydrogéologie by Belgian [1] and French workers [2] and hidrogeología by Latin American [3] researchers. German scientists [4, 5] on the other hand defined the term hydrology as that part of the general science of water which relates specifically to the water below the surface. The tendency to restrict the term hydrology to the study of subsurface waters and to use terms such as hydrography and hydrometry to denote the study of surface waters existed also in the United States [5]. As late as 1938, the Executive Committee of the International Association of Scientific Hydrology (IASH) recognized the use of the name hydrology for the branch dealing with underground waters (eaux souterraines) to distinguish from potamology (dealing with streams), limnology (study of lakes), and cryology (dealing with snow and ice).

It was Meinzer [6] who first suggested the term of "geohydrology" for the study of ground water, at a meeting of the IASH in 1939. Before him, Mead [7], hydraulic engineer and a past president of the American Society of Civil Engineers, had defined hydrogeology as the study of the laws of the occurrence and movement of subterranean waters. Engineers and geologists alike agree with him that "this must presuppose or include a sufficient study of general geology to give a comprehensive knowledge of the geological limitations which must be expected in hydrographic conditions and of the modifications due to geological changes." Mead stressed the special character of the study of "ground water as a geological agent, the understanding of which would contribute to attain a comprehension of the birth and growth of rivers and drainage systems." Meinzer subdivided the science of hydrology, which according to his definition dealt specifically with water completing the hydrologic cycle from the time it is precipitated upon the land until it is discharged into the sea or returned to the atmosphere, into surface hydrology and subterranean hydrology or geohydrology.

Obviously, some controversy has arisen as to the proper meaning of the words hydrogeology and geohydrology. Mead and Meinzer, authors of classical textbooks on hydrology, have used both terms to describe the same subdivision of hydrology. Moreover, many American geologists have followed the literal meaning of the word hydrogeology and have extended its usage to cover all studies, both of surface and subsurface water, which include a substantial amount of geologic orientation. The subject matter of this book, nevertheless, is confined to subsurface water; the title of the book has been chosen in conformity with Meinzer's definition and emphasizes the hydrology of ground water rather than its geological nature, according to the preface of the book. The knowledge of geohydrology is a basic requirement in the training of ground-water geologists and ground-water hydraulicians.

The best scientific research and professional work is often accomplished through combined efforts by geologists, hydraulicians, agronomists, chemists, and physicists specialized in the earth sciences. Cooperation has become imperative with the increasing complexity of the ground-water problems that remain to be solved and has been fostered in recent times by a better dissemination of scientific literature [8], avoiding thereby unnecessary duplication of research. Thus there is a common interest in the fundamentals of ground-water flow held by the hydrogeologist concerned with the evaluation of the safe yield of a ground-water basin and by the engineer in charge of drainage and irrigation projects. Principles of dispersion and diffusion in porous media applied by petroleum engineers in the study of the migration of gas and oil are also used in the analysis of salt water intrusion in coastal aquifers. The domain of fluid flow through porous media is not confined to that of rock and earth materials: mechanical engineers are interested in the heat exchange process associated with the movement of a gas through porous media; chemical engineers study the mass transfer of a gas from a mixture to a liquid solvent flowing in opposite sense through packed towers. Consequently, a vast body of technical literature dealing with the physics of the flow through porous media has become available to both hydrogeologists and geohydrologists [9, 10].

1.2 *History of Hydrogeology and Geohydrology* [37]

GROUND-WATER UTILIZATION

In the dry regions of Asia, the universal scarcity of water, the locally dense population, and the dominance of agriculture resulted in an early development of the art of constructing wells and infiltration galleries [13]. Accounts of well water and well construction abound in ancient literature and are specially well known from the Biblical record of Genesis.

Well construction in the Near East was by man and animal power aided by hoists and primitive hand tools, despite great difficulties. A number of large-diameter wells, some large enough to accommodate paths for donkeys, attest both to the industry of the people and to the scarcity of water. These wells rarely exceed a depth of 50 meters. There is little evidence of technological advances in well drilling during historic time in this region despite the fact that Egyptians had perfected core drilling in rock as early as 3,000 B.C. [11]. This drilling was confined to stone quarry operations. Ancient Chinese, prolific in many inventions, were also responsible for the development of a churn drill for water wells, which, in principle, was almost identical to modern machines [12]. The early machinery was largely of wood and powered by human hands. Through a slow drilling rate sustained for years, and even decades, these ancient people were able to achieve wells of amazing depths. Bowman [13] reports a depth of 1,200 meters and Tolman [14] a depth of 1,500 meters. The deepest holes, however, were drilled for brine and gas rather than for potable water. The same methods, only slightly modified during the past 1,500 years, are still used today in rural areas of Laos, Cambodia, Thailand, Burma, and China.

The greatest achievement in ground-water utilization by ancient peoples was in the construction of long infiltration galleries, or kanats, which collected water from alluvial fan deposits and soft sedimentary rock. These structures, commonly several kilometers long, collected water for both agricultural and municipal purposes. Kanats were probably used first more than 2,500 years ago in Iran; however, the technique of construction spread rapidly eastward to Afghanistan and westward to Egypt. One extensive kanat system built about the year 500 B.C. in Egypt is said to have irrigated 3,500 square kilometers of fertile land west of the Nile [14]. Many kanats are still in use today in Iran and Afghanistan, the best known of which are in Iran on the alluvial fans of the Elburz Mountains.

Owing to a lack of early cultural contact with China, modern percussion methods of well drilling were developed more or less independently in Western Europe. The impetus for this development came largely from the discovery of flowing wells, first in Flanders about A.D. 1100, then a few decades later in eastern England and in northern Italy. One of the first wells was dug in A.D. 1126 by Carthusian monks from a convent near the village of Lillers. In Gonnehem, Flanders [15], near Bethune, four wells were drilled and were cased $11\frac{1}{3}$ ft above ground level so that they were able to deliver water at sufficient height to drive a water mill [16].

The wells were several hundred feet deep and tapped water under pressure from a formation consisting of fractured chalk that had its outcrop area in the higher plateaus of the Province of Artois. These and other similar wells in the region of Artois became so famous that flowing wells were eventually called artesian wells after the name of the region [12, 17].

The widespread search for artesian water stimulated a rapid development of drilling techniques. Popular interest was so great in France that for a number of years the Royal and Central Society of Agriculture of France distributed annual medals and prizes to workers in the field, to authors, inventors, well-drillers, and to those who introduced these wells in new areas [17]. Although drilling methods were more rapid and efficient in Europe than in China during the late eighteenth century, the depths of the wells rarely exceeded 300 meters. It was not until the end of the nineteenth century that the depths of water wells drilled by modern machinery exceeded the depths of the more primitive Chinese wells.

The methods of drilling for water have improved rapidly during the past 100 years, partly owing to knowledge borrowed from oil and gas drilling. The most significant single advance in drilling techniques has been the development of hydraulic rotary methods. Early rotary drilling was done with the aid of an outer casing; however, in about 1890 thick mud was found to be sufficient for holding up the walls of the hole, and the outer casing was no longer used. With this new efficiency and with the successful drilling of the Spindle Top oil field in Texas in 1901 by rotary methods, rotary drilling has steadily gained in popularity during the past 50 years [11, 13].

The perfection of the deep-well turbine pump in the years between 1910 and 1930 added a further stimulus to the well-drilling industry. Prior to this time, deep wells were fitted with low-capacity piston pumps of poor efficiency. The new turbine pumps made irrigation by wells feasible in many areas hitherto underdeveloped for agriculture. The large production of these wells has placed a greater demand on the well-drilling industry for bigger and more permanent wells.

Although technology is still being borrowed from the oil industry, many innovations such as reverse rotary drilling, gravel envelope wells, and water-well cameras have come directly from the water-well industry itself.

THE ORIGIN OF SUBSURFACE WATER—EARLY THEORIES

When one considers the importance of ground water in the Oriental civilizations, it is strange that there is little record left concerning theories as to its origin. It remained for the inquisitive Greeks to speculate about such matters. Even though considerable thought was given by the Greeks to the origin of ground water, their contributions were surprisingly sterile, particularly in light of the amazing progress they made in philosophy and mathematics. One probable reason for the lack of progress in ground-water theory was the insistence of Plato and others that philosophy and science be more or less separated from their important contact with experimentation, field observation, and practical applications. Thus a wide gap was created between practice and theory. The significance of Greek thought is in the scientific dogma that it created for almost 2,000 years. The authority of Greek writings in the earth sciences reached a peak in the scholasticism of Albertus Magnus (A.D. 1206–1280) and Thomas Aquinas (A.D. 1225–1274) during the Middle Ages (roughly A.D. 1250–1450), but still persisted with considerable strength until a scant 200 years ago.

The Greeks were impressed with the large size of rivers in comparison with the observed runoff from heavy rains. They were also impressed by caves, sinks, and large springs characteristic of the limestone terrain which covers much of the Balkan Peninsula. The most common explanation for the origin of rivers was that they were fed from large springs, which in turn were fed by underground rivers or lakes nourished directly or indirectly by the ocean. Two problems confronted the early natural philosopher. (1) How did the ocean loose its salt? (2) How did ocean water rise from the level of the sea to springs high in the mountains?

Thales (640–546 B.C.), who was an Ionian philosopher of the School of Miletus, has been called the first true scientist. He taught that water was driven into the rocks by wind and that it was forced to the surface by rock pressure, whence it emerged as springs. Plato (427–347 B.C.), the great Athenian philosopher, conceived of one large underground cavern which is the source of all river water. Water was returned to the cavern from the ocean by various subsurface passages. The driving mechanism for this circulation was not fully explained. Krynine [18] has pointed out that the foregoing interpretation of Plato's ideas which influenced medieval science was probably not a correct interpretation of his more serious thoughts on the subject. Plato's "Critias" contains a description of the hydrologic cycle which is quite accurate. Although Aristotle (384–322 B.C.) was a student of Plato, he modified considerably Plato's concepts of the origin of ground water. Aristotle taught that ground water occurred in an intricate sponge-like system of underground openings and that water was discharged from these openings into springs. Water vapor which emanated from the interior of the earth contributes the greatest part of the spring water. Aristotle did, however, recognize that some cavern water originated from rainwater which had infiltrated into the ground and had entered the cavern in liquid form rather than as vapor [19, 20].

The Romans generally followed Greek teachings in the sciences. Marcus Vitruvius, however, who lived about 15 B.C., made several original contributions to engineering and science. He is, perhaps, best known for his contributions to architecture, particularly in the acoustics of buildings. He was also one of the first persons to have a correct grasp of the hydrologic cycle. He taught that water from melting snow seeped into the ground in mountainous areas and appeared again at lower elevations as springs. In contrast, the famous Stoic philosopher, Lucius Annaeus Seneca (4 B.C.–A.D. 65) held essentially the same theory as Aristotle but denied the reality of infiltrating rainwater. The conclusions of Seneca were taken as positive proof for more than 1,500 years that rainfall was an insufficient source for spring water [19, 20].

Bernard Palissy (1509–1589) [36], French natural philosopher, was perhaps the first to have thorough modern views concerning the hydrologic cycle as reflected in the dialogue between "Theory" and "Practice" of his chapter "Des eaux et fontaines." Nevertheless many of the ideas of the Greeks and Romans prevailed until the end of the seventeenth century. Two of the most influential scientists of their time, Johannes Kepler (1571–1630), German astronomer, and Athanasius Kircher (1602–1680), German mathematician, elaborated greatly on the earlier ideas of Seneca and Aristotle [19]. Kepler taught that the earth was similar to a large animal and that sea water was digested and that fresh water from springs was the end product of the earth's metabolism. The ideas of Kircher were exposed in his "Mundus Subterraneus," which was first printed in 1664 and soon became the standard reference work on geology for scholars of the seventeenth century. The work was ambitious in scope and unsurpassed in a display of spectacular imagination. Springs, which were fed by subterranean channels connected to the sea, were thought to issue from large caverns in mountains. Whirlpools, particularly the somewhat mystical Maelström off the coast of Norway, were thought to mark the positions of openings to the caverns in the sea bottom.

Between the dawn of scientific thinking and the end of the Renaissance, about A.D. 1600, little advance was made in hydrogeology and geohydrology. Five main facts were missed by all but a few early philosophers and scientists. (1) The earth does not contain a network of large interior caverns. (2) Although suction of the wind, capillary attraction, the forces of the waves, and other natural mechanisms exist to raise water against gravity, these mechanisms are insufficient to lift vast quantities of water in the earth's interior. (3) Sea water does not loose all its salt by infiltrating through soil. (4) Rainfall is sufficient to account for all water discharged by rivers and springs. (5) Rainfall infiltrates into the ground in large quantities.

THE FOUNDERS OF HYDROGEOLOGY AND GEOHYDROLOGY

The true source of river water was proved by two French scientists, Pierre Perrault (1608–1680), and Edmé Mariotte (1620–1684) [5, 19, 20]. Perrault [37] measured rainfall in the Seine River basin for the years 1668, 1669, and 1670 and found the average to be 520 mm per year. He then estimated runoff from the basin and concluded that it was only one sixth of the total volume of rain, thus proving that rainfall was more than sufficient to account for all stream water. Studies of evaporation and capillary rise were also made by Perrault. He proved that capillary rise could never form a free body of water above the water table and that the height of capillary rise in sand was less than one meter. Mariotte measured the amount of infiltration of rainwater into a cellar at the Paris Observatory. He noted that this infiltration, as well as spring flow at other places, varied with the rainfall. He concluded, therefore, that springs were fed by rainwater which infiltrated into the ground. Mariotte's important contributions were published in Paris in 1690, after his death, and also as collected works in Leiden in 1717 [21]. The latter contain Mariotte's essay "Du mouvement des eaux" (pp. 326–353), dealing with the properties of fluids, the origin of flowing wells, winds, storms and hurricanes, and other topics. Using the float method, Mariotte estimated the flow of the Seine River at the Pont Royal in Paris at 200,000 ft^3 per minute, or 1.05×10^{11} ft^3 per year, less than one sixth of the total annual precipitation on the basin that provides the runoff to the Seine upstream of Paris. "It was therefore evident, if one third of the precipitation evaporated from the ground," as Mariotte assumed, "and if one third remained in the earth, that there would be enough water left to sustain the flow of wells and rivers in the basin." Thus Mariotte verified Perrault's conclusions concerning the source of water for runoff. Several years later Edmund Halley (1656–1742), the famous British astronomer, published studies of evaporation from the Mediterranean Sea and concluded that this evaporation was able to account for all the water flowing into this sea by rivers, thus adding important data in support of the two French scientists [5].

Artesian wells have excited speculation since the days of the early Greeks, but correct explanations were not widely published until the first part of the eighteenth century. The first explanation which was mechanically correct was made by the brilliant Arabian philosopher al-Biruni (973–1048) [22]. The best documented explanation came much later and was by Antonio Vallisnieri, president of the University of Padua, Italy, who published a paper in 1715 on the artesian water in northern Italy. He illustrated his paper with some of the earliest geologic cross sections, which were drawn for him by Johann Scheuchzer [16].

Although his work was somewhat anticipated by Hagen and Poiseuille, Henri Darcy (1803–1858) [23] was the first person to state clearly the mathematical law which governs the flow of ground water. Darcy was a well-known French engineer whose main achievement was to develop a water supply for the city of Dijon, France. The development of his formula was the result of experimentation with filter sands and was presented in 1856 in an appendix of a report on the municipal water supply of Dijon. His report, however, resembled a scientific monograph on hydraulic engineering more than a present-day engineering report.

MODERN HYDROGEOLOGY AND GEOHYDROLOGY

Developments during the past century have been along three more or less separate lines: (1) elaboration of the relation between geology and ground-water occurrences, (2) development of mathematical equations to describe the movement of water through rocks and unconsolidated sediments, and (3) the study of the chemistry of ground water, or hydrogeochemistry.

The development of the relation between geology and ground-water occurrence is difficult to associate with individual names. In general, many geologists have contributed to specific problems. For example, the occurrence of ground water in areas of perennially frozen ground has been studied by a large number of Russian geologists. Many Dutch geologists have contributed to the understanding of ground water in coastal sand dunes. Japanese geologists and geophysicists have made numerous contributions to the understanding of hot springs. The English have made several contributions, one of which was an early application of geology by William Smith in 1827 to increase the water supply of Scarborough, England [24]. After a study of the local geology, he recommended that ground-water storage be increased by partially damming a spring. A. Daubrée [25] of France wrote one of the earliest general treatises on the geological aspects of ground water. For a specific example of modern contributions of an individual geologist the work of H. T. Stearns in the Hawaiian Islands can be cited. This work gives an excellent description of the relation between volcanic rocks and the occurrence of ground water. The work of W. M. Davis and J. H. Bretz on the formation of limestone caverns is another good example. Still another example is the work of DuToit on the consolidated rocks of the Union of South Africa. Despite the great number of geologists that could be cited, one man, O. C. Meinzer, stands out as the most important. Although he contributed to methods of making ground-water inventories and to the theory of artesian flow, and stressed the importance of phreatophytes, his main contribution was organizing the science of ground water [38]. He analyzed, defined, and welded together the various facets of the new branch of earth science largely between the years of 1920 and 1940, when he was a member of the United States Geological Survey.

Advances in ground-water hydraulics can be more easily identified with individual people because specific formulas are commonly published rather than generalized concepts which are so important in classical geology. Jules Dupuit [26], of France, was the first scientist to develop a formula for the flow of water into a well. This work was published only seven years after Darcy's monograph, yet successfully utilized Darcy's law. In 1870 Adolph Thiem [27], of Germany, modified Dupuit's formula so that he could actually compute the hydraulic characteristics of an aquifer by pumping a well and observing the effects in other wells in the vicinity. For many years Thiem continued to perfect his method and to apply it to various field situations. Modern methods of higher mathematics were first applied extensively to ground-water flow by Philip Forchheimer [28], of Austria, in 1886. He introduced the concept of equipotential surfaces and their relation to streamlines. He was also the first to apply Laplace's equation and the method of images. In the United States, C. S. Slichter [29] published similar work thirteen years later. Slichter developed his ideas independently of Forchheimer [30].

A great advance was made in the quantitative analysis of ground-water flow in 1935 when C. V. Theis introduced an equation for nonsteady state flow to a well. A formula had been developed seven years earlier in Germany by Herman Weber [30]; nevertheless, the formula of Theis has proved to be of much greater utility. Theis' equation was based on an analogy with heat flow, but a few years later C. E. Jacob derived the same expression through hydraulic considerations alone. In the past years Jacob and many other workers have improved the usefulness of Theis' basic equation by modifying it for a large number of boundary conditions.

The most significant contribution in recent times to the mathematical analysis of the movement of ground water has been made by Morris Muskat [31], whose book on fluid flow through porous media remained unchallenged and unmatched for years after it was originally published in 1937. In 1964 it is still a valuable reference work in teaching, and it will remain an indispensable tool in research for many years to come. Muskat wrote numerous original papers on the flow of subterranean fluids, providing a base upon which later research was founded and leading to stimulating discussions. Equally valuable was the work by M. King Hubbert [32] whose lucid thoughts on fundamental concepts of fluid flow in the ground reflect a deep understanding of the physics of the phenomenon. He derived Darcy's law from the Navier-Stokes equations and introduced the concept of force potential in his mathematical derivation, a concept more general and useful than that of velocity potential, which, strictly spoken, applies only when the fluid is water with constant physicochemical properties. The concept of force potential is at the base of Hubbert's work on immiscible displacement in multiphase systems and on hydrodynamical entrapment of petroleum.

The subject of ground water has attracted the attention of many Russian researchers [8, 33]; among them the names and contributions of Zhukovsky, pioneer of the air foil theory, and even more of Pavlovsky [34] are outstanding. Unfortunately, except for Polubarinova's book [8], little of this work has become available to the Western scientist in the form of unabridged translations into English. In the field of hydrogeology, Selin-Bekchurin [8] has been very prominent.

Chemical analyses of water have been routine for more than a century; however, the successful correlation of water chemistry with the hydrologic and geologic environments, or hydrogeochemistry, is a more recent development. Early attempts at geochemical interpretations were made by B. M. Lersch of Germany in 1864 and T. S. Hunt of Canada in 1865. Modern hydrogeochemical studies in North America started with the work of F. W. Clarke, done from about 1910 to 1925, and included a large number of chemical analyses of water with geochemical interpretations. Another early geochemist who made detailed studies of specific areas in the United States was Herman Stabler. His regional studies of water chemistry in the western United States have been excelled only during the past decade. Modern trends in hydrogeochemistry include exhaustive studies of chemical ratios, largely by Russian and French workers, the use of trace elements (to prospect for mineral deposits) in many countries, and detailed studies of various isotopes by workers in Japan, the United States, Russia, and many other countries. Hydrogeochemistry today is a subject of research and teaching, especially in some French and German Universities [35].

1.3 *Hydrology and Human Affairs*

Many aspects of hydrogeology and geohydrology are as yet of special interest only to scientists. Nevertheless it is a branch of the earth sciences which has in a large measure been born out of practical considerations. Indeed, many of the most important advances in hydrogeology and geohydrology today have been stimulated by studies designed to solve problems of great economic importance. This trend will probably continue, as the demand for water will undoubtedly increase with growing population and industrialization.

The development of modern household equipment, new industrial processes, and recent farming techniques have all increased the need for water.

Water is used for washing, irrigation, cooling, to extinguish fires, as a solvent, and for countless other needs of modern society. The demands of an undeveloped agricultural society are only for a minimum amount of water. Without modern plumbing, one man can have sufficient water for drinking, for cooking, and for washing even if he is limited to only 20 liters (5 to 6 gal) per day. In a modern home, in contrast, a single bath can easily consume twice this quantity. With automatic washing machines, lawn sprinklers, and other water-consuming devices of modern living, many communities have an average consumption of more than 600 liters (165 gal) per person per day. If modern industry is taken into account, the water-production problems of even small cities become tremendous. For example, it commonly takes more than 10 liters of water to process one kilogram of meat, 100 liters to produce one kilogram of paper, and almost 200 liters to produce one kilogram of steel. Fortunately, a great amount of industrial water is not consumed and can be reused after proper treatment.

Modern agriculture consumes the largest amount of water. Many irrigated crops take more than 400 liters of water to produce one kilogram of product. Alfalfa hay, for example, commonly requires more than 1,000 liters of water for each kilogram of dry hay produced. In contrast, in nonirrigated farming only a small amount of water is used. A sheep will use about 6 liters of water each day, a horse about 40 liters of water a day. A cow uses only about 10 liters of water to produce one liter of milk. A laying hen uses between one half and one liter of water per day. Consequently, in areas where water is very expensive or scarce, livestock or poultry industries can be developed more economically than water consuming types of industries.

Although most large cities and irrigation projects use water from surface streams, the economic importance of subsurface water can hardly be emphasized too strongly. Subsurface water is more desirable than surface water for at least six reasons. (1) It is commonly free of pathogenic organisms and needs no purification for domestic or industrial uses. (2) The temperature is nearly constant, which is a great advantage if the water is used for heat exchange. (3) Turbidity and color are generally absent. (4) Chemical composition is commonly constant. (5) Ground-water storage is generally greater than surface-water storage, so that ground-water supplies are not seriously affected by short droughts. (6) Radiochemical and biological contamination of most ground water is difficult. Ground water, which has been stored by nature through many years of recharge, is available in many areas which do not have dependable surface water supplies.

Three common disadvantages discourage ground-water development in some areas. (1) Most important is the fact that many regions are underlain by rocks with insufficient porosity or permeability to yield much water to wells. (2) Usually, but not always, ground water has a greater dissolved solids content than surface water in the same region. (3) The cost of developing wells is commonly greater than the cost of developing small streams. This is particularly true in regions of moderate to high precipitation.

REFERENCES

1. Fourmarier, P., *Hydrogéologie*, Masson, Paris, 1958 (294 pp.).
2. Schoeller, H., *Les eaux souterraines*, Masson, Paris, 1962 (642 pp.).
3. Lopez, C. J., "Informe preliminar sobre irrigación con agua subterranea del municipio de Codazzi, Departmento del Magdalena," Servicio Geologico Nacional (Colombia), Boletín Geologica, Vol. 9, pp. 47–90 (1960).
4. Prinz, E., *Handbuch der Hydrogeologie*, 1st Ed., Springer, Berlin, 1919; 2nd Ed., 1923.
5. Meinzer, O., "Introduction," *Hydrology*, pp. 1–31, Dover, New York, 1942 (712 pp.).
6. Meinzer, O., *Hydrology*, p. 4, Dover, New York, 1942.
7. Mead, D., *Hydrology, the Fundamental Basis of Hydraulic Engineering*, 1st Ed., McGraw-Hill, New York, 1919 (626 pp.); 2nd Ed., McGraw-Hill, New York, 1950 (717 pp.).
8. De Wiest, R. J. M., "Translator's Remarks," *The Theory of Ground-Water Movement*, p. ix, Princeton University Press, Princeton, New Jersey, 1962. (Original work by P. Ya. Polubarinova-Kochina.)
9. Scheidegger, A. E., *The Physics of Flow through Porous Media*, Macmillan, New York, 1960 (313 pp.).
10. Bird, R. B., W. E. Stewart, and E. N. Lightfoot, *Transport Phenomena*, Wiley, New York, 1960 (780 pp.).
11. Brantly, J. E., "Hydraulic Rotary-Drilling System," pp. 271–452, in *History of Petroleum Engineering*, D. V. Carter (Ed.), American Petroleum Institute, New York, 1961.
12. Brantley, J. E., "Percussion-Drilling System," pp. 133–269, in *History of Petroleum Engineering*, D. V. Carter (Ed.), American Petroleum Institute, 1961.
13. Bowman, Isaiah, "Well-Drilling Methods," *U.S. Geographical Survey Water-Supply Paper* 257, pp. 23–30 (1911).
14. Tolman, C. F., *Ground Water*, Chapter I, pp. 1–25, McGraw-Hill, New York, 1937.
15. Blanchard, R., *La Flandre, étude géographique de la plaine flamande en France, Belgique et Hollande*, p. 8, Danel, Lille, 1906.
16. Hagen, G., *Handbuch der Wasserbaukunst*, Vol. I, p. 87, Bornträger, Koenigsberg, 1853.
17. Norton, W. H., "Artesian Wells of Iowa," *Iowa Geological Survey*, Vol. 6, pp. 122–134 (1897).
18. Krynine, P. D., "On the Antiquity of Sedimentation and Hydrology (with Some Moral Conclusions)," *Bulletin Geological Society of America*, Vol. 71, pp. 1721–1726 (1960).
19. Adams, F. D., *The Birth and Development of the Geological Sciences*, Chapter 12, pp. 426–460, Dover, New York, 1938.

20. Baker, M. N. and R. E. Horton, "Historical Development of Ideas Regarding the Origin of Springs and Ground Water, *Transactions American Geophysical Union*, Vol. 17, pp. 395–400 (1936).
21. Mariotte, E., *Oeuvres de Mr. Mariotte*, 2 Vols. (701 pp.). Edited by P. Van der Aa, Leiden, 1717.
22. Dampier-Whetham, W. C. D., *A History of Science and Its Relations with Philosophy and Religions*, Macmillan, New York, 1931 (514 pp.).
23. Darcy, H., *Les fontaines publiques de la ville de Dijon*, V. Dalmont, Paris, 1856 (674 pp.).
24. Sheppard, T., "William Smith, His Maps and Memoirs," *Proceedings Yorkshire Geological Society*, Vol. 19, new series, pp. 75–253 (1917).
25. Daubrée, A., *Les eaux souterraines, aux époques anciennes et à l'époque actuelle*, Dunod, Paris, 3 Vols., 1887.
26. Dupuit, J., *Etudes théoriques et pratiques sur le mouvement des eaux dans les canaux découverts et à travers les terrains perméables*, 2nd ed., Dunod, Paris, 1863 (304 pp.).
27. Thiem, A., *Hydrologische Methoden*, Gebhardt, Leipzig, 1906 (56 pp.).
28. Forchheimer, Ph., *Uber die Ergebigkeit von Brunnen Anlagen und Sickerschlitzen, Zeitschrift des Architekten- und Ingenieur Vereins zu Hannover*, Vol. 32, pp. 539–564 (1886).
29. Slichter C. S., "Theoretical Investigation of the Motion of Ground Water," U.S. Geological Survey, *19th Annual Report*, pp. 2, 295–384 (1899).
30. Hall, H. P., "A Historical Review of Investigations of Seepage Toward Wells," *Journal Boston Society Civil Engineers*, pp. 251–311 (July 1954).
31. Muskat, M., *The Flow of Homogeneous Fluids through Porous Media*, McGraw-Hill, New York, 1937 (763 pp.); Second Printing, J. W. Edwards, Ann Arbor, Michigan, 1946.
32. Hubbert, M. King, "The Theory of Ground-Water Motion," *Journal of Geology*, November, December, 1940.
33. Harr, M. E., *Ground Water and Seepage*, 315 pp., McGraw-Hill, New York, 1962.
34. Pavlovsky, N. N., *Collected Works*, 2 Vols., Akad. Nauk, USSR, Leningrad, 1956.
35. De Wiest, R. J. M., Educational Facilities in Ground-Water Hydrology and Geology, "*Ground Water* (Journal of the NWWA), Vol. 2, pp. 18–24 (April 1964).
36. Cap, P. A., *Les oeuvres complètes de Bernard Palissy—Des eaux et fontaines*, pp. 436–483, Albert Blanchard, Paris, 1961.
37. Jones, P. B., Y. D. Walker, R. W. Harden, and L. L. McDaniels, "The Development of the Science of Hydrology," Texas Water Commission, *Circular* 63–03, p. 14, April 1963 (35 pp.).
38. Hackett, O. M., "The Father of Modern Ground-Water Hydrology," *Ground Water* (Journal of the NWWA), Vol. 2, pp. 2–5 (April 1964).

CHAPTER TWO

Elements of Surface Hydrology

The Hydrologic Cycle. Relationship between Surface Water and Ground Water

Geohydrology, the subject of this textbook, is part of the general science of hydrology, and a student should not only be able to identify it as such and situate it in the science as a whole, but also relate it to the other parts of hydrology. Therefore he should understand how the hydrologic cycle is completed and how its various components are correlated in nature. The broad picture [1] of the hydrologic cycle as represented in Fig. 2.1 seems so self-explanatory nowadays that an ignorance of it for so many centuries defies our imagination. Man as a poet was early attracted by the cyclic variations in nature, diurnal and seasonal,[1] yet his Muse failed to sing of the never ending journey of the water of the earth. It remained for modern times to describe in prose a great phenomenon of nature that did not find its place in the immortal verses of the classics.

The oceans are the immense reservoirs from which all water originates and to which all water returns. This statement is somewhat simplistic because not all water particles are in the process of completing the entire hydrologic cycle at all times. There are built-in loops, e.g., when water evaporates from land and returns to land as precipitation, only to evaporate again, etc. But in its most elaborate cycle, water evaporates from the ocean and forms clouds which move inland, condense, and fall to the earth as precipitation. From the earth, through rivers and underground, water runs off to the ocean. So far there is no evidence that water decreases in quantity at a global level. No water is destroyed but none is generated either, to paraphrase a well-known principle of physics.[2]

[1] Titus Lucretius Carus, *De Rerum Natura* (libri sex), 2nd Edition, Deighton, Cambridge, England, 1866.

[2] Antoine Laurent, Lavoisier, "Rien ne se perd, rien ne se crée."

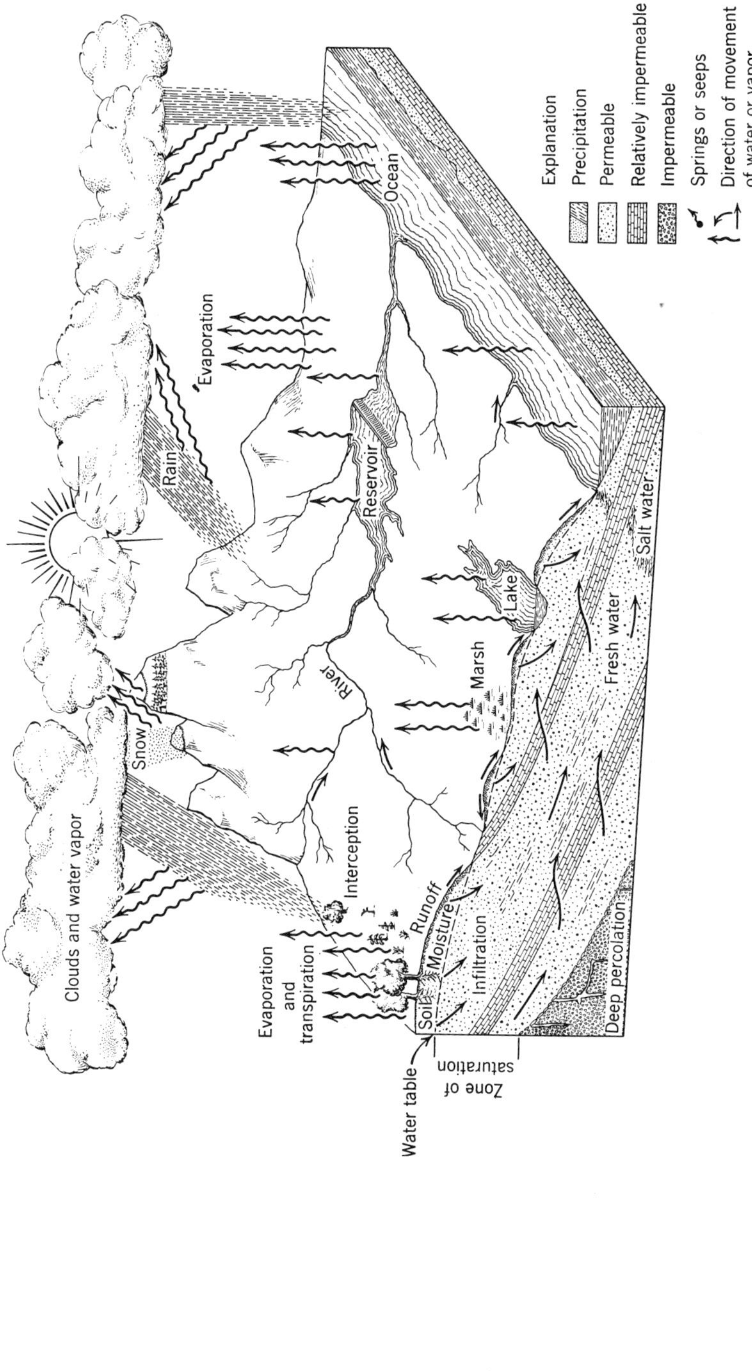

Fig. 2-1 The hydrologic cycle. (Courtesy of Texas Water Commission.)

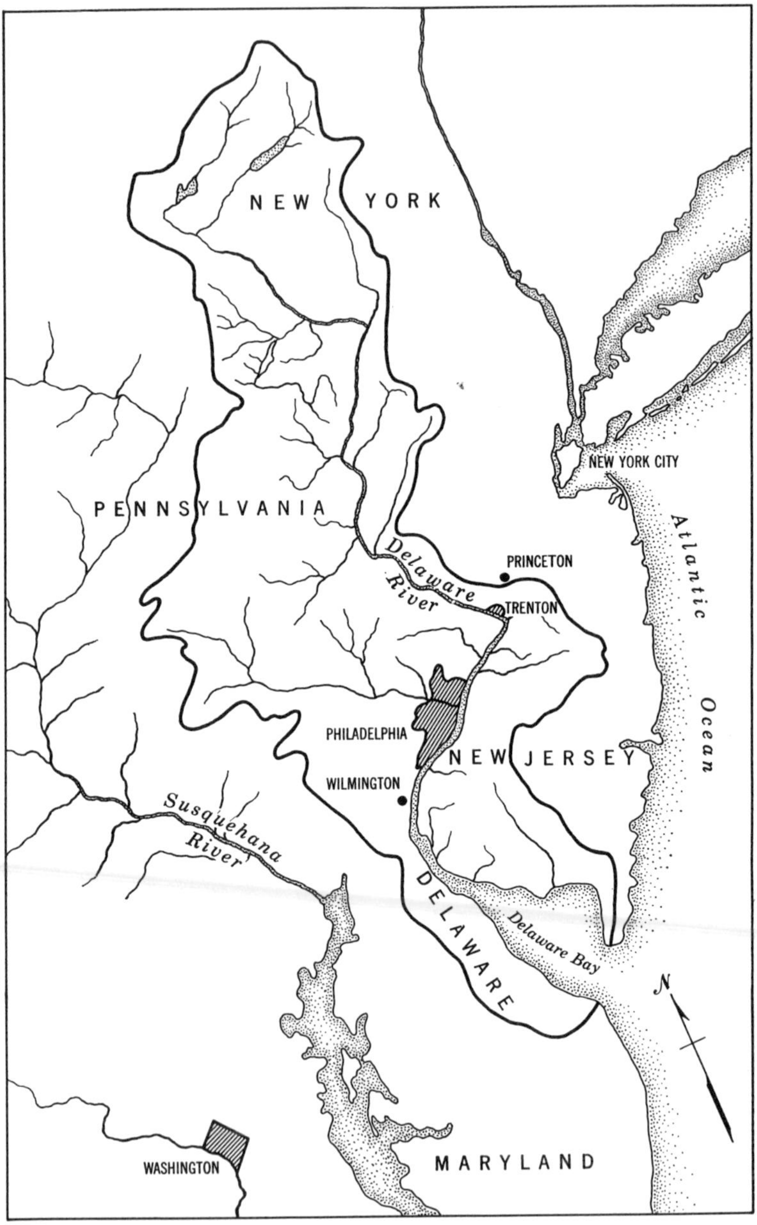

Fig. 2-2 The Delaware River Basin. (After a map by A. F. Loeben and K. L. Pyle.)

For human usage, however, the physical state of water (i.e., liquid versus solid or gaseous) is important, and so is its quality. While its available quantity is limited, the need for water is ever increasing, and consumption is bound to hit the ceiling of supply in the not so distant future. Therefore water conservation and pollution abatement have become very important in today's economic life. Until recently, the new science of hydrology did not enjoy a sufficiently large growth rate to cope with the problems created by a rational exploitation of the world's water resources. Most of the hydrologists came to this discipline more by vocation than by formation, and even now very few specific training centers exist at the university level. Therefore, in order to give more impetus to the science of hydrology and to foster research in this domain, UNESCO has sponsored the Decade of International Hydrology [2]. It is for this reason also, as a modest contribution to stimulate the interest of the hydrogeologist, that we summarize in this chapter those topics of surface hydrology which are pertinent to the understanding of the relationship between surface water and ground water.

◆ The Drainage Basin

A drainage basin of a water course is the entire area contributing to the runoff and sustaining part or all of the flow of the main stream and its tributaries [3]. This definition is compatible with the fact that the boundaries of a drainage basin and of its underlying ground-water basin do not necessarily have the same horizontal projections. The Mississippi has the largest drainage basin in the United States. It reaches from the Rocky Mountains to the Appalachians and includes countless other basins, from that of the big muddy Missouri to the smallest watersheds feeding the numerous tributaries. Every stream, even the smallest brook or rivulet, has its own drainage basin. Large drainage basins are often fan-shaped and display a drainage pattern reminiscent of a tree leaf and its veins, but small basins may assume very irregular shapes. The boundaries of the basin consist of divides or ridges which separate it from adjacent basins and which may be identified on topographic maps. A basin has a single outlet where the stream cuts through the divide, as in the case of a tributary basin, or where the stream flows into the sea, as in the case of a major basin, e.g., the Delaware River Basin (Fig. 2.2). The basin is the subject of many specific hydrologic studies, e.g., relationships between precipitation and runoff, flood routing, and flood forecasting, for which knowledge of the physiography of the basin is very important.

2.1 *Physiography of the Basin—Quantitative Study*

The morphology or study of the physical features and shapes of a basin is treated in detail in textbooks of physical geology [3, 4] and is only briefly illustrated here. A few of the various parameters used in this study are mentioned here.

The drainage area of a basin is the area of its horizontal projection, as

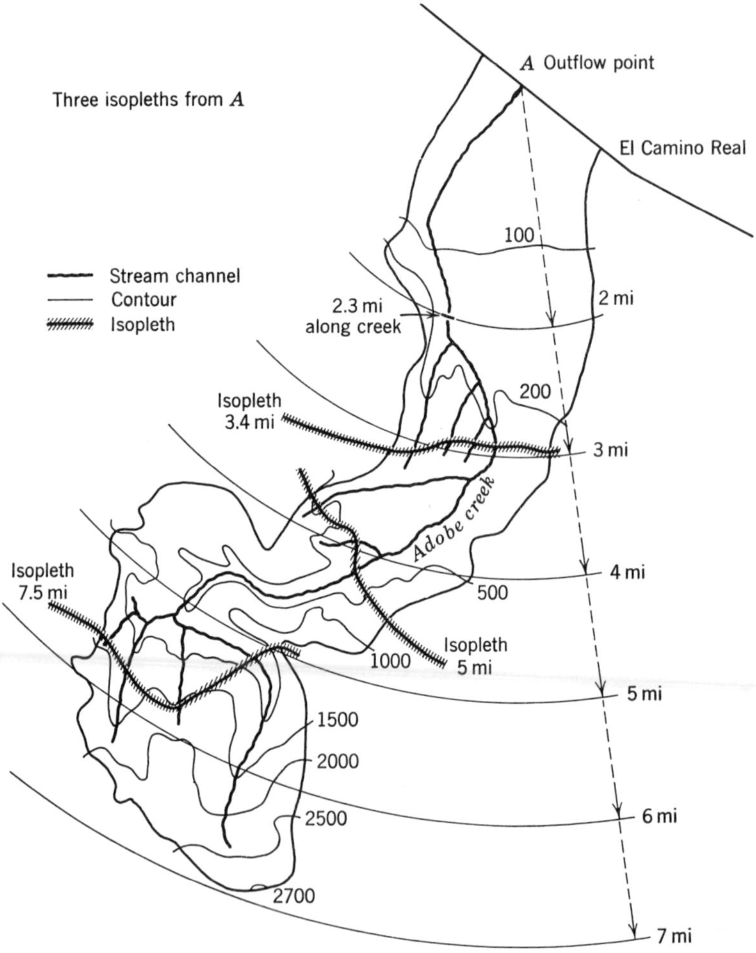

Fig. 2-3 Isopleths for the Adobe Creek Basin, near Palo Alto, Calif. Area: 11.0 sq mi; Drainage density: $D_d = \dfrac{2.18 \text{ mi}}{\text{sq mi}}$.

determined by planimetry, less the areas occupied by parts noncontributing to streamflow, i.e., areas in which stagnant runoff collects.

The drainage density, D_d, is the average length of streams per unit area within the basin, or

$$D_d = \frac{\Sigma L}{A_d} \tag{2.1}$$

in which ΣL is the total length of streams in the basin and A_d is the drainage area. Streams are generally classified according to their flow

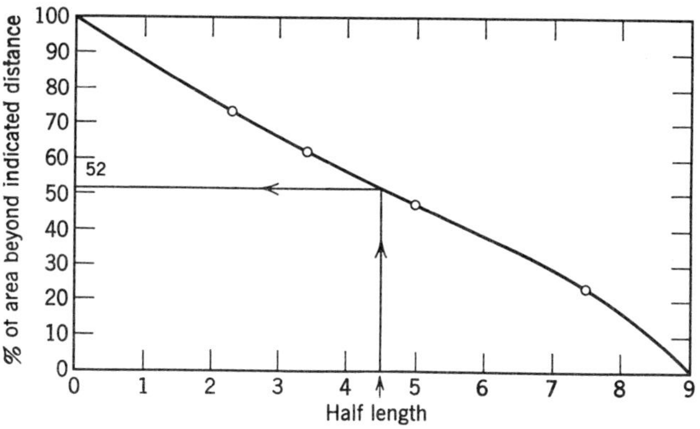

Fig. 2-4 Isopleths for the Adobe Creek Basin, near Palo Alto, Calif. Distribution of area as a function of distance from outflow.

constancy. Perennial streams flow at all times above surface, except in periods of extreme drought. Intermittent streams carry water above surface most of the time but appear dry occasionally because of evaporation and bank storage. Ephemeral streams carry water only as a result of flash runoff or snowmelt. All types of streams should be included in the computation of D_d, since this factor may be used in connection with flood flows. An accurate computation of D_d would be wieldy, and, in most cases, knowledge of the order of magnitude of D_d, varying from less than 1 mi/sq mi for a poorly drained basin to about 5 mi/sq mi for well-drained basins, is satisfactory. The drainage density of Adobe Creek basin, a small basin of 11.0 sq mi drainage area in the foothills of the Santa Clara Mountains near Stanford University, Palo Alto, California was computed to be 2.18 mi/sq mi (Fig. 2.3). This figure shows also three isopleths, or lines of equal distance from the outflow point A of the basin. The areas of the basin beyond these isopleths were planimetered and plotted as percentages of the total drainage area versus the values of the isopleths in miles (Fig. 2.4). This distribution of area with distance from the outflow station is used in the computation of the time distribution of runoff.

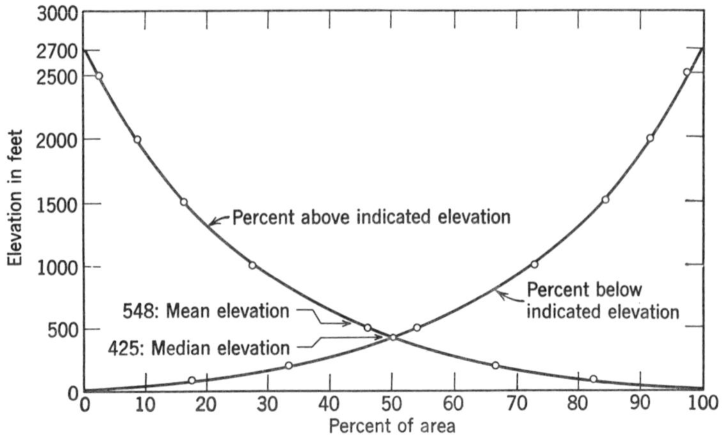

Fig. 2-5 Area-elevation distribution curve, Adobe Creek Basin.

The Adobe Creek basin has 52% of its drainage area located beyond one half the length of the major water course.

Another distribution curve, elevation versus area, Fig. 2.5, represents the basin profile and is useful in comparing drainage basins. Such a curve may be constructed by planimetering on a topographic map the areas enclosed between adjacent contours and the basin divide and by plotting the average elevations versus the areas in percentage of the total area (Fig. 2.3). A fair approximation may also be obtained, as indicated in Fig. 2.6, by superimposing a sufficiently dense grid drawn on transparent paper on a topographic map and by counting the number of grid intersections falling within various elevation intervals. The latter technique is also useful in the computation of the average precipitation on a drainage basin, when data of the variation of precipitation with elevation are available. The median elevation of the basin is determined from the area-elevation curve of Fig. 2.5 as the elevation above and below which half of the basin area extends. The mean elevation is determined as a weighted average in which the average elevation between adjacent contours is multiplied by the corresponding area and the sum of the products is divided by the total area. The stream profile (Fig. 2.7) shows the elevation of the bed of the main stream as a function of its distance from the basin outflow. The gross slope of a stream between any two points is the total fall between the points divided by the stream length (Fig. 2.7). The mean slope may be constructed by drawing a straight line (Fig. 2.7) such that the areas enclosed above and below the stream profile are equal. Stream profile and mean slope are specific characteristics of the stream channel and are not well suited as parameters for the overall shape of the basin.

Horton [5] described a method of determining the land slope of a basin, as illustrated in Fig. 2.6. A grid on transparent paper is superimposed on a topographic map of the basin. The length of each line of the grid between its intersection with the basin boundaries is measured, and the number of crossings or contacts of each line of the grid with the contours

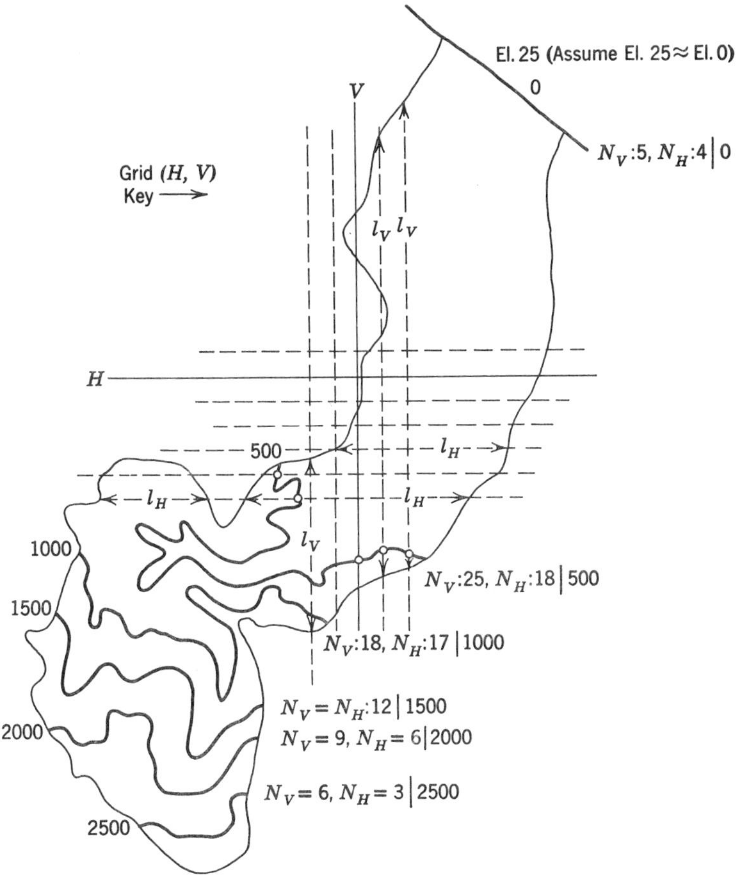

Fig. 2-6 Average landslope, Adobe Creek Basin.

V-direction

$\Sigma N_V = 75$

$\Sigma l_V = 53.2 \text{ mi}$

$\Delta Z = 500 \text{ ft}$

$$S_V = \frac{\Sigma N_V \Delta Z}{\Sigma l_V}$$

$$S_V = \frac{75 \times 500 \text{ ft}}{53.2 \times 5280 \text{ ft}} = 0.133 \text{ ft/ft}$$

H-direction

$\Sigma N_H = 60$

$\Sigma l_H = 54.1 \text{ mi}$

$\Delta Z = 500 \text{ ft}$

$$S_H = \frac{60 \times 500 \text{ ft}}{54.1 \times 5280 \text{ ft}} = 0.105 \text{ ft/ft}$$

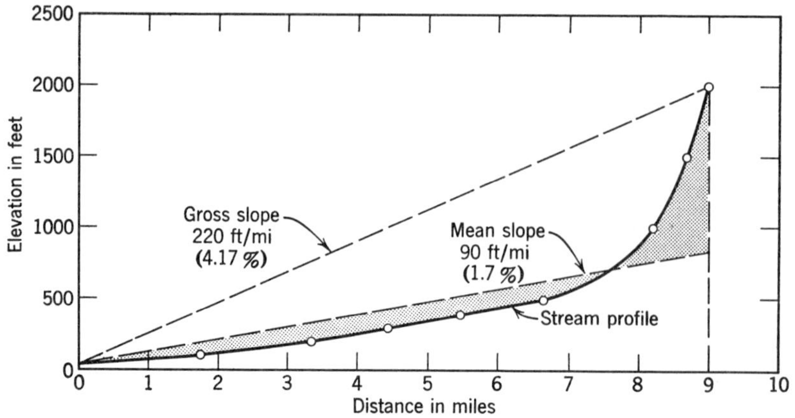

Fig. 2-7 Profile of major watercourse, Adobe Creek Basin.

is counted. The land slope s in either direction is then

$$s = \frac{N \sec \theta}{l} \Delta Z \tag{2.2}$$

in which N is the total number of contour crossings in both directions, l is the total length of line segments in both directions, ΔZ is the contour interval, and θ is the angle between the contours and the grid lines. The occurrence of $\sec \theta$ stems from an attempt to compute the slope along a normal to the contours. In practice the determination of $\sec \theta$ is tedious and uncertain and is often ignored, as in Fig. 2.6, where the land slope in the vertical and horizontal directions was computed separately and where the arithmetic average of both slopes was adopted as the land slope.

2.2 *Physiography of the Basin—Descriptive Features*

For a detailed study, the reader is again referred to references 3 and 4. It was recognized more than a century ago by Playfair that rivers were capable of carving the deepest valleys. With other elements of nature such as weather, glaciers, and waves rivers were able, by their destructive force to shape and change the landscape, given sufficient time. We have evidence of what has become known as Playfair's law when we observe land erosion and the formation of small gullies after heavy rain storms and when we estimate the enormous volumes of silt that are deposited by rivers in natural lakes, in manmade reservoirs, and in deltas. Although exceptions to Playfair's law may be found, it is likely that most valleys have been formed by stream activity.

The concept of the cycle of erosion [3] provides us with a qualitative picture of river valleys and regions. The cycle of a stream valley is divided into three main stages, referred to as young, mature, and old, in comparison with the age of man. A river is young when it is still actively and rapidly eroding its channel. Therefore it must flow at high, supercritical velocities, and it transports all sediments coming from lateral inflow. The river banks are steep, forming a V-shaped cross section, and there is no flood plain. Young streams typically are mountain streams with falls and rapids along the water course.

A stream attains maturity when its flow velocity is slowed down to the point where the stream no longer erodes its channel in depth and starts to develop a narrow flood plain. The transition from mature to old is largely a matter of degree. Old streams are characterized by a slow, tranquil flow, wide flood plains, broad meander belts, and the formation of deltas where the streams flow into the ocean. Typical old streams are the Mississippi, Rhine, and Nile in their lower reaches. There are, of course, streams and valleys that have reached different stages of development at different points along their courses—maturity in their lower reaches and youth toward their head waters [3].

Stream patterns may offer valuable information on the possibility of ground-water recharge of geologic formations underlying the drainage basin. A dendritic or tree-like pattern develops when the bedrock is uniform in its resistance to erosion and is made up either of flat-lying sedimentary rocks or massive igneous or metamorphic rocks. A rectangular pattern is indicative of underlying bedrock criss-crossed by fractures, which forms zones of weakness that are particularly vulnerable to erosion and may provide outcrop areas well-suited for intake of water. Likewise a trellis pattern usually indicates that the region is underlain by alternate bands of resistant and nonresistant rock. The possibility of recharge is shown by the cross section of the trellis pattern of Fig. 2.8 [3].

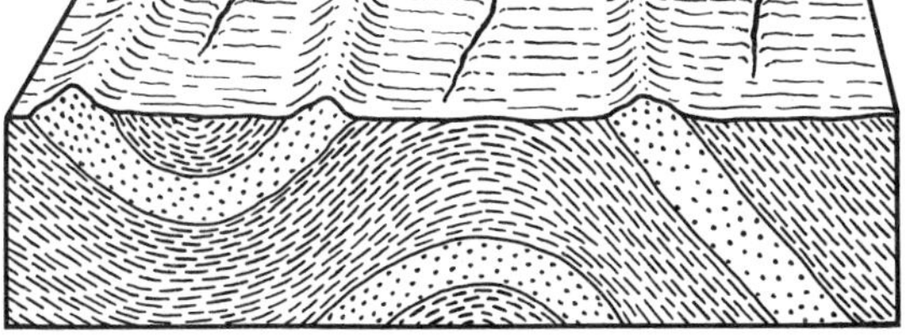

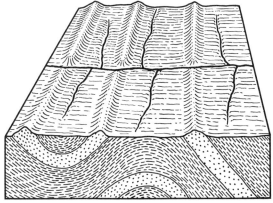

Fig. 2-8 Trellis pattern. (After Leet and Judson.) Possibility of ground-water recharge.

◆ Precipitation

Hydrometeoric precipitation as a result of condensation of atmospheric water vapor includes rain, snow, and hail under various forms, and is measured for its total amount, intensity, and duration.

2.3 *Types*

Three types of precipitation are characteristic.

CYCLONIC PRECIPITATION

It is associated with areas of low pressure moving across the earth as a result of planetary motion and solar effects. Usually this type leads to frontal precipitation, so-called because of the intersection with the ground of a sloping interface between masses of warm and cold air (Fig. 2.9).

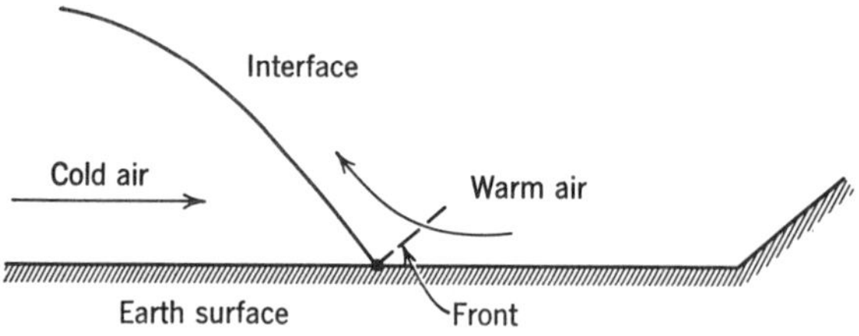

Fig. 2-9 Cyclonic precipitation. Vertical cross section through interface.

Warm-front precipitation results when the wedge of cold air is forced back by the oncoming warm air and the front moves to the left. Precipitation is light to moderate and covers a large area ahead of the moving front. Cold-front precipitation, on the other hand, occurs when cold air displaces warm air at the earth's surface; it is showery and more localized in the vicinity of the front. Cyclonic precipitation is the most common type and takes place all over the earth.

CONVECTIVE PRECIPITATION

It results from moist air that rises from the earth's surface where it is heated by radiation and reflection. As this air rises, it expands, and in this process does work against its surroundings and loses its heat. After cooling it condenses and precipitates in the form of showers. This type of precipitation is typical of the tropics and large continental regions, e.g., the Great Plains in the United States (summer thunderstorms).

OROGRAPHIC PRECIPITATION

It results from the mechanical lifting of air, caused by the presence of mountain barriers, to a certain altitude. Actually mechanical lifting has the same effect as the lifting due to thermal gradients in convective precipitation; the orographic effect also influences the intensity and distribution of cyclonic rainfall.

2.4 *Measurement of Precipitation*

Because it requires a minimum of apparatus, precipitation was probably the first hydrological phenomenon to be recorded by man (e.g., in India as early as in the fourth century B.C.). Existing rainfall data have proved to be very helpful in estimating runoff when no stream flow measurements were on record. In the United States rainfall data are accumulated and published primarily by the U.S. Weather Bureau but also by the Water Resources Departments of the individual states, by the Bureau of Reclamation, and by the U.S. Army Corps of Engineers. Boston has the oldest continuous rainfall record, which dates from 1750, whereas Croton River has the longest runoff record, dating from 1870. The study of the precipitation runoff relationship has been one of the primary aims of the science of hydrology. Many parameters influence this cause and sequence relationship which may be investigated for short and long durations, as will be seen in section 2.15. They add to the uncertainty inherent to the extension in the past of existing runoff records by means of rainfall data available in the past.

Rainfall data are studied for their intensity to determine the probability of occurrence of flood-producing storms. Precipitation is measured by rain gages, which may be of the nonrecording or recording type. The nonrecording type has been standardized by the U.S. Weather Bureau in the form of a circular cylinder having an 8-in. diameter (Fig. 2.10*a*), which serves as an overflow can. A funnel-shaped receiver of the same diameter is connected to a measuring tube with a cross section equal to one tenth that of the receiver. When 1 in. of rain falls into the receiver, the measuring tube is filled to a depth of 10 in. Rainfall is measured with a graduated rod to the nearest hundredth of an inch. About 11,000 nonrecording gages of various types were used in the United States [6] in 1958. At that time there were also about 3,500 recording gages (Fig. 2.10*b*) which kept a continuous record of precipitation and therefore also showed the intensity of rainfall.

Precipitation data over a long period of years must be analyzed for consistency before they may be used in any hydrologic project. Indeed,

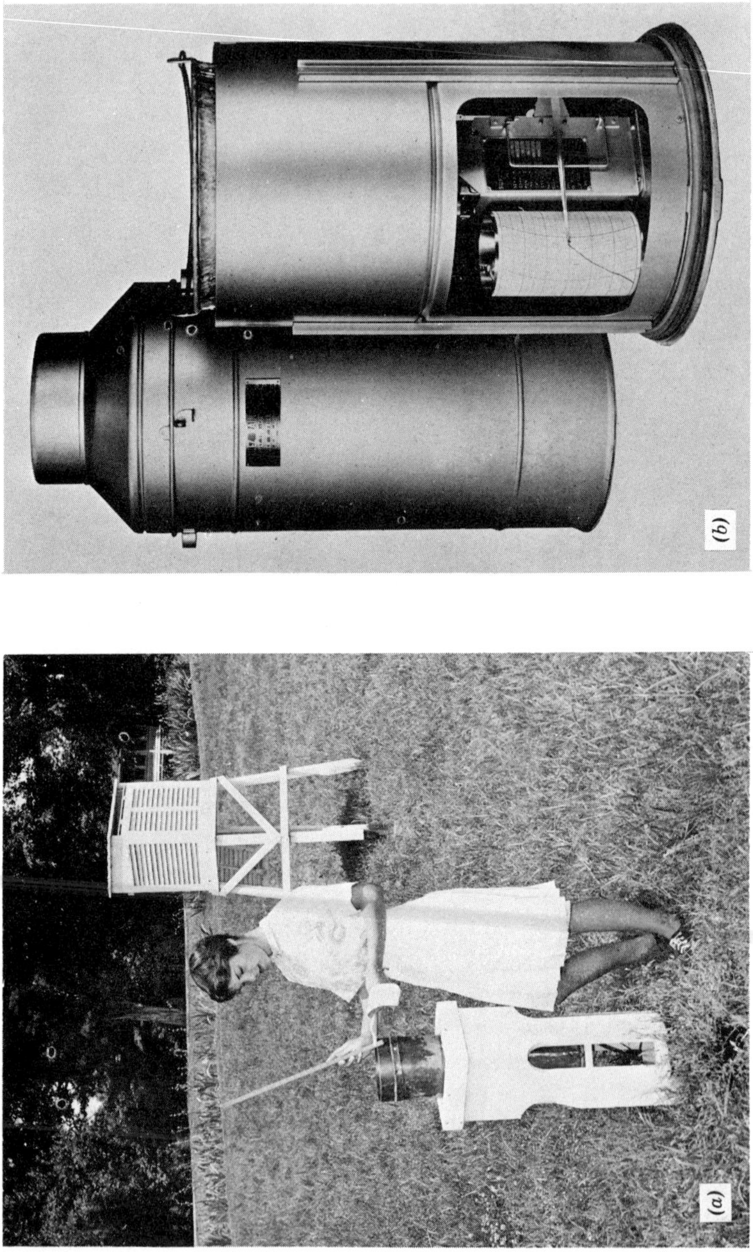

Fig. 2-10 (*a*) Standard 8-in. precipitation gage. (*b*) Recording rain gage. (Both courtesy of U.S. Weather Bureau.)

alterations such as changes in gage location, gage type, method of observation, and environment may be reflected in the data. These changes may be detected by the double-mass-curve technique, in which the accumulated annual or seasonal rainfall at the station to be tested is plotted against the concurrent accumulated values of the mean precipitation of a series of

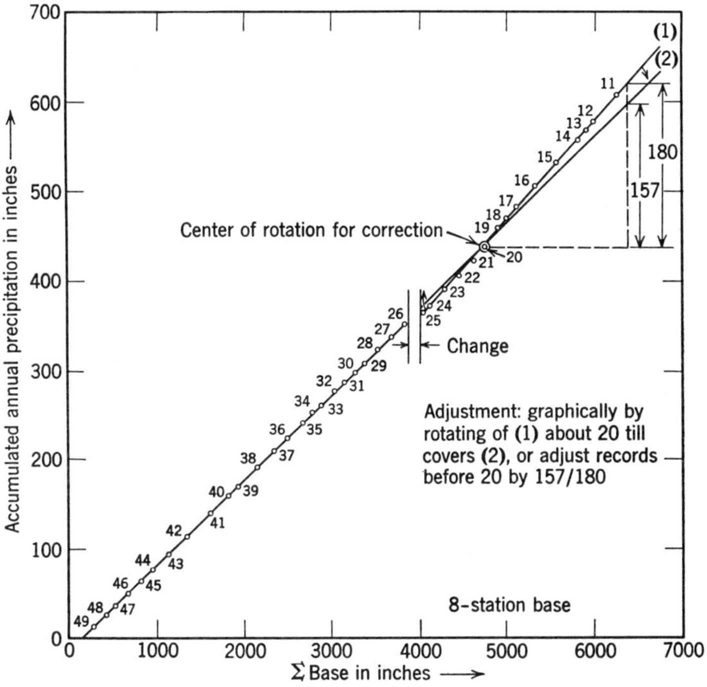

Fig. 2-11 Double-mass curve for annual precipitation in inches, Palo Alto, Calif.

surrounding base stations. An abrupt change in the slope of the double-mass curve marks the time when the alteration of the station became effective. Data obtained before that time are adjusted to be compatible with recent observations by prorating their value, as indicated on Fig. 2.11. A change in the regional weather conditions would not have had any influence on the slope of the double-mass curve because the station and its surroundings would have been equally affected.

2.5 *Average Precipitation over Area*

Point rainfall data as measured by gages are used to compute average depths of precipitation over a specific area either for an individual storm

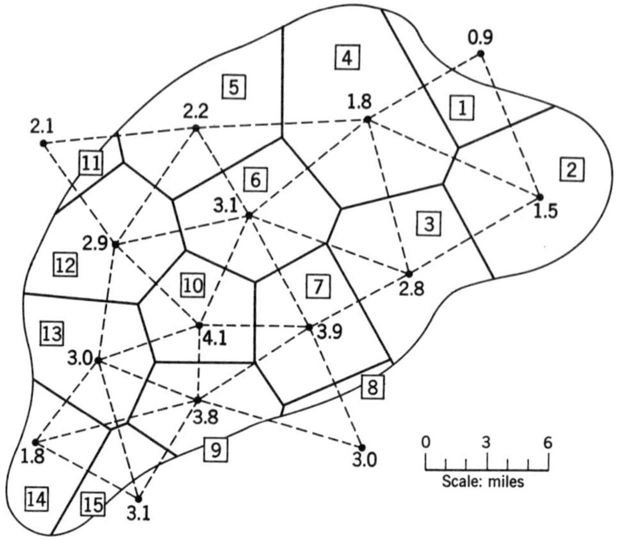

Fig. 2-12 Thiessen's method for average precipitation.

Polygon	Area (sq mi)	Precipitation Station (in.)	$P \times A$
1	15.48	0.9	13.93
2	45.36	1.5	68.04
3	38.16	2.8	106.85
4	57.96	1.8	104.33
5	36.72	2.2	80.78
6	33.12	3.1	102.67
7	33.12	3.9	129.17
8	1.80	3.0	5.40
9	23.04	3.8	87.55
10	23.04	4.1	94.46
11	2.88	2.1	6.05
12	33.12	2.9	96.05
13	27.00	3.0	81.00
14	20.16	1.8	36.29
15	9.72	3.1	30.13
	$A_d = 400.68$		$\Sigma P \times A = 1042.70$

Average precipitation on basin

$$\frac{\Sigma P \times A}{A_d} = 2.60 \text{ in.}$$

or on a seasonal or annual basis. The density of the gage network in the first place affects the value of these averages and their derivation from the true average precipitation as it fell on the area. Other important factors, which may or may not show up in the computations, depending upon the method used, are orographic effects and storm patterns. The simplest method, which, however, does not consider the latter factors nor the distribution of the gages, is to take the arithmetic mean of the point measurements. Nonuniform distribution of gages is accounted for in Thiessen's weighting method [7], whereby each gage is given a weighting factor expressed as the ratio of the area surrounding the gage over the total area. The average precipitation is the sum of the gage data, each multiplied with its weighting factor. The polygons surrounding each gage are obtained by erecting bisectors orthogonal to the segments joining adjacent stations. These stations are joined in such a manner that a simple grid of triangles results (Fig. 2.12). Orographic effects are neglected in this method, but they may be accounted for in another weighting method, the isohyetal method [6]. Contours of equal precipitation or isohyets are drawn on a map with station locations and recorded rainfall (Fig. 2.13). The average precipitation between two adjacent isohyets is then weighted by the ratio of the area between these isohyets over the total area. The isohyetal method is the most subjective method, and therefore its accuracy will depend upon the skill of its user and his knowledge of the topography of the terrain and the storm characteristics.

For small areas measuring up to 200 sq mi and reasonably uniform spacing of the rain gages, the arithmetic mean will suffice. For intermediate areas, of 200 to 2,000 sq mi, with small orographic effects, Thiessen's method may be used, although for large areas the isohyetal method is most suitable.

The orographic effect on average precipitation of a basin may easily be taken into account by the use of a grid on transparent paper, as explained before in the area-elevation computations (Fig. 2.14). Analysis of rainfall records for the drainage basin shown leads to a variation of mean annual precipitation as a function of elevation, given by the curve of Fig. 2.14. A grid placed over the contour map gave the results summarized in Table 2.1 (a tally is made by a node of the grid falling between two contour lines).

2.6 *Depth-Area-Duration Analysis* [6, 9]

Knowledge of the maximum amount of precipitation occurring on areas of various sizes for storms of different durations is of interest in many hydrologic problems, such as in the design of culverts and in the

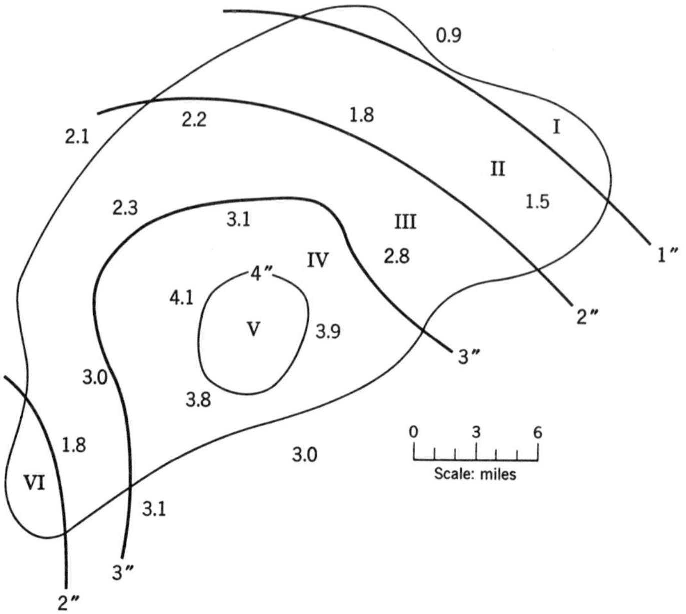

Fig. 2-13 Isohyetal method for average precipitation.

Zone	Area (sq mi)	P_{ave} (in.)	$P_{ave} \times A$
I	21.24	0.95	20.18
II	84.96	1.50	127.44
III	150.12	2.50	375.30
IV	107.28	3.50	375.48
V	24.48	4.10	100.37
VI	12.60	1.80	22.68
	$A_d = 400.68$		$\Sigma P_{ave}A = 1{,}021.45$

Average precipitation on basin

$$\frac{\Sigma P_{ave} \times A}{A_d} = 2.55 \text{ in.}$$

study of time effect of varying river stages on ground-water flow. To obtain this knowledge a series of storms with continuous records are analyzed. The analysis of a typical storm with precipitation over the entire basin follows.

The maximum precipitation at various stations for different durations is determined from the data given by recording rain gages at these stations. The basin is subdivided into zones by isohyets and the mean accumulative precipitation for each zone is determined. Next the maximum average depths of rainfall for different durations are determined for accumulative

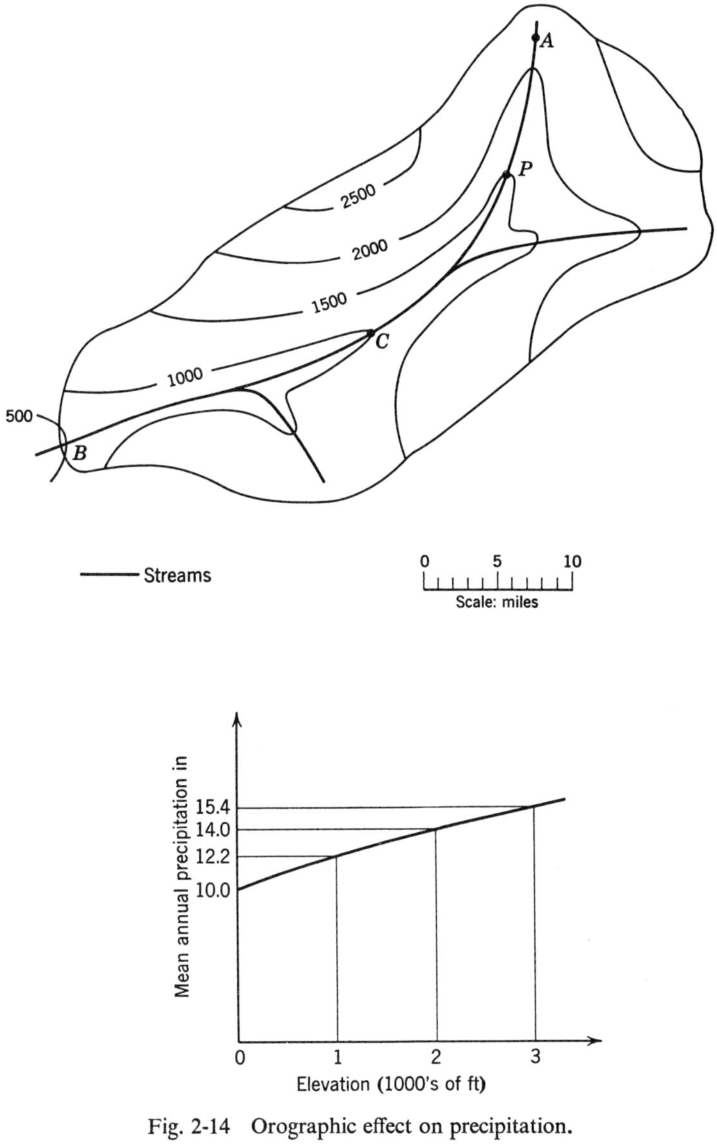

Fig. 2-14 Orographic effect on precipitation.

Table 2.1 Mean Annual Precipitation

Elevation (in 100's of ft)	Tally, t	Mean Annual Precipitation, in. Elevation (in 100's of ft)	Precipitation, p	Tally, t
5	66	7.5	11.6	66
10	225	12.5	12.8	225
15	184	17.5	13.6	184
20	212	22.5	14.4	212
25	43	27.5	15.2	43
30	$\Sigma t = 730$			

Mean annual precipitation over drainage basin:

$$P = \frac{\Sigma p \times t}{\Sigma t} = 13.5 \text{ in.}$$

areas of the basin. The results are plotted on semilogarithmic paper, i.e., for each duration the maximum average depths are plotted against the logarithms of the areas. A curve may be fitted through the results, or for each duration a depth-area formula as proposed by Horton [8] may be determined, so that previously analyzed storm data may be extrapolated. An example will illustrate this method [6].

EXAMPLE

The isohyets for 1.0 in., 1.5 in., 2.0 in., and 2.5 in. are constructed for the river basin sketched in Fig. 2.15. They divide the basin into five zones, I to V, with areas as indicated. Only the basins I to IV are used in the analysis, and the isohyetal method gives the result shown in Table 2.2.

Table 2.2 Isohyetal Method

Zone	Area, sq mi	Ave. Precip., in.	Volume Precip. (sq mi) × in.	Accum. Area, sq mi	Accum. Volume (sq mi) × in.	Depth Rainfall, in.
I	100.0	2.85	285	100.0	285	2.85
II	125.0	2.25	281	225.0	566	2.52
III	112.5	1.75	197	337.5	763	2.26
IV	65.0	1.25	81	402.5	844	2.10
V	17.5	0.80	14	420.0	858	2.04

The data from the recording rain gages at the five stations A, B, D, E, and F inside the basin and the two stations C and G outside it are given in Table 2.3.

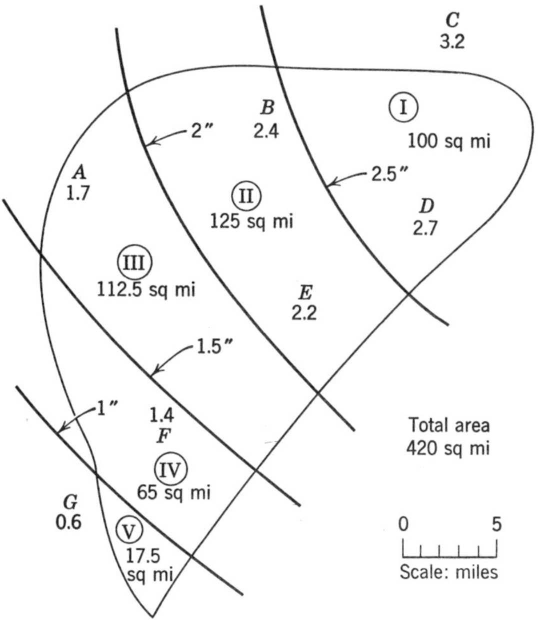

Fig. 2-15 Sketch of drainage basin for area-depth-duration analysis.

From an inspection of Table 2.3, we obtain the maximum precipitation at the various stations for three chosen durations, say 4 hr, 8 hr, and 12 hr. For example, the maximum precipitation in A for 8 hr is 1.2 in. (i.e., between 4 A.M. and noon). (See Table 2.4.)

Table 2.3 Total Station Accumulative Precipitation in Inches

Time	A	B	C	D	E	F	G
4 A.M.	0.0	0.0	0.0	0.0	0.0	0.0	0.0
6 A.M.	0.4	0.0	0.0	0.0	0.0	0.0	0.0
8 A.M.	0.6	0.5	0.0	0.0	0.0	0.2	0.0
10 A.M.	0.9	0.8	0.0	0.0	0.3	0.5	0.2
noon	1.2	1.2	0.6	0.2	0.8	0.8	0.3
2 P.M.	1.4	1.5	1.1	0.6	1.2	1.1	0.5
4 P.M.	1.7	1.7	1.7	1.2	1.4	1.2	0.6
6 P.M.	1.7	2.1	2.2	1.7	2.0	1.3	0.6
8 P.M.	1.7	2.4	2.9	2.2	2.2	1.4	0.6
10 P.M.	1.7	2.4	3.2	2.7	2.2	1.4	0.6
midnight	1.7	2.4	3.2	2.7	2.2	1.4	0.6

Table 2.4 Maximum Station Precipitation for Given Durations

Duration	A	B	C	D	E	F	G
4 hr	0.6	0.8	1.2	1.1	0.9	0.6	0.3
8 hr	1.2	1.5	2.3	2.1	1.7	1.1	0.6
12 hr	1.7	2.1	3.2	2.7	2.2	1.3	0.6

In the following step, the mean accumulative precipitation for each zone during the storm is determined. This is accomplished by means of the data of Table 2.3 and by expressing the precipitation for every zone as a function of the weighted precipitations of nearby stations. One possible way to estimate the mean accumulative precipitation for the different zones is

$$\text{I} = \frac{C + 2D}{3} \qquad \text{II} = \frac{E + 1.5B}{2.5}$$

$$\text{III} = \frac{A + F + E}{3} \qquad \text{IV} = \frac{A + G + 7F}{9}$$

This leads to the results of Table 2.5.

Table 2.5 Total Zonal Accumulative Precipitation in Inches

Time	Zone I	Zone II	Zone III	Zone IV
4 A.M.	0.0	0.0	0.0	0.0
6 A.M.	0.0	0.0	0.13	0.04
8 A.M.	0.0	0.3	0.26	0.22
10 A.M.	0.0	0.60	0.56	0.51
noon	0.3	1.04	0.93	0.80
2 P.M.	0.8	1.38	1.23	1.07
4 P.M.	1.4	1.58	1.43	1.20
6 P.M.	1.9	2.06	1.66	1.27
8 P.M.	2.4	2.32	1.76	1.35
10 P.M.	2.9	2.32	1.76	1.35
midnight	2.9	2.32	1.76	1.35

From the last row (midnight) of Table 2.5, we may see that the assumptions about the mean accumulative zonal precipitation were indeed reasonable and that the results do not differ very much from the simple arithmetic average precipitation for the zones as given in the third column of Table 2.2.

The maximum average depth of rainfall for durations of 4, 8, and 12 hr for the accumulative areas I, I + II, I + II + III, I + II + III + IV is now determined by means of Table 2.6.

Table 2.6 Total Accumulative Precipitation in Inches

Time	I	I + II	I + II + III	I + II + III + IV
4 A.M.	0.0	0.0	0.0	0.0
6 A.M.	0.0	0.0	0.04	0.04
8 A.M.	0.0	0.17	0.17	0.16
10 A.M.	0.0	0.33	0.41	0.40
noon	0.3	0.71	0.78	0.75
2 P.M.	0.8	1.12	1.16	1.10
4 P.M.	1.4	1.50	1.47	1.36
6 P.M.	1.9	2.00	1.88	1.71
8 P.M.	2.4	2.35	2.16	1.94
10 P.M.	2.9	2.58	2.30	2.06
midnight	2.9	2.58	2.30	2.06

Table 2.6 may be reconstructed without difficulty by noticing that

$$\text{I} + \text{II} = \frac{\text{I} \times 100 + \text{II} \times 125}{225}$$

$$\text{I} + \text{II} + \text{III} = \frac{\text{I} \times 100 + \text{II} \times 125 + \text{III} \times 112.5}{337.5}$$

$$\text{I} + \text{II} + \text{III} + \text{IV} = \frac{\text{I} \times 100 + \text{II} \times 125 + \text{III} \times 112.5 + \text{IV} \times 65}{402.5}$$

It is evident that the midnight row of Table 2.6 should be in close agreement with the last column of Table 2.2. From an inspection of Table 2.6 we determine the maximum average precipitation. For example, the maximum average precipitation for item I of Table 2.7 for a duration of 4 hr is 1.1 in. (e.g., between noon and 4. P.M., or between 2 P.M. and 6 P.M.)

Table 2.7 Maximum Average Precipitation in Inches

	AREA			
Duration	I (100 sq mi)	I + II (225 sq mi)	I + II + III (337.5 sq mi)	I + II + III + IV (402.5 sq mi)
4	1.1	0.88	0.75	0.70
8	2.1	1.67	1.47	1.31
12	2.9	2.25	1.99	1.78

The results of Table 2.7 are plotted on semilog paper in Fig. 2.16 and very closely fit straight lines. It is not desirable, however, to extrapolate the curves linearly for small areas. Indeed, if such an extrapolation were

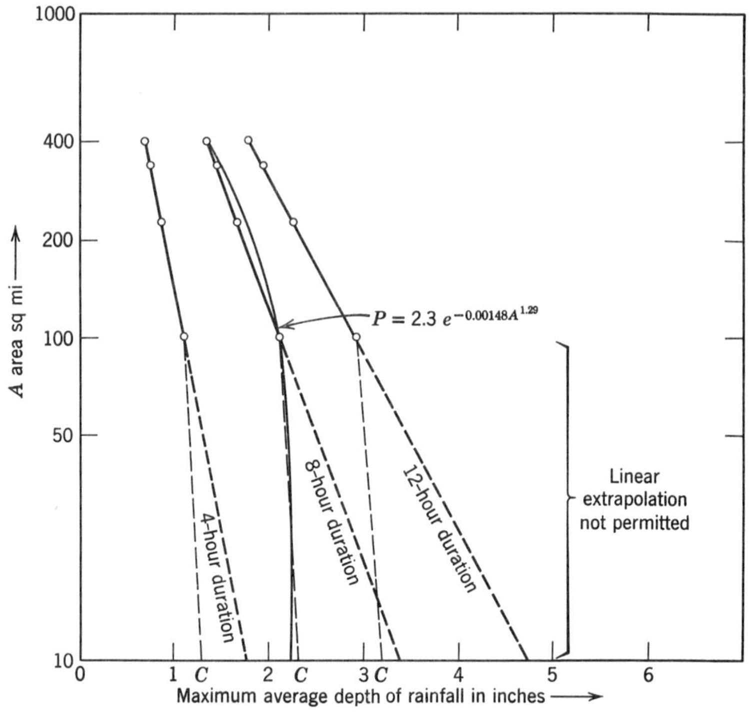

Fig. 2-16 Maximum depth-area-duration curves.

made, it would give maximum depths of rainfall on a 10 sq mi area much in excess of the maximum precipitation of station C, which is unacceptable. Therefore it is better to fit a depth-area curve to the results, as proposed by Horton [8]. Horton proposed the formula

$$\bar{P} = P_0 e^{-kA^n} \tag{2.3}$$

in which $\bar{P}$ is the average depth of rainfall in inches for a given duration over an area A in square miles, P_0 is the highest precipitation at the center of the storm and k and n are constants for a given storm. These constants may be determined if, for a given P_0 in Eq. 2.3, the values of $\bar{P}$ and A are known for two storms of the same duration, so that substitution in Eq. 2.3 leads to two equations with two unknowns, k and n.

In the case of the 8-hr duration storm, we select two pairs of values:

$$\bar{P}_1;\ A_1 = 2.1 \text{ in.};\ 100 \text{ sq mi} \qquad \bar{P}_2;\ A_2 = 1.3 \text{ in.};\ 400 \text{ sq mi}$$

and assume $P_0 = 2.3$ in. equal to the maximum station precipitation for 8 hr. Insertion of these values in Eq. 2.3 and solution of the two equations leads to $n = 1.29$ and $k = 0.00148$ and a depth-area formula for the 8-hr duration storm

$$\bar{P} = 2.3e^{-0.00148A^{1.29}} \tag{2.4}$$

For $A = 10$ sq mi, Eq. 2.4 gives $\bar{P} = 2.23$ in., a reasonable value.

The technique explained in the foregoing paragraph has been applied to a series of storms with precipitation on the Passaic River basin, New Jersey, selected from data given by the U.S. Army Corps of Engineers [9] and by the U.S. Weather Bureau.

2.7 *Variation in Precipitation*

Precipitation varies with geographical situation and with time of year. Equatorial regions have the highest rainfall and precipitation decreases with increasing latitude. Other regional effects, however, such as the presence of mountains, large lakes, and oceans, have a greater influence on the precipitation pattern than does the distance to the equator. In many regions precipitation has a seasonal character, that has not much changed over the period for which a record of rainfall exists. A typical monthly distribution of precipitation for the Northwest Coast of the United States, for example, shows heavy precipitation averaging around 6 in. per month during the winter and lighter precipitation averaging less than 1 in. during the summer. This pattern is reversed for the Great Plains, showing heavy summer precipitation around 3 in. per month, and tapering off to monthly winter precipitation below 1 in. Figure 2.17 [10] shows the average or mean annual precipitation in the United States for the period 1899–1938.

◆ Evapo-Transpiration

A large amount of the water that is precipitated upon the earth is returned to the atmosphere as vapor, through the combined actions of evaporation, transpiration, and sublimation. The last three actions essentially are three modifications of a single process [11], owing to the energy of the solar engine that keeps the hydrologic cycle running. Evaporation or vaporization is the process by which molecules of water at the surface of water or moist soil acquire enough energy through sun radiation to escape the liquid and pass into the gaseous state. Sublimation differs from this phenomenon only in that water molecules are converted directly from the solid state (snow or ice) into vapor, without passing through the liquid state. Billions of tons of snow, either from the top of the snow surface that covers the earth or from the snow intercepted by trees, roofs, etc., are converted annually into vapor. Water vapor is also breathed into the air from the lungs of animals or is transpirated by plants. Some of the water requirements necessary to the growth of crops have already been mentioned in Chapter 1. Most of the water absorbed by plants is passed into the atmosphere and some plantations therefore release almost as much water into the air as would rise from a comparable area of water surface.

Evapo-transpiration is considered a regional water loss, although in the strict sense of the word, loss applies only to evaporation and sublimation. Plants use water in much the same way as do human beings and industries, but without polluting their waste waters. It is still a matter of speculation as to how important some weeds, herbs, trees, etc., are. Evapo-transpiration is a regional loss because on a global basis its total amount must be equal to the total amount of precipitation, both estimated at 95,000 cu mi per year. Wüst [12] estimated that some 80,000 cu mi of water are evaporated annually from the oceans, while only about 15,000 cu mi are evaporated from lakes and land, and that the total annual precipitation over the continents is about 24,000 cu mi. These estimates, combined with the knowledge of the areas of the continents and the earth's total surface, allow us to compute precipitation and evaporation in inches and also to compute total runoff from the continents (streamflow to the oceans) in inches or in acre-feet per acre. Let the areas of the continents be as follows:

Asia	17,000,000 sq mi
Africa	11,500,000 sq mi
North America	8,300,000 sq mi
South America	6,800,000 sq mi
Antarctia	6,000,000 sq mi
Oceania	4,000,000 sq mi
Europe	3,800,000 sq mi
Total	57,400,000 sq mi

Based on an equatorial circumference of 24,902 mi and a meridional circumference of 24,860 mi, the earth's surface computed as the average of two spheres would be 197.4×10^6 sq mi. (Krick [11] quotes 197.0×10^6 sq mi of surface area.) The oceans therefore would occupy roughly 140×10^6 sq mi, or 70.8 % of the earth's surface.

E_o = Annual evaporation from oceans

$$= \frac{80{,}000 \times 5{,}280 \times 12}{140 \times 10^6} = 36.25 \text{ in.}$$

E_c = Annual evapo-transpiration from continents

$$= \frac{15{,}000 \times 5{,}280 \times 12}{57.4 \times 10^6} = 16.55 \text{ in.}$$

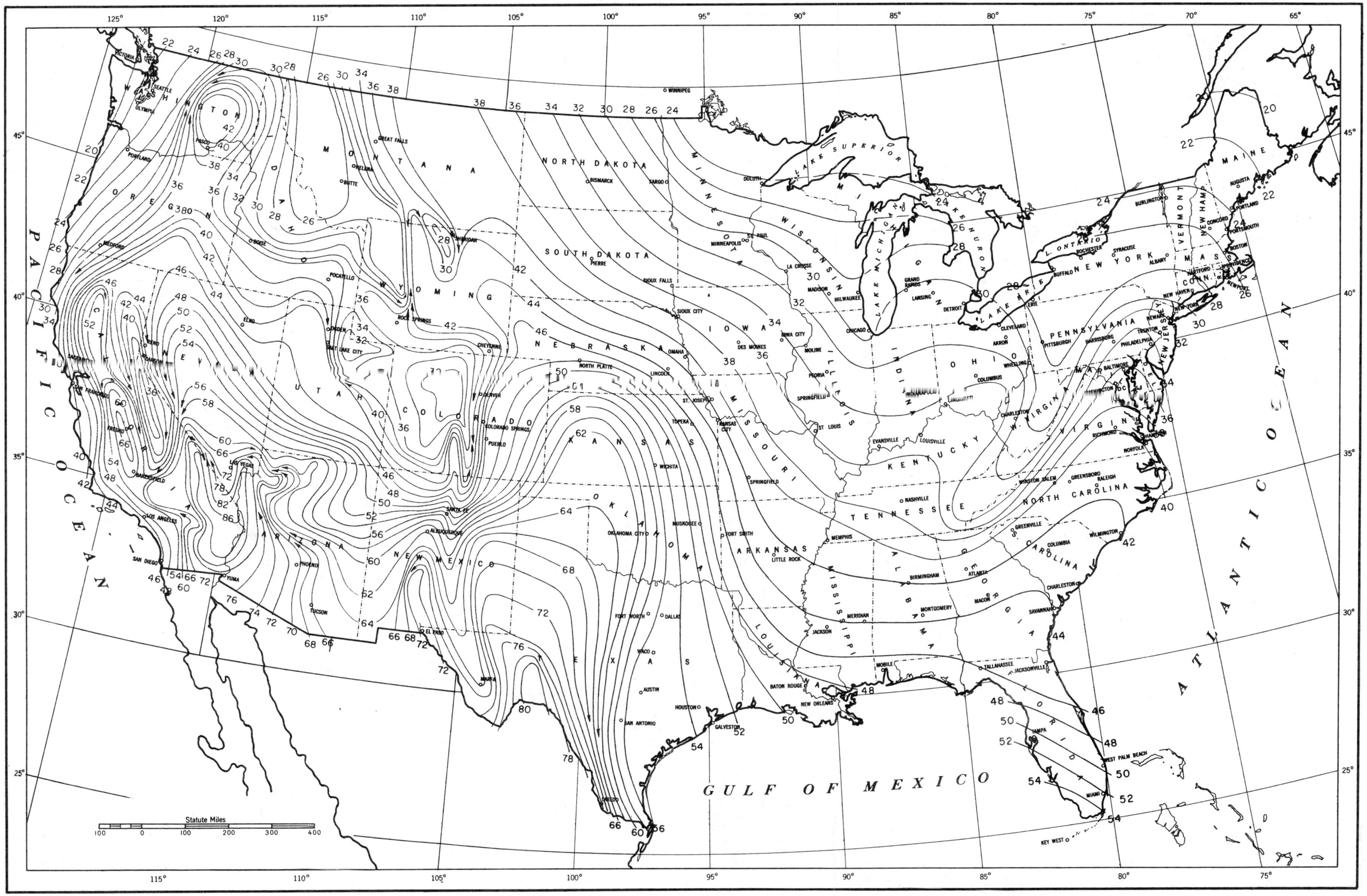

Fig. 2-18 Mean annual lake evaporation for the United States. (Courtesy Water Information Center Inc.) Lines show mean annual lake (free-water) evaporation in inches based on period 1946–1955. (After *U.S. Weather Bureau Technical Paper 37.*)

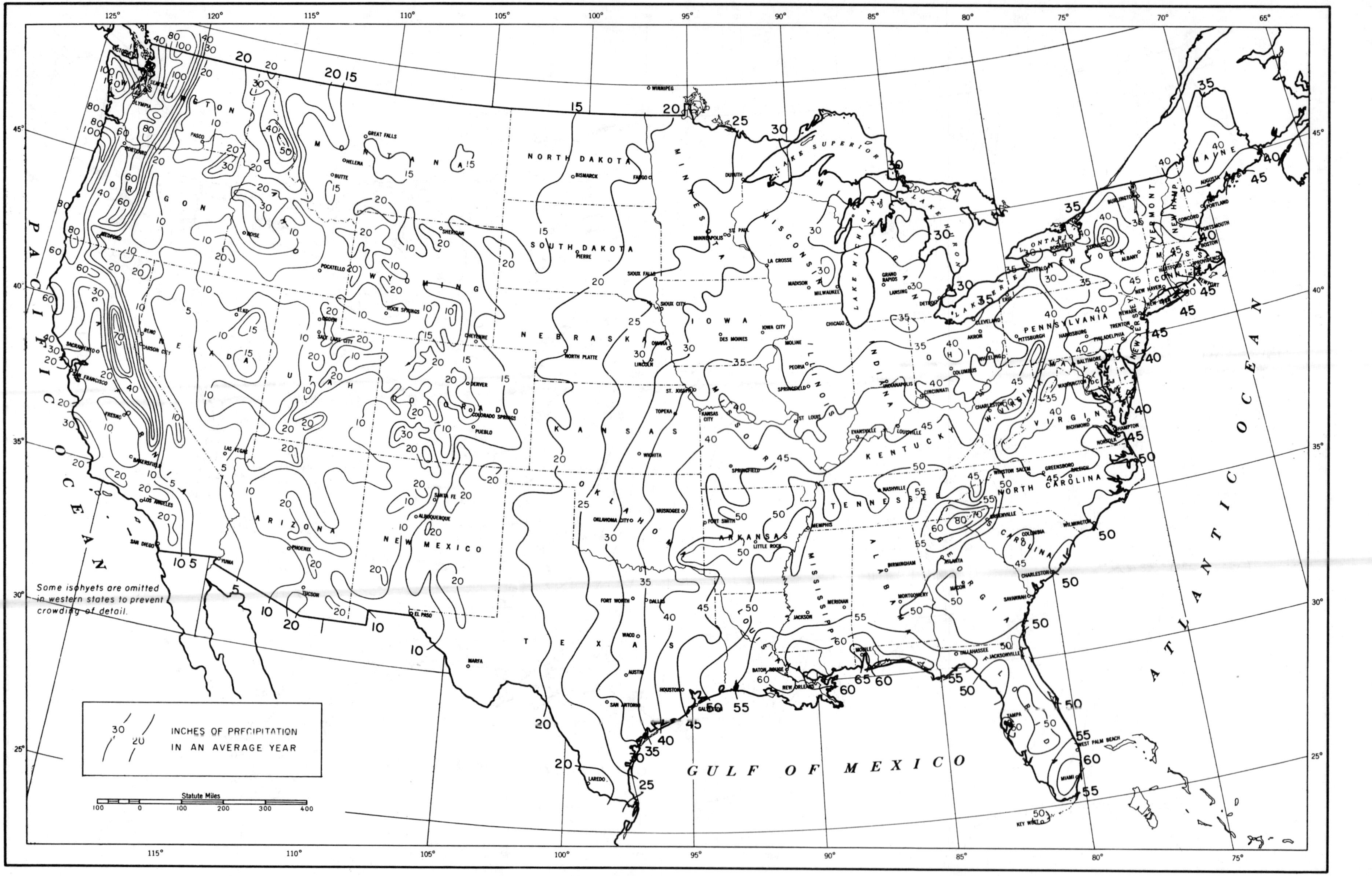

Fig. 2-17 Average annual precipitation for the United States. (Courtesy Water Information Center Inc.) Based on 40-year period. (After U.S. Department of Agriculture, "Climates of the United States.")

P_o = Annual precipitation over oceans

$$= \frac{71{,}000 \times 5{,}280 \times 12}{140 \times 10^6} = 32.25 \text{ in.}$$

P_c = Annual precipitation over continents

$$= \frac{24{,}000 \times 5{,}280 \times 12}{57.4 \times 10^6} = 26.55 \text{ in.}$$

On a global basis, the annual change of water storage in the upper layers of the earth's crust is negligible and therefore the total annual runoff from the continents is about 24,000 cu mi − 15,000 cu mi = 9,000 cu mi, or about 10.0 in. (i.e., 0.83 acre-feet per acre).

The previous computations were based on the following relationships:

$$E_o + E_c = P_o + P_c \tag{2.5}$$
(on a global basis)

$$P_c = E_c + \text{runoff} \tag{2.6}$$
(on a continental basis)

In the United States Eq. 2.6 is numerically expressed as [13]:

30 in. precipitation = 21 in. evaporation + 9 in. runoff

In the United States, the atmosphere makes up the difference between precipitation and evapo-transpiration by providing a net transport of moisture from the oceans to rainfall over the land in the amount of 9 in., to balance the discharge of rivers to the sea.

Finally, we obtain a better idea of the energy involved in the evapo-transpiration process when we realize that on the average the sun raises about 2.2 million tons of moisture annually from each square mile of the earth, as may be derived from the previous computations.

Evaporation also affects the numerical analysis of ground-water flow problems [14], especially in unsteady state computations for periods of drought. A map with the average annual lake evaporation of the United States is given in Fig. 2.18.

2.8 *Factors Affecting Evaporation*

The rate of evaporation depends on the continuous relative ease with which water molecules leave the liquid. This, in turn, depends on many factors such as temperature, relative humidity of the ambient atmosphere, wind velocity, barometric pressure, and pollution of the water. Evaporation is perhaps the most observed phenomenon in daily life and is explained in elementary physics courses. When water molecules escape from the surface of heated water in a closed container, they mix intimately with the air molecules. According to Dalton's law the total pressure of the mixture is equal to the sum of the partial pressure of the water, called vapor pressure, and the partial pressure of the dry air. When the mixture is saturated (with water), the vapor pressure is maximum and is called saturation vapor pressure. Saturation vapor pressure increases with temperature and has been tabulated for water in psychrometric tables by the U.S. Weather Bureau [15]. A graphical representation of saturation vapor pressure in millibars (from 0 to 50°F) and in inches of mercury (from 50°F to 100°F) is also given in Fig. 2.20. In the case of evaporation from moist surfaces in nature, the atmosphere plays the role of an expansible container lid. Evaporation can take place as long as the mixture, air-water vapor, is not saturated and the rate of evaporation, dE/dt is proportional to the difference between the saturation vapor pressure e_s and the existing vapor pressure e_a in the air above the water. The latter is also the saturation vapor pressure at the dew point of the air. Dew point is the temperature at which air becomes saturated when cooled under constant pressure and with constant water vapor content. Hence the maximum possible evaporation rate is

$$\frac{dE}{dt} = K(e_s - e_a) \tag{2.7}$$

in which K is a proportionality factor. Equation 2.7 is valid when water and overlying air are at the same temperature. When the temperature of the water is lower than that of the air, evaporation will continue until e_a reaches a value e_w, which is the saturation vapor pressure corresponding to the temperature of the water, and is smaller than e_s. However, when the water is at a higher temperature than the air, e_w is larger than e_s and condensation would occur if evaporation should take place at a rate larger than the one given by Eq. 2.7. The superfluous vapor usually condenses into fog and increases the relative humidity of the atmosphere. The relative humidity is the ratio of water vapor present in the atmosphere to the quantity that would saturate the air at the same temperature. It is proportional to the vapor pressure at the same temperature. Hence,

$$\frac{dE}{dt} = K\left(\frac{100\% - \text{relative humidity}}{100\%}\right) \tag{2.8}$$

As the temperature falls, the relative humidity of a given volume of air and water vapor increases and the rate of evaporation decreases.

Wind, as is well known from experience, promotes evaporation simply because it sweeps the vaporized water molecules away from the surface of evaporation and thereby decreases e_a. The effect of wind velocity on evaporation is expressed empirically by the expression

$$\frac{dE}{dt} = K'\left(1 + \frac{W}{10}\right) \tag{2.9}$$

in which W is the wind velocity in mph at about 25 ft above the surface. Meyer [16] combined the effects of Eqs. 2.7 and 2.9 in the formula named after him:

$$\frac{dE}{dt} = C(e_s - e_a)\left(1 + \frac{W}{10}\right) \tag{2.10}$$

in which the rate of evaporation is expressed in inches per day; C is a coefficient varying from 0.36 for ordinary lakes of about 25-ft depth to 0.50 for wet surfaces of soil, small puddles, and shallow evaporation pans; e_s, e_a, and W are as defined previously.

The effect of every day variations in barometric pressure on evaporation is only minor. An increase in evaporation of the order of 20% is accomplished only by a reduction of barometric pressure of mercury from 30 in. to 20 in.

Water pollution reduces evaporation approximately in proportion to the percentage of solids in solution. Sea water with 35,000 parts per million (ppm) of total solids (i.e., 96.5% water) will evaporate 96.5% as rapidly as fresh water, all other factors being equal.

Finally, evaporation from land and other surfaces exposed to precipitation differs from evaporation over water surfaces in the degree of opportunity, based on availability of water at all times. Thus the opportunity of evaporation is 100% for lakes and streams, while for land it attains that value only immediately after precipitation.

2.9 *Reservoir Evaporation*

DETERMINATION FROM WATER BUDGET

The principle of conservation of mass may be applied to determine reservoir evaporation. The water budget for a reservoir over a given period of time is expressed as

$$P + I = E + \frac{\Delta S}{A_{\text{ave}}} + 0 \tag{2.11}$$

in which P is the precipitation on the reservoir, I is the inflow, surface, and subsurface into the reservoir, O is the outflow, surface and subsurface from the reservoir, $\Delta S/A_{\text{ave}}$ is the change in reservoir storage per unit area taken as the average of the surface areas at the beginning and the end of the given time period, and E is the evaporation during the period of observation. Theoretically all quantities of Eq. 2.11 may be measured directly, except E, which would then be computed from Eq. 2.11. However, although it may be possible to measure P, $\Delta S/A_{\text{ave}}$ and the surface components of O and I with relative ease, all errors in measurement are reflected in the final value of E. Moreover, it may be very difficult to evaluate the subsurface components of O and I, as should become clear from the study of this book. If these components are large and difficult to determine, values of E resulting from Eq. 2.11 may not be reliable. The method therefore is more suitable for computation of annual or monthly values than for short period, daily values.

Reservoir evaporation may also be determined by the energy-budget approach [17] and by the use of the mass-transfer theory [18].

MEASUREMENT BY MEANS OF EVAPORATION PANS

The most common method of estimating reservoir evaporation is to establish a relationship between evaporation from a pan near the reservoir and between evaporation from the reservoir determined in the most accurate way possible. Thus the value of a pan coefficient, expressing the ratio of lake-to-pan evaporation may be determined on a monthly and annual basis. Future reservoir evaporation may then be determined from pan measurements. A variety of pans is used, differing in exposure, such as the sunken, floating, and surface type. The sunken pan is buried in the soil adjacent to the reservoir while the floating pan floats on the reservoir. Both types have shortcomings and advantages, but they are now outnumbered by the standard Weather Bureau Class A pan. This pan is of unpainted galvanized iron 4 ft in diameter and 10 in deep, set on a wooden frame 12 in. above ground level. To convert measurements made with this pan to equivalent reservoir evaporation, multiply by a pan coefficient of 0.70 to 0.75.

Evaporation records are made available by the Bureau of Reclamation, the U.S. Department of Agriculture (Bureau of Plant Industry), and the U.S. Weather Bureau.

Figure 2.19 shows a class A U.S. Weather Bureau evaporation station.

INFLUENCE ON RUNOFF

Evaporation from manmade reservoirs may have a significant influence on another phenomenon that is part of the hydrologic cycle, namely runoff or streamflow, which will be treated in section 2.13. If net evaporation is larger than precipitation, streamflow always decreases as a result of the construction of a reservoir. If precipitation is higher than net evaporation, streamflow will generally increase by the construction of a reservoir, except when the runoff coefficient of the area before construction is relatively high, in which case there may be a decrease of streamflow. The runoff coefficient is an empirically determined factor that, when multiplied by the total precipitation, gives the total streamflow. The estimates of the runoff coefficient may vary widely, from 0.05 for a basin in a semiarid region where conditions for runoff are unfavorable, to 0.35 for a basin in a region

Fig. 2-19 Class A evaporation station at Landisville, Pa. showing sheltered thermometer, evaporation pan with anemometer, recording rain gage, and standard 8-in. rain gage in foreground. (Courtesy of U.S. Weather Bureau.)

with continuous and abundant rainfall. The runoff coefficient for a lake, of course, is one. As an example, we can compute the net increase or decrease in streamflow as a result of the construction of each reservoir for which the data are tabulated in Table 2.8.

2.10 *Monthly Evaporation*

A formula for monthly evaporation similar to Meyer's equation for daily evaporation may be derived for a standard Class A evaporation station, by means of a graphical correlation procedure described by Linsley et al. [12]. Given are data on temperature, dewpoint, wind velocity, and evaporation as observed at the station, averaged on a monthly basis over a period of years. It is necessary to establish a graphical relationship between the parameters so that monthly evaporation may be estimated from temperature, dewpoint, and wind data, and to determine the equation of this graphical relationship.

We assume that Dalton's law applies to the data and that the water temperature is equal to the air temperature. The family of curves in the upper quadrant I has been constructed from the graphs of e_s (Fig. 2.20), using mean temperature as ordinates, vapor pressure difference $(e_s - e_a)$ as abscissas, and dewpoint as parameter. The family of curves in the lower

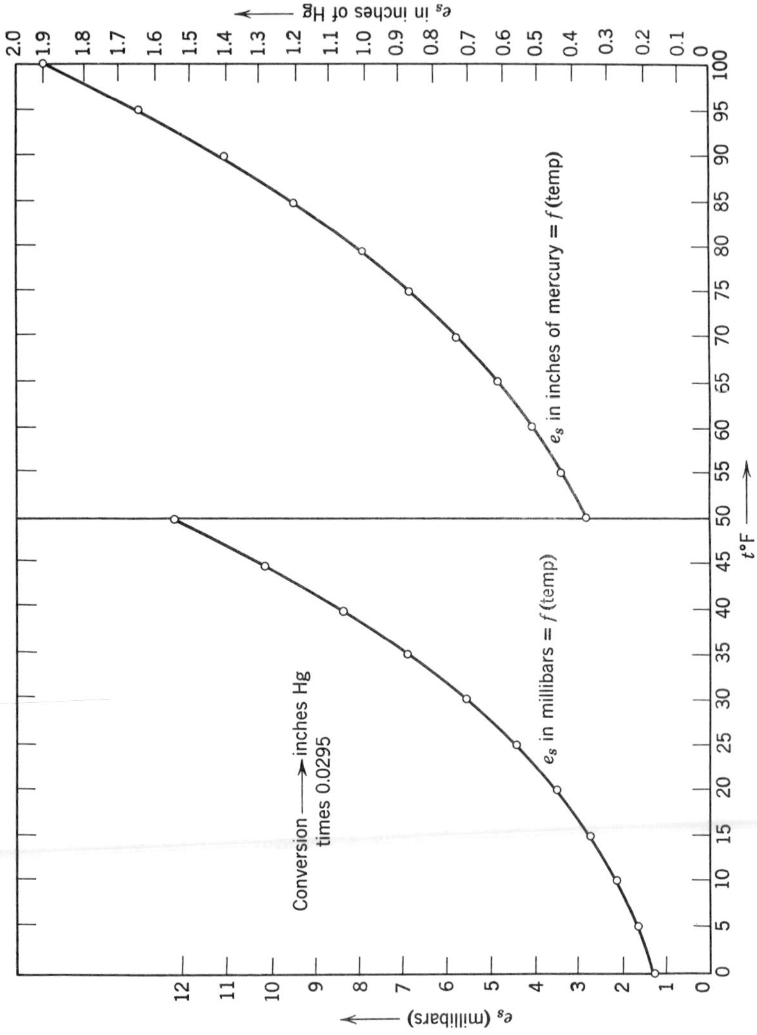

Fig. 2-20 Saturation vapor pressure versus temperature.

Table 2.8 Effect of Construction of Reservoir on Runoff

Reservoir	Lake Mead	Shasta	Norris	Ross
Location	(Ariz.-Nev.)	(No. Calif.)	(Tenn.)	(Wash.)
Area (acres)	146,500	26,000	40,000	1,895
Annual precipitation	7 in.	40 in.	50 in.	80 in.
Annual pan evaporation	120 in.	65 in.	44 in.	32 in.
Runoff coefficient	0.05	0.20	0.30	0.35
Pan coefficient	0.7	0.7	0.7	0.7
Results	944,275 AF/yr loss	29,750 AF/yr loss	14,000 AF/yr increase	4,640 AF/yr increase

Note: AF = acre-foot, amount of water which would cover 1 acre at a depth of 1 ft, equivalent to 326,000 gal.

Sample Computations:

Lake Mead: (a) Streamflow before construction of reservoir: = Precipitation on area of reservoir × runoff coefficient

$= \frac{7}{12} \times 146{,}500 \times 0.05 = 4{,}275$ AF

(b) Streamflow after construction of reservoir = Precipitation on area of reservoir − evaporation from reservoir

$$= (7 - 0.7 \times 120)\frac{146{,}500}{12} = -940{,}000 \text{ AF}$$

Therefore:

Change in streamflow = end value − begin value

$= -940{,}000\text{AF} - 4{,}275 \text{ AF} = -944{,}275 \text{ AF}$

Change is a net decrease.

Lake Norris: (a) Streamflow before construction of reservoir

$$= \frac{50}{12} \times 40{,}000 \times 0.3 = 50{,}000 \text{ AF}$$

(b) Streamflow after construction of reservoir

$$= (50 - 0.7 \times 44) \times \frac{40{,}000}{12} = 64{,}000 \text{ AF}$$

Therefore the change is a net increase of 64,000 AF − 50,000 AF = 14,000 AF.

quadrant II has been constructed, using evaporation as ordinates, vapor pressure difference $(e_s - e_a)$ as abscissas, and wind velocity as parameter. The curves of quadrant II have been drawn approximately as straight lines passing through the origin of coordinates, in anticipation of their equation derived in the sequel. The parameters have been indicated for both families of curves by writing their numerical values beside each point. Subsequently the curves have been labeled with the corresponding value of the parameters. After the two families of curves have been drawn, the accuracy

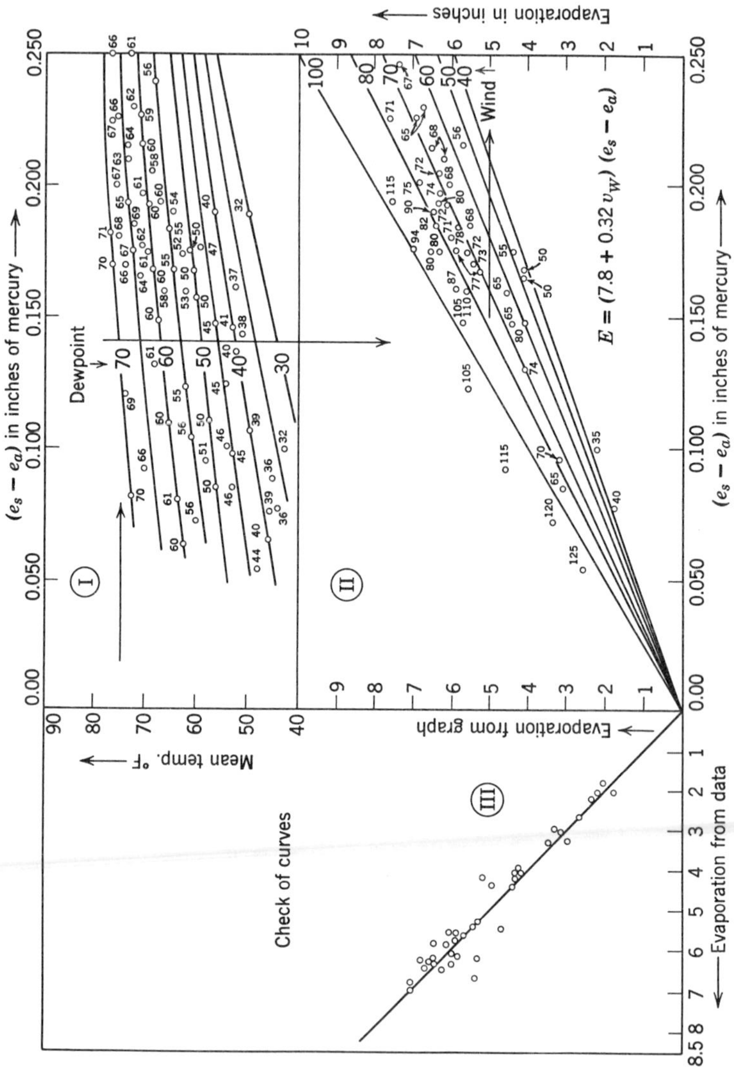

Fig. 2-21 Coaxial relation for monthly evaporation as a function of mean temperature, dew point, and wind velocity.

of the data fitting may be checked by plotting in quadrant III the evaporation values resulting from the curves for mean temperature, dewpoint, and wind velocity as given by the data versus the evaporation values given by the data. For a good fitting, the points in quadrant III should be evenly distributed about the bisector of the angle between abscissa axis and ordinate axis, a perfect fitting for a particular set of data giving a point exactly on that line. In Fig. 2.21, a fairly accurate result was obtained on the first trial to draw the curves. If this were not the case, the distribution of points in quadrant III should indicate how to correct the position of the curves in quadrants II and III. To determine the equation which would give the same results as the graph, let, in an analogy with Eq. 2.10,

$$E = (a + bv_w)(e_s - e_a) \tag{2.12}$$

in which E is the monthly evaporation in inches, v_w is the wind velocity in miles per hour measured at the station, e_s, e_a are as defined before and are expressed here in inches of mercury, and a and b are two constants which are determined by picking out two of the six curves which have been plotted in quadrant II, i.e. the curves for $v_w = 60$ mph and 80 mph.

$$E = 7 \text{ in. for } e_s - e_a = 0.25 \text{ in. Hg and } v_w = 60 \text{ mph}$$
$$E = 8.6 \text{ in. for } e_s - e_a = 0.25 \text{ in. Hg and } v_w = 80 \text{ mph}$$

Insert in Eq. 2.12 and solve for a and b. Finally, the equation becomes

$$E = (7.8 + 0.32v_w)(e_s - e_a) \tag{2.13}$$

2.11 *Transpiration*

Transpiration, the process by which water from plants is discharged into the atmosphere as vapor, depends essentially on the same factors as those which control evaporation, namely, air temperature, wind velocity, and solar radiation. Transpiration also varies with the species and density of plants and to a certain extent with the moisture content of the soil, in that a certain minimum amount of water must be available to the plant roots. This amount is determined by the wilting coefficient, a gravimetric moisture content (see section 3.5) below which plants wilt and do not recover again in a humid atmosphere. Wilting coefficients vary from 0.9% in coarse sand to over 16% in clay loam [19]. Moisture in excess of the specific retention of a soil (see section 3.7) is often detrimental, but in the range from wilting coefficient to specific retention, transpiration is relatively independent of soil moisture. Most plants do not grow in saturated soil or open water.

Transpiration is measured in the laboratory by means of pots or vessels in which plants are grown while a wax seal prevents evaporation from the

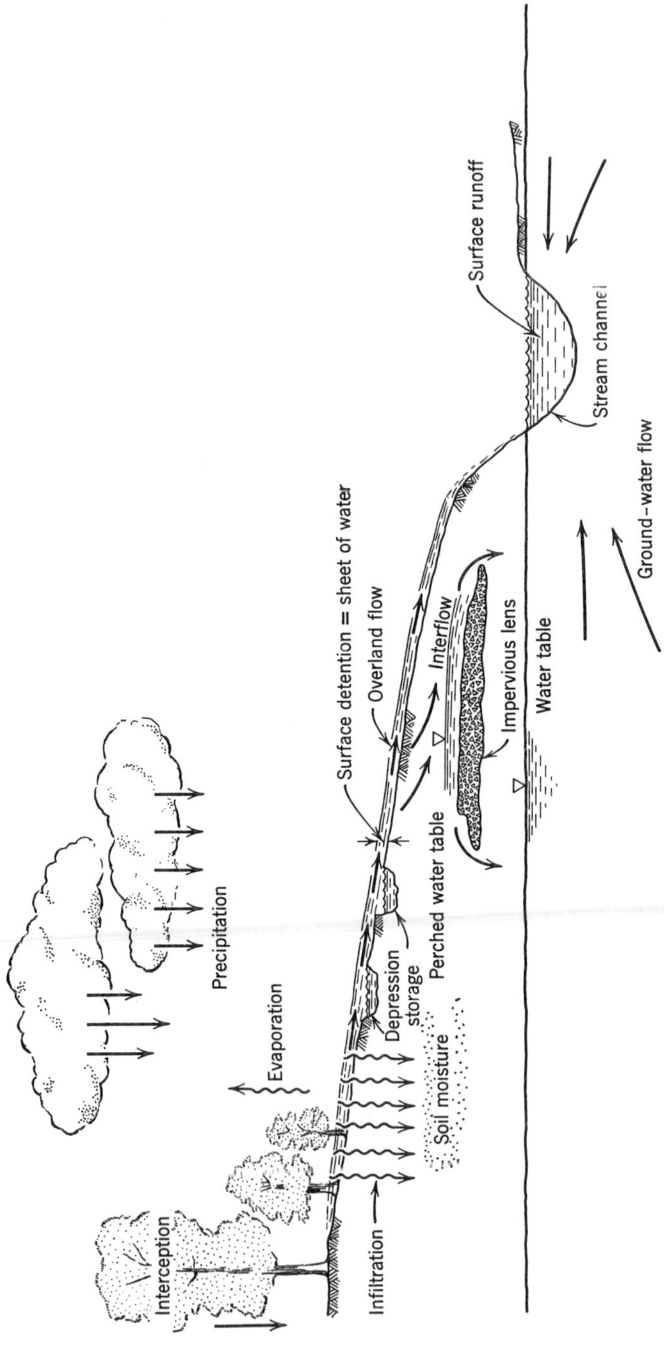

Fig. 2-22 Simple picture of runoff cycle.

ground surface. The change in weight due to loss of transpired water is measured. As in the case of evaporation pans, a coefficient must be applied to convert laboratory data into field data. The use of pots is limited to relatively shallow-rooted plants.

Measurements of evapo-transpiration, or consumptive use, i.e., the total amount of water absorbed by vegetation for transpiration or building of plant tissue, plus evaporation from the soil, are made in deep tanks. The quantity of water required to maintain constant moisture conditions in the tank during plant growth is a measure of the consumptive use.

◆ Runoff

The term runoff is usually considered synonymous with streamflow and is the sum of surface runoff and ground-water flow that reaches the streams. Surface runoff equals precipitation minus surface retention and infiltration. Infiltration is the passage or movement of water through the surface of the soil and is to be distinguished from ground-water flow. For the study of precipitation-runoff relationships which are of vital importance in many hydrologic projects, a clear picture of the runoff cycle is of great help.

2.12 *Runoff Cycle*

The runoff cycle describes the distribution of water and the path it follows after it precipitates on land until it reaches stream channels or returns directly to the atmosphere through evapo-transpiration. The relative magnitude of the various components into which the total amount of precipitation of a given storm may be broken down depends on the physical features and conditions, natural and manmade, of the land as well as on the characteristics of the storm. At the beginning of a storm, a large amount of precipitation is intercepted by trees and vegetation. Water stored on vegetation is usually well exposed to wind and offers large areas to evaporation. Precipitation from storms of mild intensity and short duration may be entirely depleted by interception and by the small amount of water that infiltrates through the soil surface and fills puddles and surface depressions. For water to infiltrate, the soil surface must be in the proper conditions. When the available interception and depression storage are completely exhausted and when the rainfall intensity at the soil surface exceeds the infiltration capacity of the soil, overland flow begins (Fig. 2.22). The soil surface is then covered with a thin sheet of water called surface detention. Once the overland flow reaches a stream channel it is called surface runoff.

Part of the water that infiltrates into the soil will continue to flow laterally at shallow depths as interflow owing to the presence of relatively impervious lenses just below the soil surface and will reach the stream channel in this capacity. Another part will percolate to the ground-water table and eventually will reach the stream channel to become the base flow of the stream. A third part will remain above the water table in the zone of unsaturated flow (see section 3.7).

The contribution to streamflow from a storm of moderate intensity and essentially constant in time may be visualized by means of Fig. 2.23. A portion of the rain falls directly into the stream channel and is indicated as channel precipitation. Initially almost all of the rain is collected on the

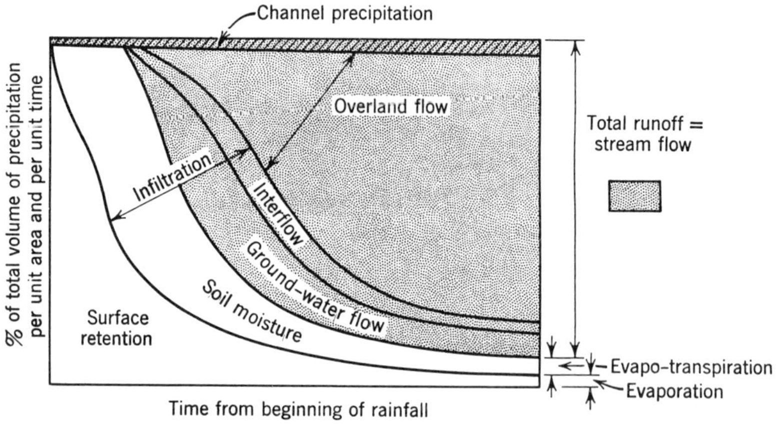

Fig. 2-23 Decomposition of a storm of uniform intensity.

earth as surface retention, the sum of interception, depression storage, and evaporation. As time goes on, the storage capacity of foliage and depressions becomes more saturated and more water infiltrates into the soil. Finally overland flow occurs and becomes surface runoff. After a sufficiently long time, surface retention approaches a steady state value, consisting exclusively of evaporation. Soil moisture levels off to evapotranspiration via plant roots and thermal gradients just below the soil surface, and ground-water flow becomes a more or less constant base flow. The phenomenon of infiltration deserves special attention because it may eventually lead to a more effective method of estimating runoff from rainfall than is the use of runoff coefficients.

INFILTRATION

Horton [20] introduced the concept of infiltration in the hydrologic cycle and defined infiltration capacity f_p as the maximum rate at which a given soil can absorb precipitation in a given condition He suggested that infiltration capacity would decrease exponentially in time from a maximum initial value to a constant rate. The actual rate of infiltration f_i is always smaller than f_p except when the rainfall intensity i equals or exceeds f_p. It also decreases exponentially with time as the soil becomes saturated and as its clay particles swell. The rate of infiltration depends upon such factors as condition of the soil surface, density of vegetation, temperature, chemical composition of water reaching the soil, physical properties of the soil (porosity, grain-size distribution, cohesion, etc.), intensity of the rainfall (high-intensity rainfalls causing compaction of the soil).

The infiltration approach to runoff is based on the use of infiltration indices such as the Φ-index, which is defined in Fig. 2.24 as the average rainfall intensity above which the observed mass of runoff equals the mass

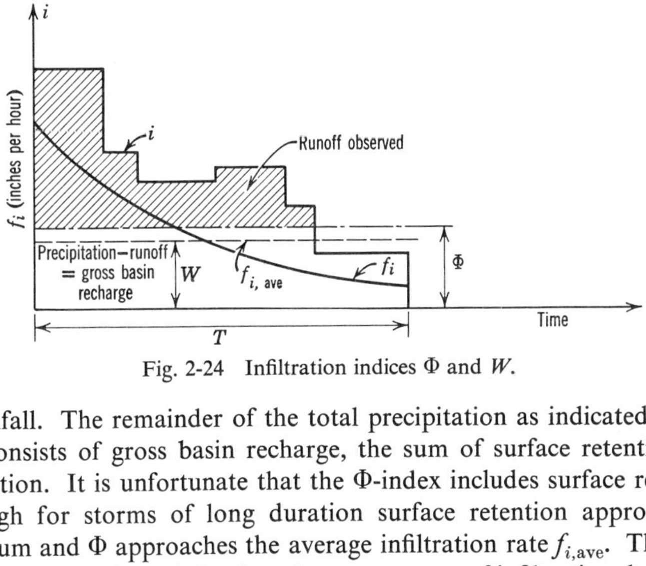

Fig. 2-24 Infiltration indices Φ and *W*.

of rainfall. The remainder of the total precipitation as indicated in Fig. 2.24 consists of gross basin recharge, the sum of surface retention and infiltration. It is unfortunate that the Φ-index includes surface retention although for storms of long duration surface retention approaches a minimum and Φ approaches the average infiltration rate $f_{i,\text{ave}}$. Therefore the *W*-index has been defined as the average rate of infiltration during the time the rainfall intensity exceeds the infiltration capacity,

$$W = \frac{F_i}{T} = \frac{1}{T}(P - Q_s - S_e) \tag{2.14}$$

in which F_i is the total amount of infiltration, T is the time during which the rainfall intensity exceeds infiltration capacity, P is the precipitation, Q_s is the observed surface runoff from the storm, and S_e is the total surface retention.

Infiltration data may be used with some success to estimate maximum flood flows which consist almost entirely of surface runoff. Such flows generally occur when the initial moisture condition of the soil is quite uniform and therefore when the infiltration capacity curve is quite stable. For a simple storm distribution of continuous rainfall with intensity at all times exceeding the infiltration capacity as shown in Fig. 2.25, the surface runoff may be determined from rainfall and infiltration data as follows. The W-index is estimated and an infiltration curve is superimposed on the rainfall plot. The surface retention is estimated and added to the infiltration so that the surface runoff is represented by the area between the precipitation curve and the infiltration plus surface retention curve.

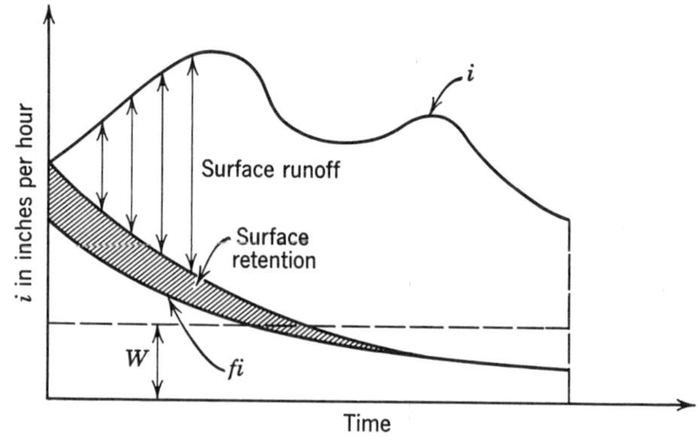

Fig. 2-25 Estimation of surface runoff from precipitation and *W*-index.

2.13 *Streamflow* [52]

The study of runoff necessarily includes analysis of streamflow records. It is therefore of interest to know how streamflow is measured and how existing stage-discharge relationships may be extended.

Streamflow is measured by recording the stage or elevation of the water surface at a given station and by entering the recorded value in a stage-discharge relationship, called a rating curve. A rating curve is established by measuring with a current meter the actual average flow velocity in the cross section at the station for a number of flow conditions at different stages, computing the discharge and plotting the data, stage as ordinate, discharge as abscissa. Figure 2.26 shows how the discharge computations are carried out. Because the velocity varies from the water surface to the river bed, as indicated in Fig. 2.26, the average velocity in any vertical is obtained by measurements made at 0.6 of the depth (from the water surface) for shallow water and by the average of measurements at 0.2 and 0.8 depth in deep water. Gaging stations have been established on all principal streams by the U.S. Geological Survey and by state agencies. River stage may be measured by simple staff gages, with scales graduated in feet and tenths. These gages consist of three sections which measure low, medium, and high flows, or are inclined along the riverbank to improve the reading of the scale. Zero datum for river stage is arbitrary. Sometimes it is set at mean sea level, but usually it is set below the lowest flow on record. A staff gage is inexpensive and easy to maintain but it requires frequent

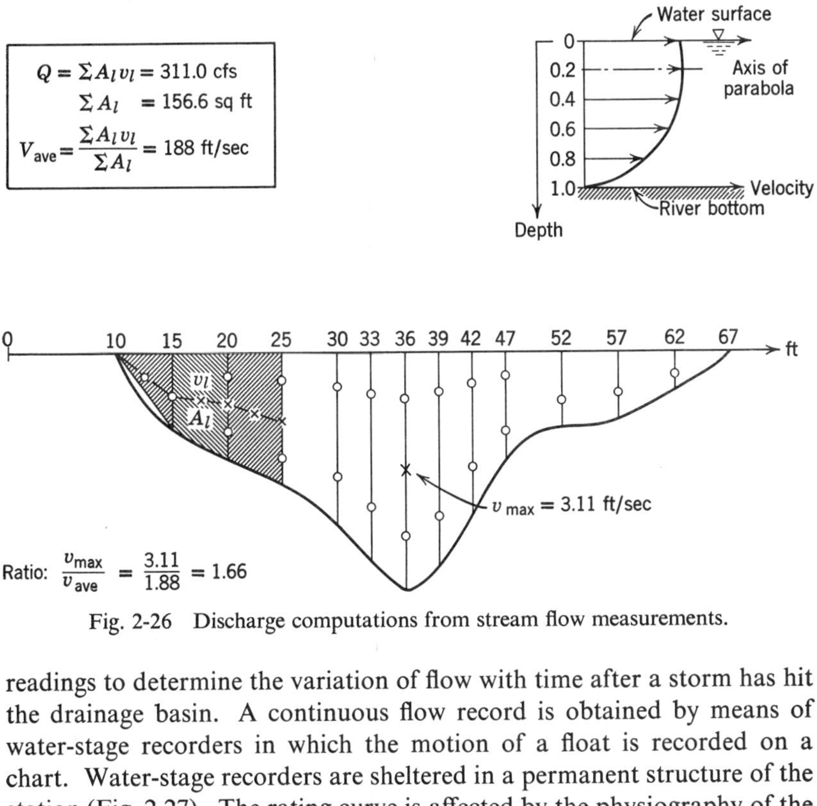

Fig. 2-26 Discharge computations from stream flow measurements.

readings to determine the variation of flow with time after a storm has hit the drainage basin. A continuous flow record is obtained by means of water-stage recorders in which the motion of a float is recorded on a chart. Water-stage recorders are sheltered in a permanent structure of the station (Fig. 2.27). The rating curve is affected by the physiography of the channel downstream from the gage. A section control prevails for the gage if the affecting features are located in a short reach of channel. A channel control is said to be effective for the station when the rating curve is affected by the channel characteristics over a considerable distance. Scour and sedimentation alter the channel characteristics so that a channel control changes with time and requires frequent recalibration of the rating curve. The station of Fig. 2.27 has a permanent artificial low-water control consisting of a concrete weir with a shallow V-notch, and at high floods the bridge contraction controls.

RATING CURVES

When the control is permanent and when the slope of the energy line at the station is reasonably constant for all occurrences of a given stage, a simple stage-discharge relation is sufficient (Fig. 2.28). Sometimes, however, the relation between stage and discharge is not uniquely determined, as in the rise and fall of a flood wave, or when back water effects are felt in the station, and the discharge will also depend on the slope of the energy

Fig. 2-27 Stony Brook gaging station in Princeton. (Courtesy of USGS, Trenton Branch.)

line, as shown in Fig. 2.29. Because the velocity head terms are small, the slope of the energy line is approximately the same as that of the water surface (Fig. 2.29), which is easier to measure. A relation for the discharge based upon the slope may now be derived with the help of Mannings formula [23].

$$Q = A\frac{1.49}{n}R^{2/3}S^{1/2} \tag{2.15}$$

in which Q is the discharge in cubic feet per second, A is the cross-sectional area of flow in square feet, R is the hydraulic radius in feet, the ratio of A and the wetted perimeter (or contact line of A with the river bed and banks), S is the slope, and n is Manning's roughness coefficient. The ratio for any two discharges Q and Q_o at a given station corresponding to the same stage but for different slopes S and S_0 is then

$$\frac{Q}{Q_0} = \left(\frac{S}{S_0}\right)^{1/2} \tag{2.16}$$

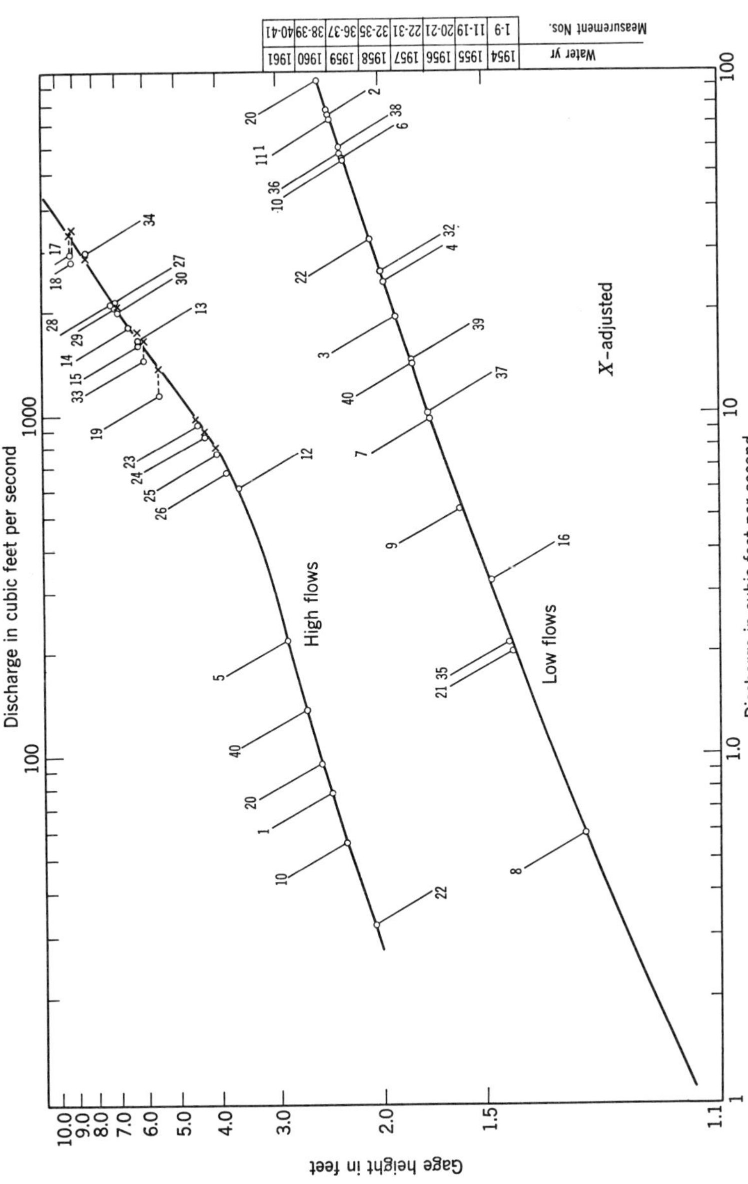

Fig. 2-28 Rating curves for Stony Brook at Princeton, N.J. Drainage Area: 44.0 sq mi. (Courtesy of USGS, Trenton Branch)

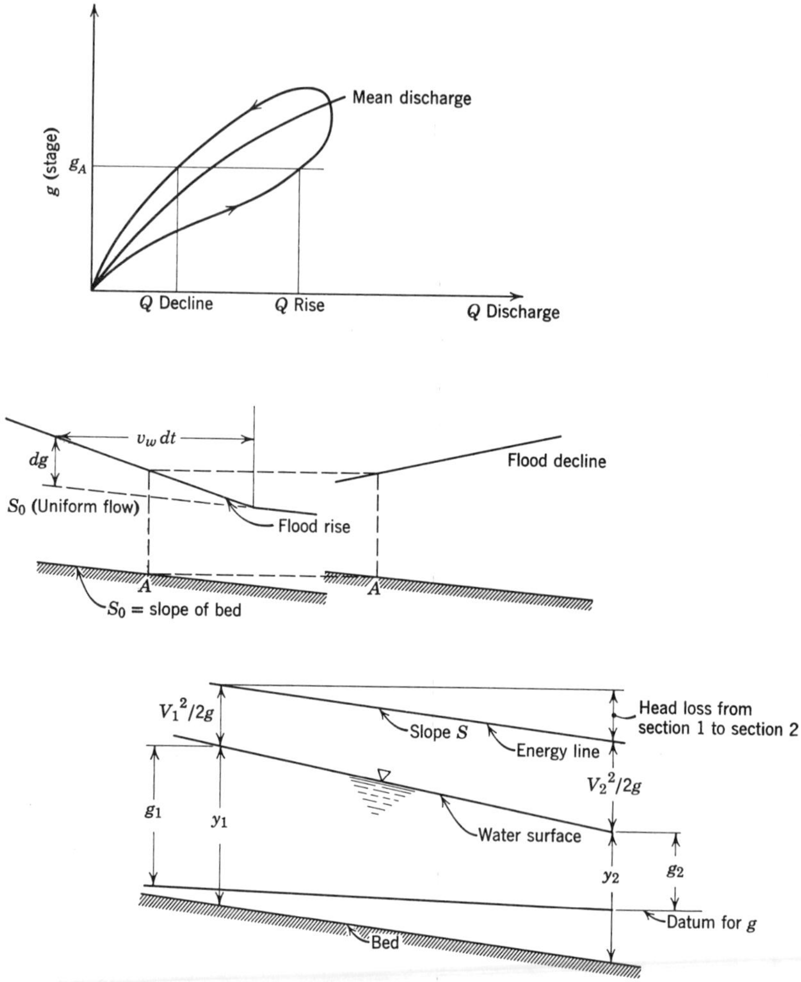

Fig. 2-29 Stage, discharge, and slope of a river.

If the slope of the energy line is replaced as a first approximation by F/L, in which F is the fall of the water surface between the main (base) gaging station and an auxiliary gage a distance L downstream, Eq. 2.16 must be replaced by

$$\frac{Q}{Q_0} = \left(\frac{F}{F_0}\right)^m \tag{2.17}$$

in which the exponent m would still have a magnitude about $\frac{1}{2}$.

A constant-fall rating curve may now be developed as follows. Assume that all Q_0 correspond to a constant fall F_0, for convenience taken as 1 ft. We can make a series of current meter measurements with corresponding stages at the gaging station and at the auxiliary gage downstream, and plot (Fig. 2.30) the base stage versus the corresponding discharge, noting the value of F/F_0 beside each point. The data will be divided into two fields, $F/F_0 > 1$ and $F/F_0 < 1$, so that the curve $F/F_0 = 1$ may easily be fitted as the boundary between the two fields. It is now possible for every base stage to find Q_0 corresponding to Q as indicated in Fig. 2.30. The

Table 2.9 Constant Fall Rating Curve

Base Stage, ft	Fall, ft	Q (by Eq. 2.17), cfs	Q (by adjustment)
34	1.5	130,000	135,000
34	0.5	68,700	66,000
26	0.8	46,800	47,200

data of Eq. 2.17 are now plotted on log-log paper (Fig. 2.31), and the exponent m is determined as the slope of the straight line fitted to the data. Once m is determined, Eq. 2.17 and the constant fall rating curve may be used to determine any Q with the help of simultaneous stage measurements at the base gage and at the auxiliary gage. Q may also be determined after an adjustment curve is constructed as in Fig. 2.30, in which values of Q/Q_0 are plotted versus values of F/F_0. For a given base stage, the constant fall rating curve gives Q_0, and the adjustment curve gives Q/Q_0 with the help of the observed fall. The two methods give close results as is illustrated for Fig. 2.30 by Table 2.9.

A normal-fall rating curve is developed when the range of the fall is large and when the normal fall is known as a function of the base gage. Let Q_n be the normal discharge corresponding to the normal fall F_n for a given base stage. A series of simultaneous measurements of Q and F are made, and the base stages are plotted versus the corresponding discharges, with the value of F/F_n noted beside each point. A normal fall rating curve is then drawn and an adjustment curve is constructed in the same way as before. With the help of these two curves, Q is determined for any given base stage and fall.

When a station is not affected by back water but by a rapidly varying stage due to a flood wave [21] (Fig. 2.29), the slope of the surface is not S_0 (uniform flow) but $S_0 + dg/(v_w\, dt)$, in which dg is the change in stage during a given period of time dt, and v_w is the velocity of the flood wave. A rating curve is first established for a series of discharges Q_0 without flood

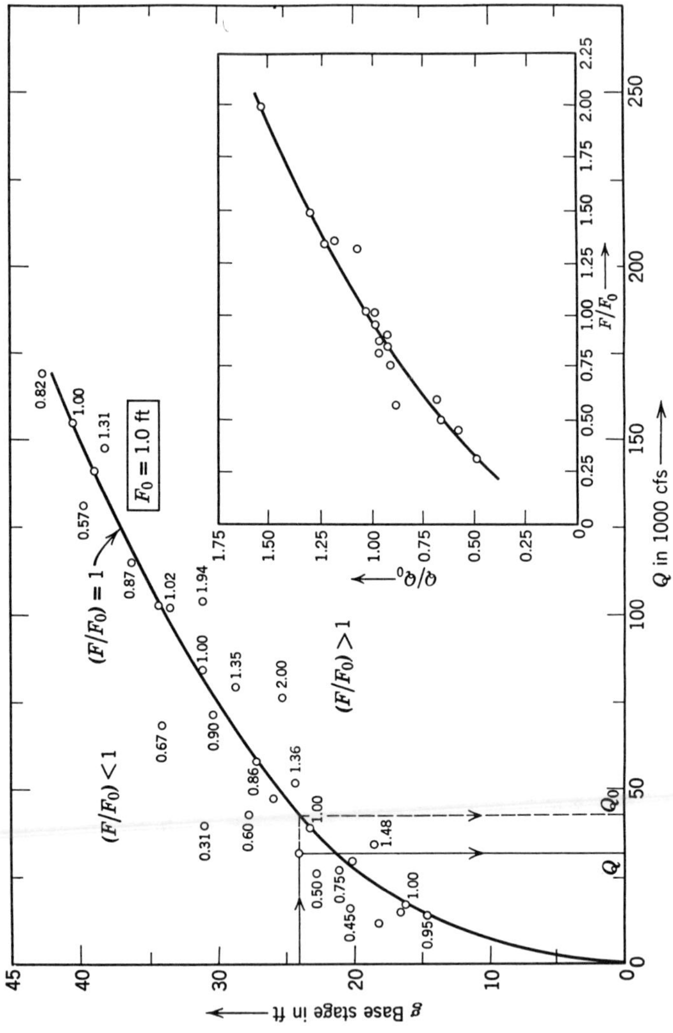

Fig. 2-30 Constant-fall rating curve. Insert is adjustment curve.

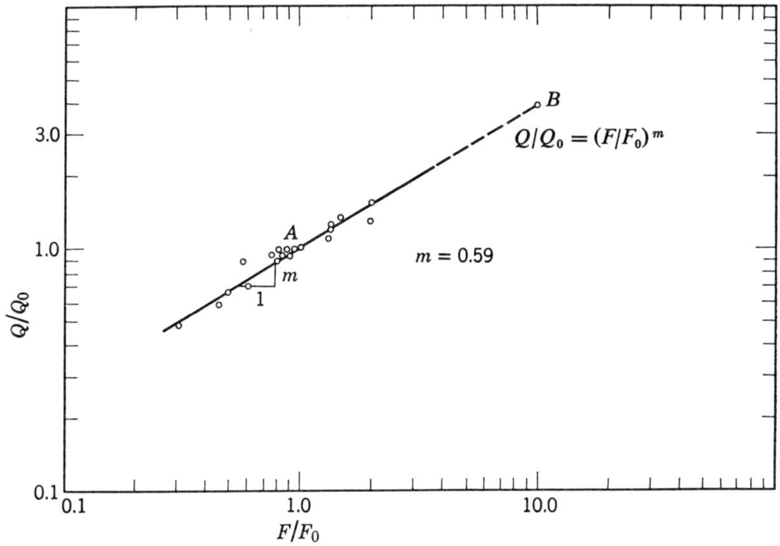

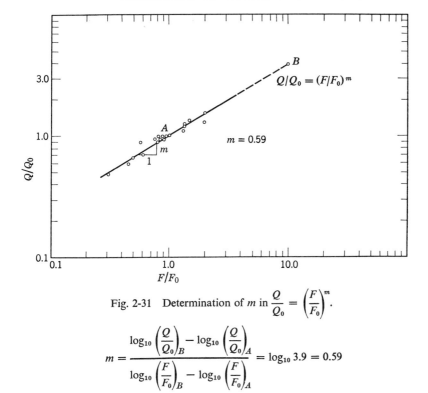

Fig. 2-31 Determination of m in $\dfrac{Q}{Q_0} = \left(\dfrac{F}{F_0}\right)^m$.

$$m = \frac{\log_{10}\left(\dfrac{Q}{Q_0}\right)_B - \log_{10}\left(\dfrac{Q}{Q_0}\right)_A}{\log_{10}\left(\dfrac{F}{F_0}\right)_B - \log_{10}\left(\dfrac{F}{F_0}\right)_A} = \log_{10} 3.9 = 0.59$$

wave effect ($dg/dt = 0$), to which a correction curve Q/Q_0 versus dg/dt is added.

EXTENSION OF RATING CURVES

Occasionally flood stages occur in the range where no flow measurements were made and the existing rating curve must be extended. The best procedure is to assume an equation for the rating curve of the form

$$Q = a(g - z)^b \tag{2.18}$$

which plots as a straight line on log-log paper (Fig. 2.32) provided a and b are constants, z is the stage value for zero flow, known approximately, and may be determined by trial and error from the graph, g is the stage, and Q is the discharge.

Another method, devised by Stevens [22], is based on Chezy's formula [23]. This formula is

$$Q = ACR^{1/2}S^{1/2} \tag{2.19}$$

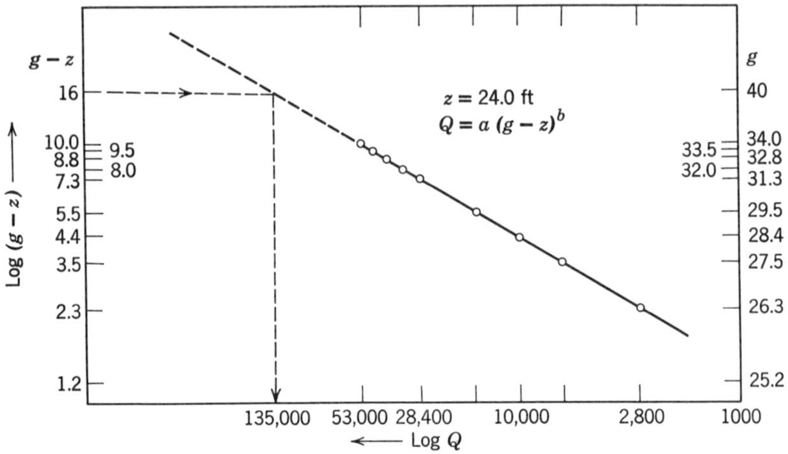

Fig. 2-32 Logarithmic extension of a rating curve.

in which Q is the discharge in cubic feet per second, A is the cross sectional area of flow in square feet, R is the hydraulic radius in feet, S is the slope of the energy line, and C is Chezy's roughness coefficient. If $S^{1/2}$ is assumed to be constant for the station and if the mean depth D_m is substituted for R, Eq. 2.19 becomes

$$Q = kA\sqrt{D_m} \tag{2.20}$$

in which k is a constant. Measured values of Q versus $A\sqrt{D_m}$ plot very closely to a straight line (Fig. 2.33). On the other hand, values of $A\sqrt{D_m}$ may be computed as a function of g and plotted on the same figure. It is then easy to eliminate $A\sqrt{D_m}$ graphically between its dependent variables Q and g. This method gives reliable results only when area and top width of the flow cross section are measured closely enough to the stage for which Q has to be extrapolated. In Fig. 2.33, where the data are missing between points P_1 and P_2, the extrapolation is quite uncertain. For a base stage of 40 ft, the logarithmic extension gives $Q = 135{,}000$ cfs as compared to $Q = 121{,}000$ cfs by Stevens' method.

SOURCES AND INTERPRETATION OF RUNOFF DATA

The U.S. Geological Survey publishes its annual Water Supply Papers in fourteen parts for various basins in the United States. It also cooperates with the Department of Water Resources (State of California) or with the Department of Conservation and Economic Development (State of New Jersey) in the publication of special reports covering the surface water supply of the various states [24]. Other sources of information are the U.S. Army Corps of Engineers, TVA, and local flood-control districts.

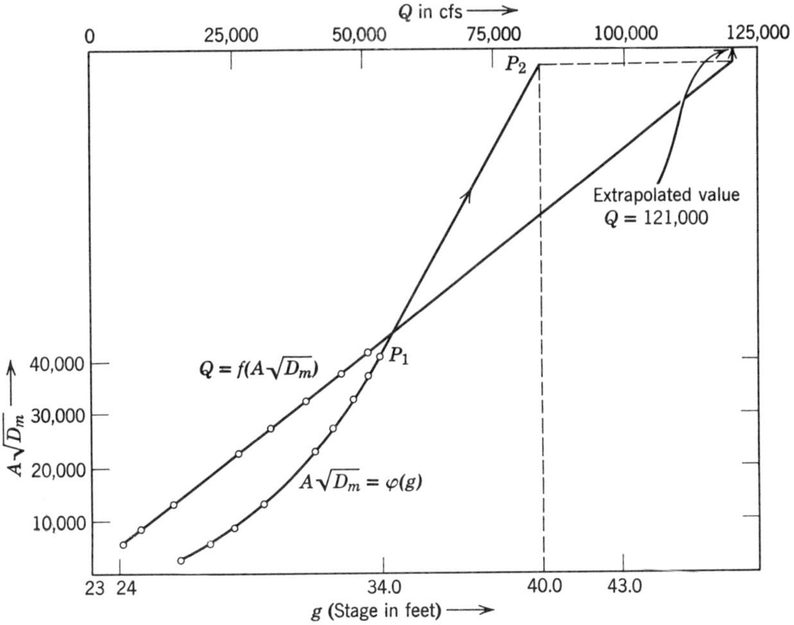

Fig. 2-33 Stevens' method for extension of a rating curve.

The U.S. Geological Survey has defined the water year as the period from October 1 to September 30, in order to have the entire high water period in one and the same year. The water year is designated by the calendar year in which it ends. Streamflow is measured in cubic feet per second (cfs), also called second-foot, and cu sec in its abbreviated form. Flow volume is usually expressed in acre-foot (quantity of water required to cover an acre to the depth of 1 ft) or in second-foot-day (volume generated by a flowrate of 1 cfs during one day). One second-foot-day (SFD) contains 86,400 cu ft whereas one acre-foot (AF) is equivalent to 43,560 cu ft, so that roughly 1 SFD = 2 AF. The measurement cubic feet per second per square mile (csm) is used less. CSM is defined as the average number of cubic feet of water flowing per second from each square mile of area drained, assuming that the runoff is distributed uniformly in time and area. The runoff in inches (in.) is the depth at which an area would be covered if all the water draining from it in a given period were uniformly distributed on its surface. The term is useful in comparing runoff with rainfall and evaporation data, and must be associated with a specific drainage basin. In the metric system, streamflow is expressed in cubic meters per second (1 m^3/sec = 35 cfs) and flow volume in cubic meters times a power of ten.

A typical USGS Water Supply Paper (WSP) gives the mean daily discharge in cfs and the monthly mean in cfs (WSP* 1,285) or in SFD per month (WSP* 1,181). The monthly runoffs, converted in inches, and other data given by the Water Supply Papers for the Arroyo Seco River near Soledad, California, are represented in Fig. 2.34.

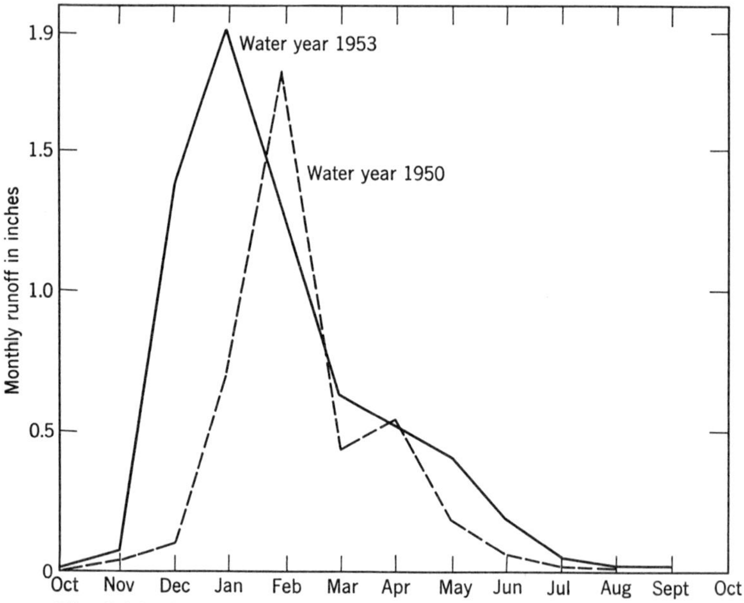

Fig. 2-34 Sample data from *U.S. Geological Survey Water Supply Papers.* Arroyo Seco near Soledad, Calif. Drainage area: 241 sq mi; Period of record: Nov. 1901–Sept. 1953; Mean annual discharge: 170 cfs; Maximum discharge of record: 22,000 cfs; Mean annual runoff: 9.55 in. (From *U.S. Geological Survey Water Supply Papers* 1181 and 1285.)

Variations in streamflow depend mainly on geographical and seasonal conditions in the United States. The mean annual runoff in the humid East is 10 to 30 in. per year, or 40 to 70% of the annual rainfall. The mean annual streamflows of eastern streams exceed 1.0 csm. On the other hand, the mean annual runoff in the arid Southwest is 0.25 to 10 in. per year, or 1 to 40% of the annual precipitation, while the mean annual streamflow in cfs per square mile is very low, often less than 0.1 csm. In the Great Plains the mean annual streamflow varies from less than 0.1 csm (Dakotas) to more than 1.0 csm (Arkansas). Seasonal variations depend primarily upon rainfall distribution and melting snows. Snow-fed streams have their peak flow during May and June, while rain-fed streams have peaks at times of greatest rainfall.

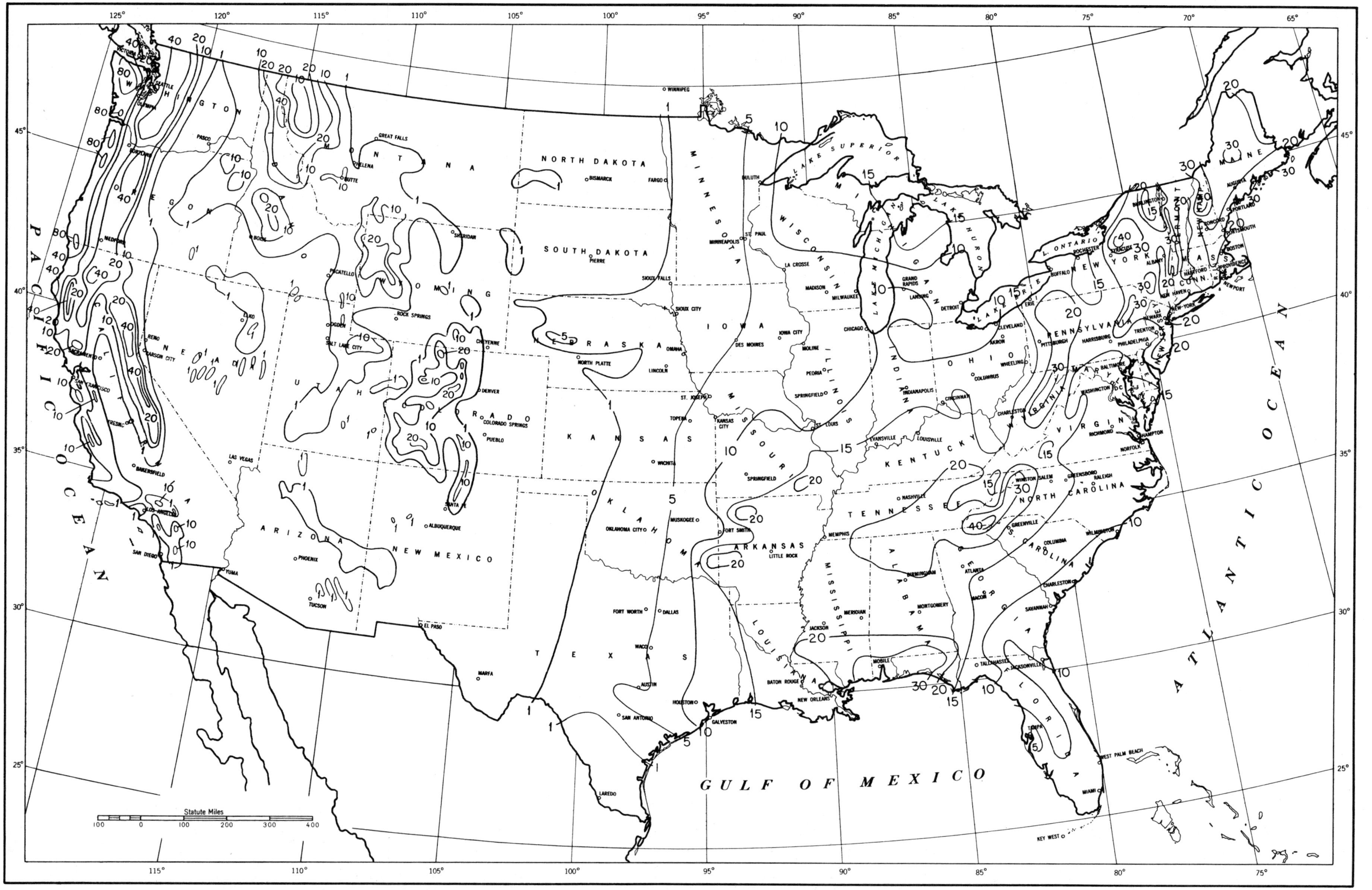

Fig. 2-35 Average annual runoff for the United States. (After U.S. Geological Survey.) Lines show average annual runoff in inches. The 5-, 15-, and 30-in. runoff lines have been omitted in western states to prevent crowding of map detail. (Courtesy of Water Information Center, Inc.)

The mean annual runoff in the United States is given in Fig. 2.35.

2.14 *Hydrographs*

The hydrograph is a plot of discharge from a hydraulic or hydrologic unit or system such as a river, drainage basin, or runoff component of storm versus time. The duration for which the distribution of flow is sought varies from a few hours for a storm hydrograph to a year for the hydrograph of annual flow of a river. The study of the hydrograph of a river before, during, and after its drainage basin is hit by a storm is particularly helpful in determining the amount of precipitation that reaches the river as direct runoff (surface runoff or storm runoff). The same analysis of a series of storm of various durations and under different conditions is therefore a first step in establishing a precipitation-runoff relation for a given basin.

HYDROGRAPH COMPOSITION

When the drainage basin of a perennial stream is hit by a storm in the dry season, and when the river discharge decreases in time as indicated in Fig. 2.36, the hydrograph of the stream will be disturbed from its smoothly leveling off curve and may assume various shapes according to the relative magnitude of rainfall intensity, rate of infiltration, volume of infiltrated water, soil moisture deficiency, rainfall duration, and other characteristics of the storm and the basin. The most relevant parameters and their influence on the four components of runoff (surface runoff, interflow, ground-water flow, and channel precipitation) are briefly compared here. Four different hydrograph pictures result from the following possibilities.

CASE 1

Rainfall intensity $i <$ rate of infiltration f_i

Volume of infiltrated water $F_i <$ soil moisture deficiency

In order to have contributions from interflow and ground-water flow due to the storm, F_i must be larger than the soil moisture deficiency. Because of our definition of field capacity (see section 3.7), soil moisture deficiency here is the volume of water required to bring the moisture content of the soil up to field capacity.

There can be surface runoff only when $i > f_i$, and therefore the only addition to stream flow is due to channel precipitation, which leads to a slight increase of discharge Q with time over the prolonged curve (Fig. 2.37*a*).

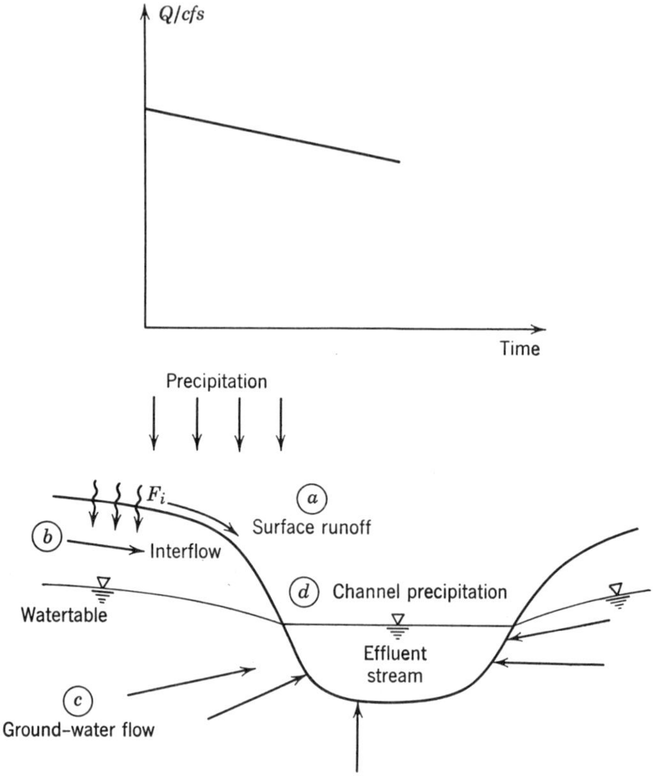

Fig. 2-36 Hydrograph of effluent stream before storm. Runoff contributions after storm.

CASE 2

Rainfall intensity $i <$ rate of infiltration f_i

Volume of infiltrated water $F_i >$ soil moisture deficiency

After the moisture content of the soil reaches field capacity, interflow and ground-water flow accretion due to the storm occur. Added to channel precipitation, these flow components give the hydrograph picture of Fig. 2.37*b*.

CASE 3

Rainfall intensity $i >$ rate of infiltration f_i

Volume of infiltrated water $F_i <$ soil moisture deficiency

In this case there are contributions from surface runoff and channel precipitation but no additional (i.e., due to the storm) ground-water flow on top of the existing base flow sustained by the ground-water basin of the river (the river is called effluent). The hydrograph picture is given in Fig. 2.37*c*.

CASE 4

Rainfall intensity $i >$ rate of infiltration f_i

Volume of infiltrated water $F_i >$ soil moisture deficiency

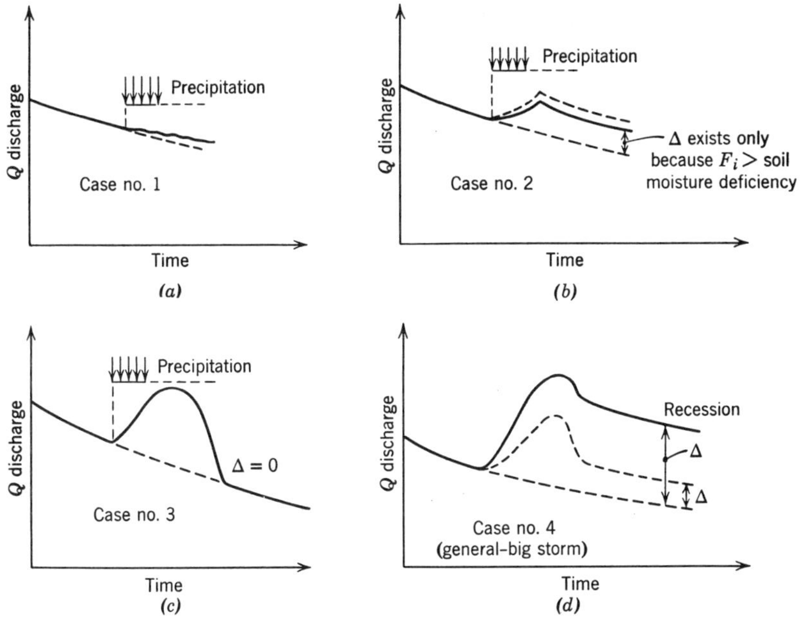

Fig. 2-37*a*, *b*, *c*, *d* Hydrograph composition.

This is what usually happens during a big storm. There is additional streamflow due to channel precipitation, surface runoff, interflow, and ground-water flow, although the contribution from ground-water flow may be negative when the river becomes influent and the ground-water flow is reversed in direction (Fig. 2.38). The hydrograph is given in Fig. 2.37*d*.

SEPARATION OF HYDROGRAPH COMPONENTS

An enlarged picture of Fig. 2.37*d* is given in Fig. 2.39 and shows the contributions of the various components: (*a*) surface runoff; (*b*) interflow; (*c*) ground-water flow; (*d*) channel precipitation. Channel precipitation ends, of course, with the rainfall. The curve indicating the ground-water component may assume a variety of positions between those of Fig. 2.38 and Fig. 2.39. In these extreme cases the surface runoff contribution differs significantly in magnitude for the same total hydrograph.

In practice the problem is not to compose a hydrograph, because the hydrograph is given by measurements in a gaging station, but is to separate its components. For reasons of simplicity, channel precipitation and interflow are included in surface runoff as a single item, designated as direct or

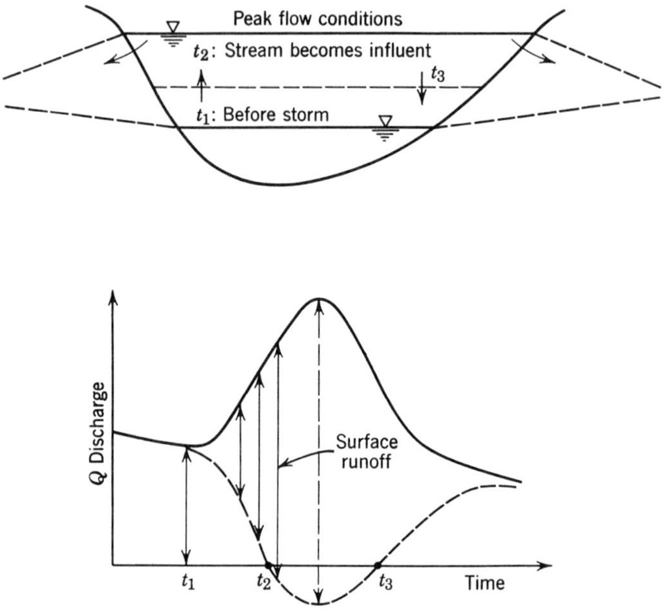

Fig. 2-38 Hydrograph for influent stream conditions.

storm runoff. The problem is to separate this item from the ground-water flow component, also called base flow. The parts of a hydrograph (Fig. 2.39) are commonly designated as rising limb, or concentration curve, crest segment, and recession or falling limb. The time base of the hydrograph may be obtained by drawing a horizontal line (Fig. 2.39) through point A where the rising limb starts and by finding its point of intersection B with the recession curve. This horizontal line may also be considered, as a first approximation, to be the boundary between direct runoff and base flow.

A more sophisticated method of separation, however, is based on an analysis of the recession curve. The recession curve for a given basin, as suggested by Barnes [25], may be represented by an equation which does

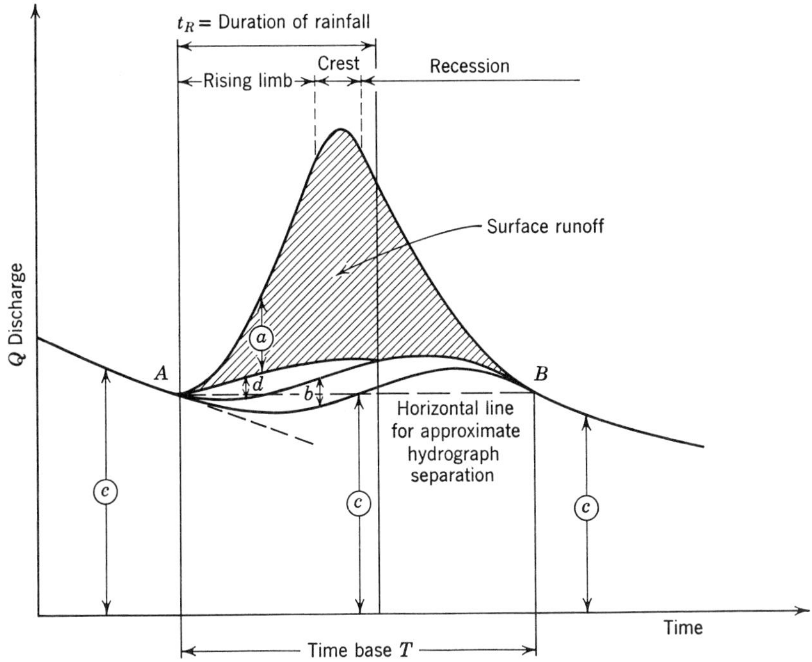

Fig. 2-39 Hydrograph parts and flow contributions.

not change in form for different storms, but only varies in the value of the recession constant K_r of

$$q_1 = q_0 K_r \tag{2.21}$$

or

$$q_t = q_0 K_r^{\,t} \tag{2.22}$$

in which t is the time between the occurrence of discharge q_0 and q_t. Variations in K_r are due to differences in areal rainfall distribution and in relative magnitude of the components of flow (direct runoff versus ground-water flow), each component contributing to the final shape of the overall recession curve. Langbein [26] devised a method of filtering out the ground-water component from the total recession. His technique is illustrated in Fig. 2.40, where the mean daily flows of one day, say q_n, are plotted versus those of the following day, say q_{n+1} (n assumes values from 0 to $t - 1$ in Eqs. 2.21 and 2.22). For high flows the data plot as a straight line, from which a constant K_r for the total flow recession may be determined according to Eq. 2.21. The value of K_r may be checked by plotting $\log q_t$ versus t. From Eq. 2.22 it follows that

$$\log q_t = \log q_0 + t \log K_r \tag{2.23}$$

which is the equation of a straight line for constant K_r.

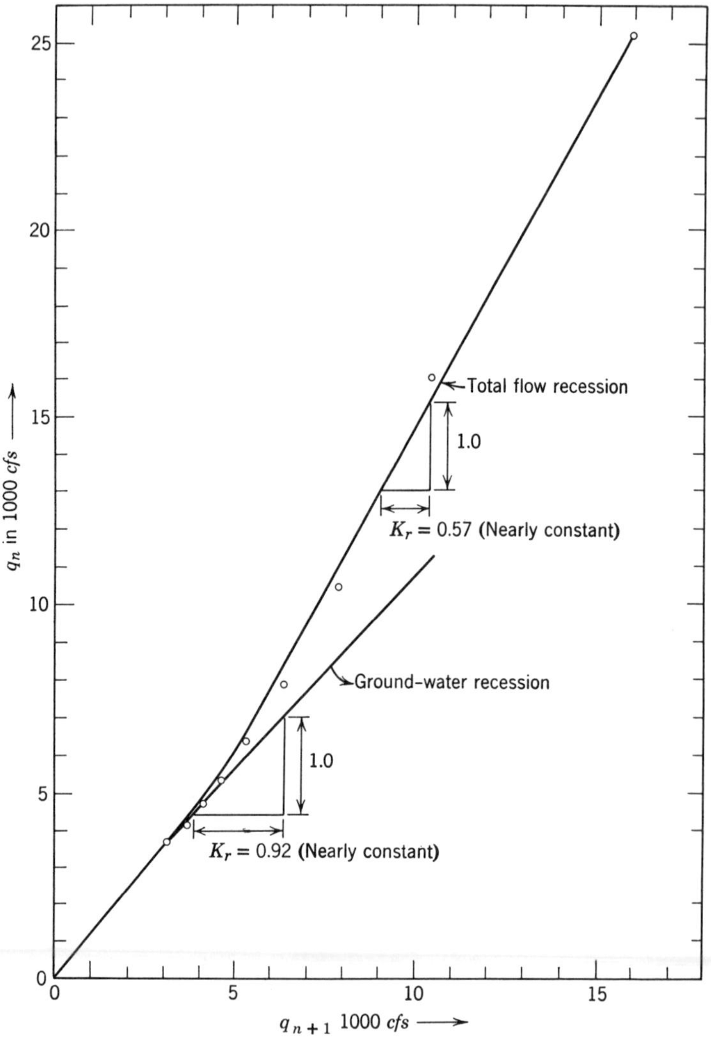

Fig. 2-40 Ground-water and total recession curves, Potomac River at Paw Paw, West Virginia. Flood of April 17, 1943.

For low flows, when the flood has clearly subsided, the q_n and q_{n+1} data again plot as a straight line from which the constant K_r, this time for the ground-water recession, may be determined. The value of the ground-water constant $K_r = 0.92$ has been used in Fig. 2.41 to construct the ground-water recession curve, prepared from data at the end of the flood when the recession is known to represent only ground-water flow. Successive earlier ordinates of the ground-water recession curve are computed

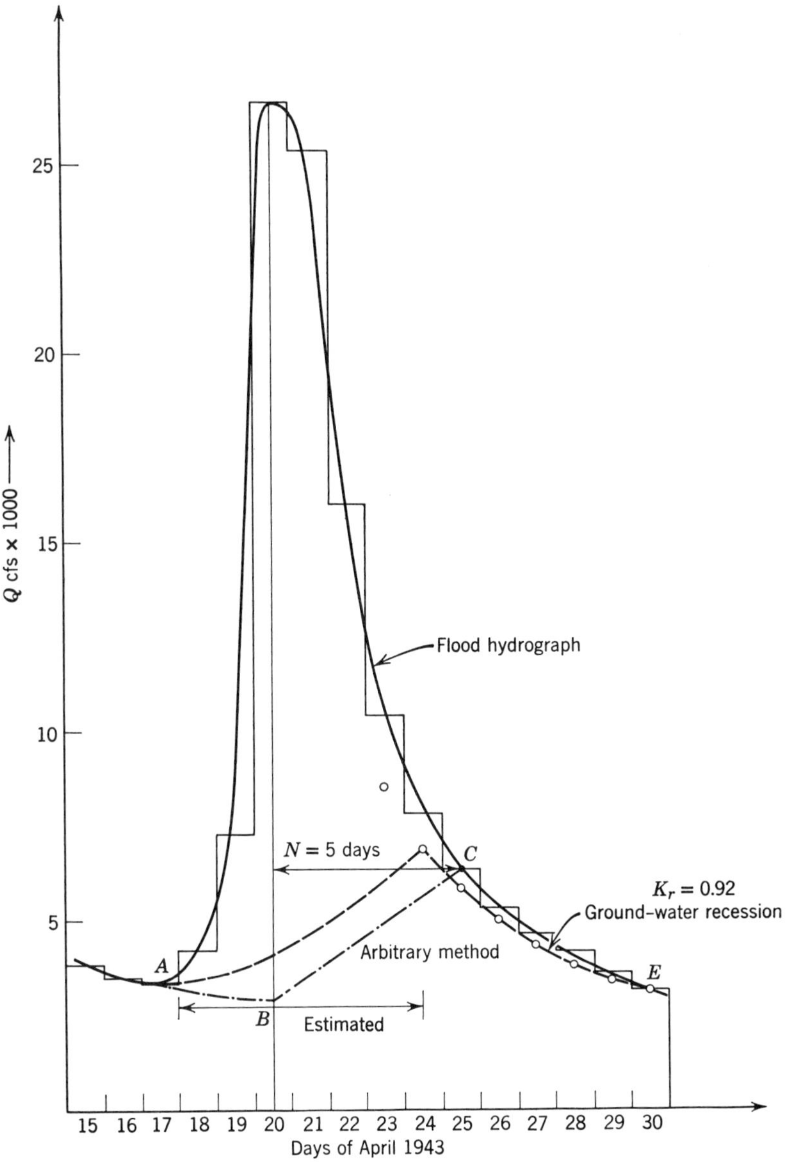

Fig. 2-41 Hydrograph separation, Potomac River at Paw Paw, West Virginia. Flood of April 17, 1943. Runoff from drainage area of 3,109 sq mi.

1. Arbitrary Method

SFD: 70.1×10^3

AF: $70.1 \times 1.98 \times 10^3 = 139 \times 10^3$

Runoff: $\dfrac{139 \text{ A ft} \times 12 \times 10^3}{3{,}109 \times 640 \text{ Acre}} = 0.84$ in.

2. Ground-water recession

SFD: 65.5×10^3

Runoff: 0.78 in.

starting from the point E of the flood hydrograph of Fig. 2.41, taken as the end of direct runoff. The time of the peak of the ground-water recession curve and the shape of the rising limb are selected arbitrarily. Although this method is classified as analytical, it is only semianalytical.

The following method is often used as a variant to the simple method of separating direct runoff and base flow by a horizontal line through the starting point of the rising limb, leading to a very long time base for the direct runoff hydrograph. The dry weather recession curve (Fig. 2.41) is extended from A to B at the time of peak flow. From this time, a value of N (days) is laid out to give the point C on the hydrograph, B and C are then joined by a straight line. This method is completely arbitrary, while for N a formula of the following form may be used:

$$N = (A_d)^{0.2} \tag{2.24}$$

in which N is the time in days and A_d is the drainage area in square miles [12].

From these considerations we can see that hydrograph separation is a rather arbitrary procedure, mainly because of the multitude of shapes which are acceptable for the ground-water hydrograph as a result of the possible combinations of effluent and influent conditions of the stream.

HYDROGRAPH SHAPE

The shape of the hydrograph, in particular the rising limb, crest segment, and early recession, are determined by surface runoff. This portion of the hydrograph may be identified with the direct runoff hydrograph, and its shape depends on such storm characteristics as the duration, areal distribution, intensity variation of rainfall, and also on the shape of the basin.

The time base of the hydrograph spans the period between the beginning of the rising limb and the time on the hydrograph when direct runoff is practically zero. Sherman [27] asserted that for a given basin the time bases of all floods caused by rainfalls of equal duration are the same, and he based the concept of his unit hydrograph (see section 2.16) on this assertion. The time base may be defined by the equation

$$\text{Time base} = t_R + t_c \tag{2.25}$$

in which t_R is the duration of the storm and t_c the time of concentration for the drainage basin. In this definition t_c is the time required for water to travel from the most remote portion of the basin to the outlet and may be estimated for a drainage channel by means of either Manning's or Chezy's formula (Eqs. 2.15 and 2.19). As a first approximation, if changes in the roughness of the channel and in other characteristics such as the area of flow and the hydraulic radius are neglected, the average velocity V is proportional to the square root of S, the slope of the energy line. The time of concentration t_c for the drainage basin of Fig. 2.14 may be computed by considering the stream channel profile of Fig. 2.42 and assuming that the time of travel from one point to the other along the profile is proportional to l/V or $l/\sqrt{S}$, in which l is the length of the path traveled by the water or, for small slopes of the profile, of its horizontal projection. S may then be approximated by $\Delta z/l$ as indicated in Fig. 2.42. If the travel

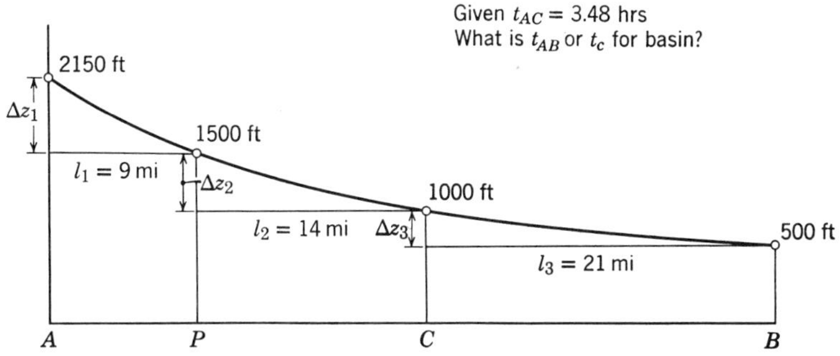

Fig. 2-42 Time of concentration t_c for a drainage basin (see Fig. 2-14).

time measured from A to C is 3 hr 29 min, the time of concentration may be computed from

$$t_c = t_{AC} \frac{\sum_1^3 \frac{l^{3/2}}{(\Delta z)^{1/2}}}{\sum_1^2 \frac{l^{3/2}}{(\Delta z)^{1/2}}} \tag{2.26}$$

With the numerical values designated in Fig. 2.42, the time of concentration for that particular basin happens to be 8 hours.

Meyer [25] defined the time of concentration as the time from the beginning of storm until the runoff becomes constant. Both definitions will give different results for the value of t_c. In this textbook Sherman's approach is used because Sherman's concept of unit hydrograph is adopted later in the text (see section 2.16).

The influence of nonuniform areal distribution of rainfall on the hydrograph is marked by a flat slope of the rising limb when most of the rainfall is concentrated in the region most remote from the basin outflow, and by a steep, rapid rise of the concentration curve when, most of the rainfall precipitates near the outflow (Fig. 2.43*a*, *b*). The shape of the basin for a uniform areal distribution of rainfall is reflected in the various hydrograph shapes of Fig. 2.43*c*, *d*, and *e*, while the double peak of Fig. 2.43*f* indicates a varying intensity. The effects of rainfall duration, distribution and intensity

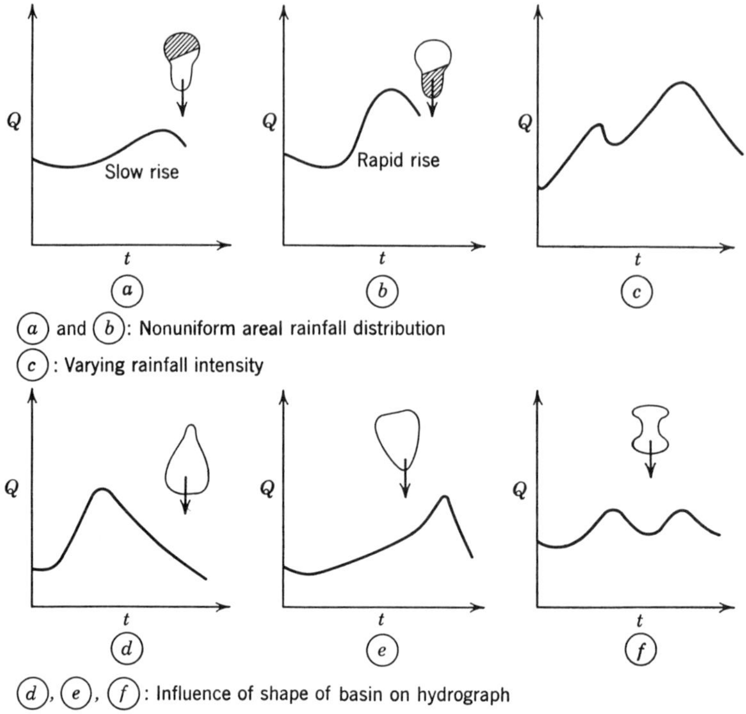

Fig. 2-43 Different hydrograph shapes.

variations on the shape of the hydrographs derived for an impermeable parking lot with dimensions and times of concentration as indicated, are shown in Fig. 2.44. The characteristics of the storms are:

A. Storms with different areal distribution

Storm 1. Storm of duration 15 min, intensity 2.0 in./hr in northern half of lot, intensity 1.0 in./hr in southern half of lot.

Storm 2. Similar to storm 1 but has reversed intensities, with the greater intensity in the southern half of the lot.

B. Storms of varying intensity in time

Storm 3. Uniformly distributed over the area—first 10 min falls at rate of 2.0 in./hr, second 10 min at rate of 1.0 in./hr.

Storm 4. Uniformly distributed over the area—first 10 min falls at rate of 1.0 in./hr, second 10 min at rate 2.0 in./hr.

C. Storms of varying duration

Storm 5. Uniformly distributed over the area with constant intensity of 1.0 in./hr lasting for 10 min.

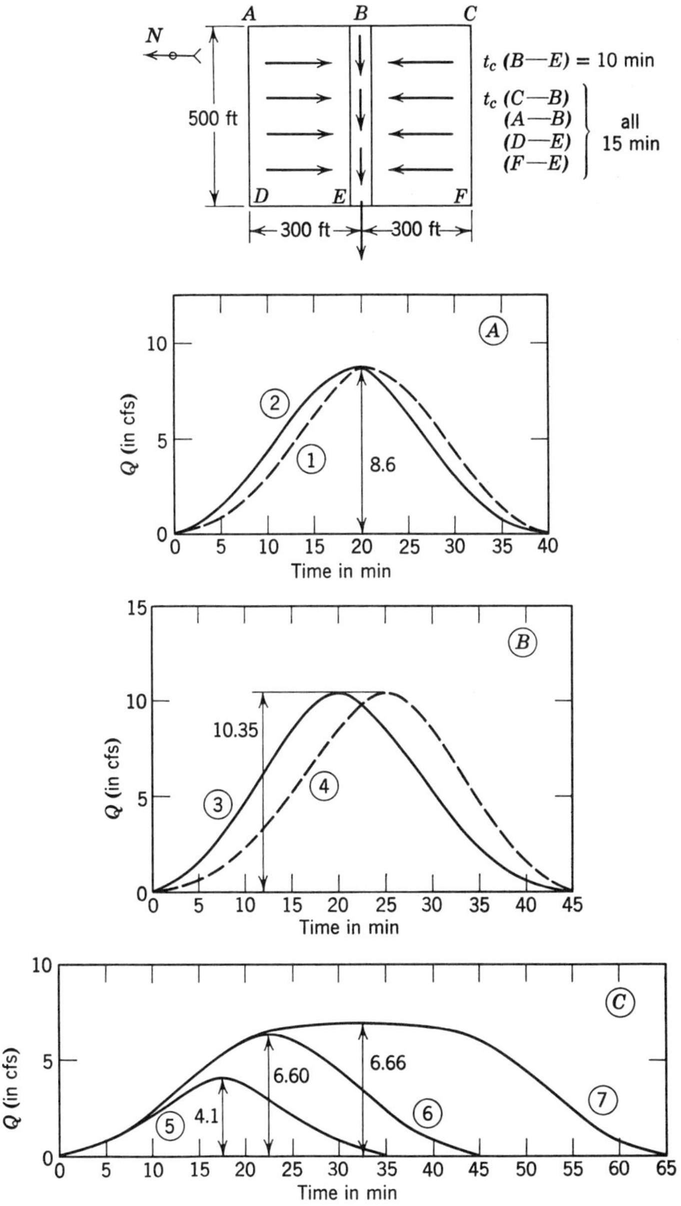

Fig. 2-44 Hydrographs for a parking lot under different storm characteristics.

Storm 6. Identical with storm 5 except for a duration of 20 min.
Storm 7. Identical with storm 6 except for a duration of 40 min.

It should be noted that actually the times of concentration vary with the velocities of runoff and therefore are not constant for the various storm conditions, although constancy was assumed here for reasons of simplicity.

2.15 *Rainfall-Runoff Relations*

The relation of runoff to rainfall is affected by many factors characteristic of rainfall, drainage basins, temperature, and season, and it is therefore obvious that no single relationship can be established to predict the runoff from a storm with known total precipitation and duration.

Nevertheless, for many years the rational approach to runoff was the use of dimensionless runoff coefficients to estimate runoff as a certain percentage of storm rainfall. Recorded runoff was plotted versus precipitation as observed in a nearby station for a number of storms, and a curve was fitted to the data, possibly a straight line (Fig. 2.45), rendering the equation

$$\begin{aligned} Q &= k(P - b) \qquad & P \geqslant b \\ &= 0 & P < b \end{aligned} \tag{2.27}$$

in which Q is the direct runoff in inches, P is the amount of storm rainfall in inches, b is a threshold of precipitation below which there is no runoff, and k is the runoff coefficient. If streamflow records are available, the ground-water flow must be extracted from the stream flow hydrograph to give the storm hydrograph.

Another example of curve fitting to precipitation-runoff data is given in Fig. 2.46, in which the method of least squares [29] has been used to determine the coefficients a, b, and c of the equation

$$Q = a + bP + cP^2 \tag{2.28}$$

in which Q is the annual runoff and P is the annual precipitation. Sometimes the correlation is better if instead of P^2 in Eq. 2.28 the annual precipitation of the previous year, designated as P_{-1}, is used.

The rational approach is still used in the planning of small hydraulic structures and has some merits in the design of storm sewers for large impervious areas such as parking lots and airport pavements, where the drainage area features are more or less constant. For natural soil conditions, however, it seems only logical to group the precipitation-runoff data according to the soil conditions preceding the storm and to establish a series of curves similar to those of Fig. 2.45.

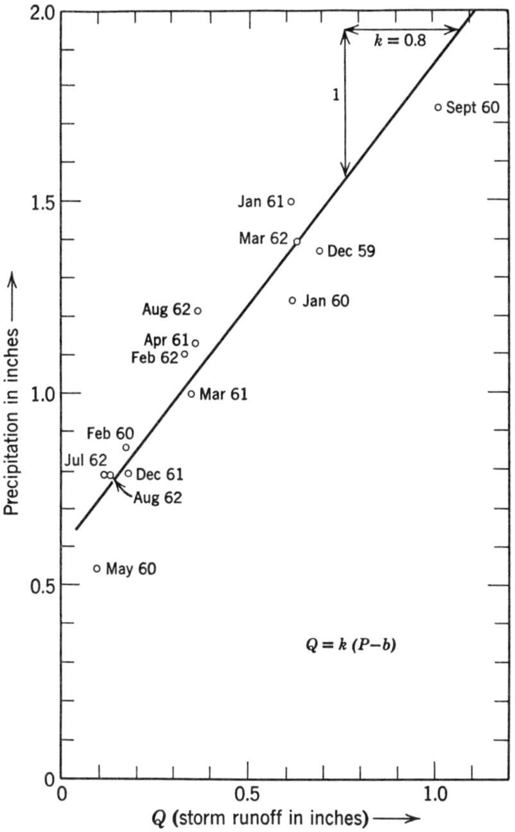

Fig. 2-45 Simple relationship between storm runoff at Spruce Run, N.J. and nearby precipitation for individual storms.

INITIAL MOISTURE CONDITIONS

Several indicators of initial moisture conditions may be used, such as initial ground-water flow, soil moisture deficiency, pan evaporation data before the storm for individual storms and total precipitation in the previous season when rainfall and runoff are related on a seasonal or annual basis.

Kohler and Linsley [18] have introduced the antecedent-precipitation index (API), which attaches a numerical value to the moisture conditions of the soil before a storm takes place, working from the idea that soil moisture should decrease logarithmically with time during periods of no

$y - y'$	$(y - y')^2$
28	784
4	16
86	7,396
19	361
14	196
16	256
5	25
72	5,184
14	196
25	625
34	1,156
10	100
70	4,900
20	400
40	1,600
19	361
16	256
7	49
36	1,296
30	900
18	324
26	676
18	324
32	1,024
28	784
34	1,156
45	2,025
42	1,764
38	1,444
15	225
0	0
Σ 861	35,803

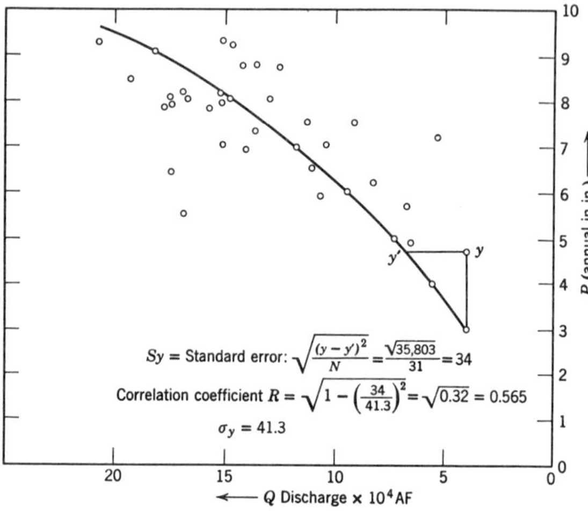

Fig. 2-46 Simple relationship between annual precipitation at Green River, Wyoming and annual runoff at Linwood, Utah. Curve fitted with help of least squares method.

precipitation. They proposed the equation

$$I_t = I_0 k^t \tag{2.29}$$

in which I_0 and I_t are expressed in inches of water, I_0 is the initial value of the antecedent-precipitation index, I_t is the reduced value t days later, and k is a recession constant varying in the range of 0.85 to 0.98. The analogy between Eqs. 2.29 and 2.22 is obvious. From Eq. 2.29 it follows that the index for any day is equal to that of the previous day multiplied by k. To this, the rainfall of the day, if any, is added.

The use of a precipitation index is not limited to runoff predictions from individual storms. It is also valuable in establishing annual runoff-precipitation relations, in which the index assumes a seasonal meaning.

COAXIAL RELATIONS FOR TOTAL STORM RUNOFF

Coaxial relations between storm rainfall and runoff for such parameters as the antecedent-precipitation index (API), duration of the storm, week of the year, and others may be established in a manner similar to the one shown in section 2.10. A simple example in which API and duration of storm are used as parameters is illustrated in Fig. 2.47.

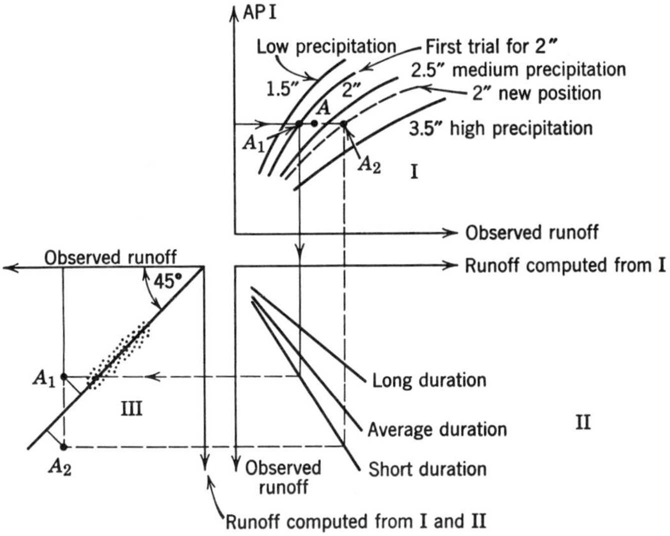

Fig. 2-47 Coaxial relation between total precipitation and runoff for given API and storm duration.

In the first quadrant the API for each storm is plotted versus the observed runoff, and the total amount of precipitation is marked beside each point. This allows the data to be stratified along curves of low, medium, and high precipitation, here assumed to be 1.5 in., 2.5 in., and 3.5 in. In the second quadrant, the runoff computed by means of the given API and the established curves of equal precipitation of quadrant I, is plotted versus the observed runoff, while the storm duration is marked beside each point. Thus curves of short, average, and long storm duration are constructed. With the help of curves I and II it is now possible to predict the runoff for any individual storm, given its total precipitation, duration, and antecedent precipitation. The question arises, however, as to how good the graphical relations are. A check on their accuracy for the available data may easily be made. For this purpose, the runoff as computed from curves I and II is plotted versus the observed

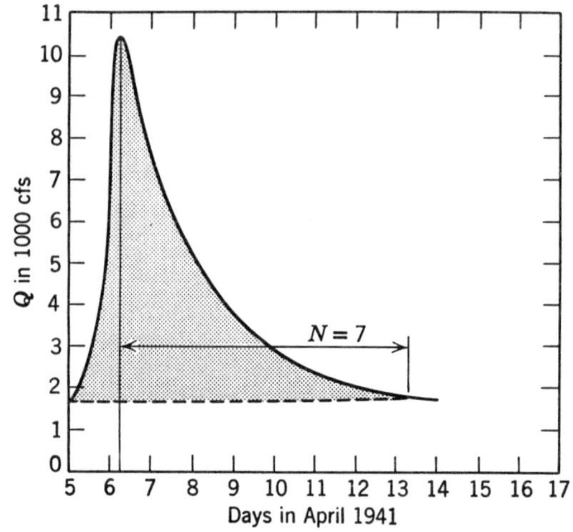

(a)

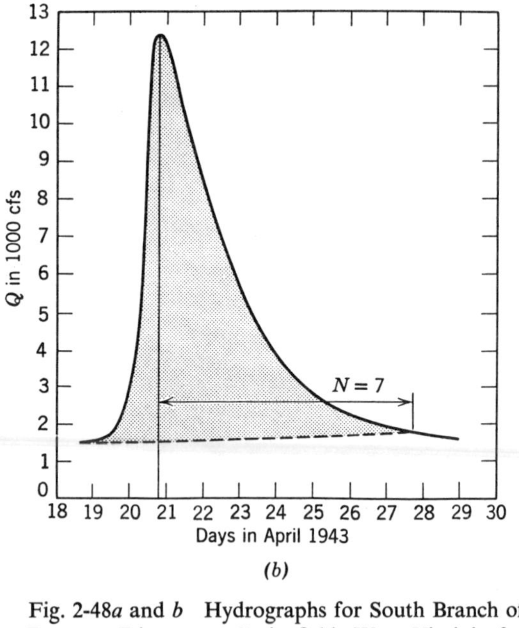

(b)

Fig. 2-48*a* and *b* Hydrographs for South Branch of Potomac River near Springfield, West Virginia for storms of April 5, 1941 and April 20, 1943, duration each 18 hr. Basin area: 1471 sq mi

(*a*) Surface runoff = 166×10^7 cu ft
= 0.485 in. over drainage area

(*b*) Surface runoff = 214×10^7 cu ft
= 0.625 in. over drainage area

runoff for all the data. The curves are well fitted if the points in quadrant III are closely and evenly distributed about the bisector of the quadrant. If this were not so curves I and II would have to be relocated, starting with those in quadrant I. This is illustrated for point A, first located by its observed coordinates, and then located as A_1 by the first position of the curve of 2-in. precipitation. Quadrant III shows that for point A the runoff computed from curves I and II is much less than the observed runoff. If a new trial location of the curve of 2-in. precipitation is established along with a corresponding position A_2 of point A, quadrant III now shows that the runoff computed from curves I and II is more than the observed runoff. It is now possible to determine the position of A in quadrant III where the error changes sign (or becomes zero) and to work back to quadrant I to find a correct position for point A. This process must be repeated a number of times for different sets of data and may also involve a shifting of the curves of quadrant II. After several efforts, a final solution should satisfy the criterion of quadrant III.

2.16 *Unit Hydrographs* (*Unit graphs*)

Sherman [27] introduced the concept of the unit hydrograph for a storm of a specific duration for any given drainage basin. It is based on the fact that the time of concentration t_c, as defined before, is constant for a basin and therefore the time base for the direct runoff hydrograph of all storms of the same duration is constant for the basin. The storm characteristics affecting the shape of the hydrograph have been discussed before. The shape of the unit hydrograph results from the average characteristics of a number of storms of the same duration. The area bounded by the unit graph and the time axis represents one inch of direct runoff from the basin. The usefulness of the unit hydrograph lies mainly in the forecasting of peak flows. Given a unit hydrograph of specific duration for a basin, the hydrograph for any other storm of the same duration may be constructed by multiplying the ordinates of the unit graph by the storm runoff. Consequently, if we can predict by any method the total direct runoff for a storm, we shall be able to predict the peak flow as well provided that we have the unit graph of the basin for the duration of the storm. Eagleson [51] developed unit hydrographs for sewered areas.

To construct a unit hydrograph a series of storms with fairly uniform rainfall intensity and of the same duration (producing significant runoff) are plotted and the base flow is extracted. The volume of direct runoff is computed (Fig. 2.48*a*, *b*) and the ordinates of the direct runoff hydrograph are divided by the measure of the direct runoff in inches. The adjusted (to 1 in. direct runoff) storm hydrographs are then replotted (Fig. 2.49) and

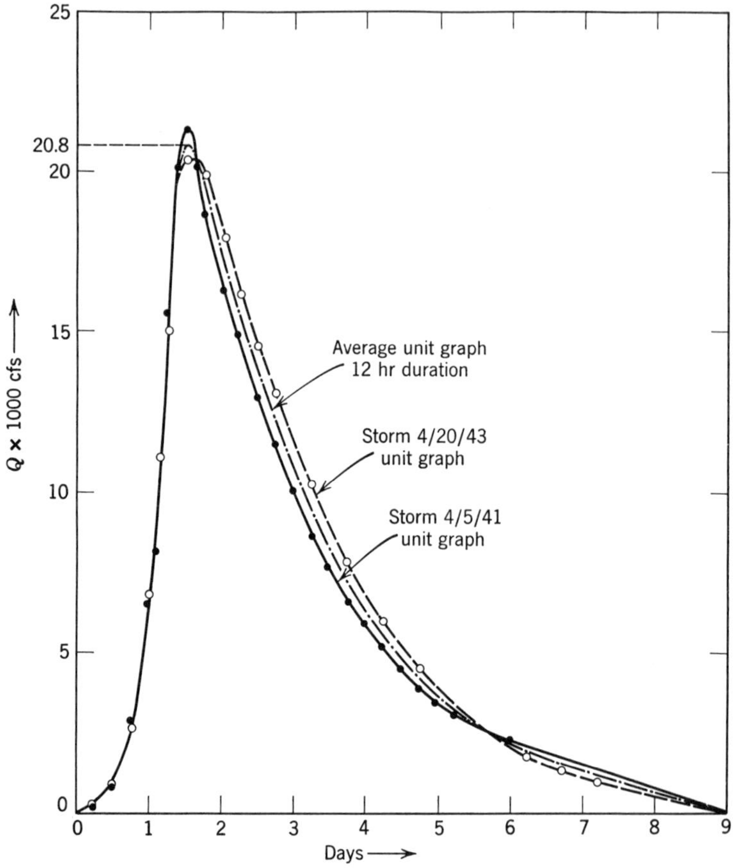

Fig. 2-49 Average unit graph for storms of 18-hr duration, South Branch, Potomac River, near Springfield, West Virginia.

averaged. In our example only two storms were used and their peaks were averaged. In practice, a series of storms are analyzed and the various peaks and times of the peaks are averaged.

This procedure is valid for simple direct runoff hydrographs having only one peak. In the case of a complex storm with several peaks in the hydrograph, it is sometimes possible to filter out individual hydrographs each connected with one peak of the storm and to apply the procedure described above. This, in fact, is equivalent to the following method illustrated in Fig. 2.50. The total duration of the storm is divided into periods of equal duration, each being the time base of a new hydrograph and the runoff Q_1, Q_2, and Q_3 for these new hydrographs is determined.

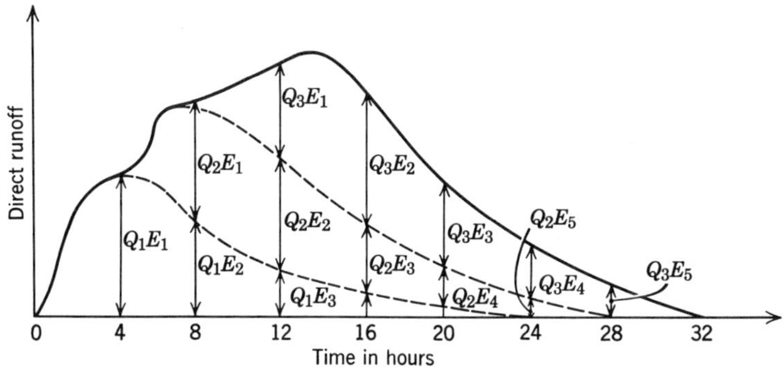

Fig. 2-50 Unit hydrograph construction from a complex storm.

Assume that Q_1, Q_2, and Q_3 are the portions of the total runoff due to the rainfall respectively in the first, second, and third periods. It is now possible to set up a series of equations of the form

$$\begin{aligned} q_1 &= Q_1E_1 \\ q_2 &= Q_1E_2 + Q_2E_1 \\ q_3 &= Q_1E_3 + Q_2E_2 + Q_3E_1 \\ &\cdots\cdots\cdots\cdots\cdots\cdots \end{aligned} \tag{2.30}$$

in which $q_1, q_2, q_3, \cdots, q_n$ are the ordinates of the complex hydrograph, a constant time fraction of the common period or base length apart, and $E_1, E_2, E_3, \cdots, E_n$ are the ordinates of the common unit graph also spaced the same constant time fraction apart. Considering all the assumptions made, the unit graphs for the three runoff portions would be the same; in other words, E_1 is constant for the three portions, and so is E_2, E_3, and so on. The ordinates of the three portions are simply:

(1) $Q_1E_1, Q_1E_2, Q_1E_3, \cdots, Q_1E_n$

(2) $Q_2E_1, Q_2E_2, Q_2E_3, \cdots, Q_2E_n$

(3) $Q_3E_1, Q_3E_2, Q_3E_3, \cdots, Q_3E_n$

UNIT HYDROGRAPHS OF DIFFERENT DURATIONS

It has already been implied that there is no specific unit graph for a drainage basin but that there is one for every storm duration. Often a unit graph associated with a given storm duration is available for a basin, and the question is to construct one for another duration. To go from a unit hydrograph for duration t_R to one for duration $2t_R$, the following simple method is used as a consequence of Eq. 2.25. A hydrograph for duration t_R is delayed a time t_R with respect to its identical graph, and the

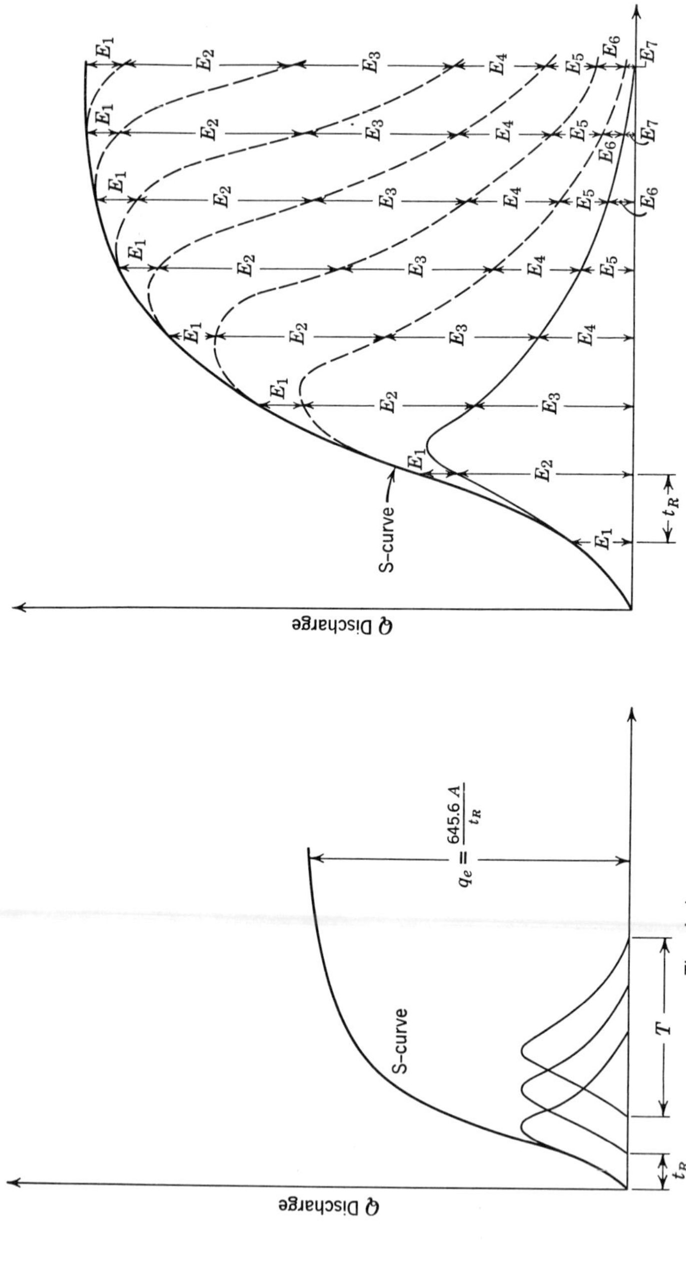

Fig. 2-51 The summation curve or S-curve. Graphical construction of S-curve.

two unit graphs are added. The complete hydrograph has a runoff of 2 in., and if all its ordinates are divided by 2 the result is a unit graph for duration $2t_R$.

To go from a unit graph for a storm of long duration to one for a storm of short duration, the technique of the S-curve or summation curve is extremely useful [12]. The S-curve associated with a certain duration t_R is the hydrograph that would result from an infinite series of unit graphs of duration t_R, each delayed t_R hours with respect to the preceding one. It is shown [12] that the limit ordinate of the S-curve is 1 in./t_R if t_R is expressed in hours, and only T/t_R unit graphs need to be combined if T is the time base of the unit graph. The S-curve reaches an equilibrium at flowrate q_e (Fig. 2.51):

$$q_e = \frac{1 \text{ in. } A \text{ sq mi}}{t_R \text{ hr}} = \frac{24A \times 26.9\ SFD}{t_R \text{ day}} = \frac{645.6A}{t_R} \tag{2.31}$$

in which A is the area of the drainage basin in square miles, t_R is the time in days, and q_e is the flowrate in cubic feet per second.

In constructing the S-curve, we notice that the S-curve additions are the ordinates of the S-curve set t_R hr ahead. Therefore it is sufficient to combine the S-curve additions to the initial unit hydrograph.

To construct the unit graph for duration t_R', the difference between 2 S curves is found, associated with a duration t_R, and lagged t_R' hr. The ordinates of the difference are multiplied by the ratio t_R/t_R'. This technique has been applied to the unit graph for 18 hr of Fig. 2.49 in the construction of the unit graphs for the same drainage basin and for storm durations of 12 hr and 24 hr as represented in Fig. 2.52.

SYNTHETIC UNIT GRAPHS [50]

In Fig. 2.52 the time t_P between the centroid of storm rainfall and the peak of runoff has been designated as time lag or basin lag. The concept of basin lag has been used by Snyder [30] in the first attempt to construct a synthetic unit hydrograph. The construction of a synthetic unit hydrograph is the only rational solution if the stream flow of the drainage basin is not measured, as is the case for most of the small watersheds in the United States. To sketch a unit graph, it is necessary to know the time of peak, the peak flow and the time base. These elements must be determined for every particular or regional location of the drainage basin. Snyder proposed the following empirical formula for basins in the Appalachian Mountains (the formula gives accurate results only in that region):

$$t_P = C_t(LL_c)^{0.3} \tag{2.32}$$

in which t_P is the basin lag in hours, L is the length of the main stream

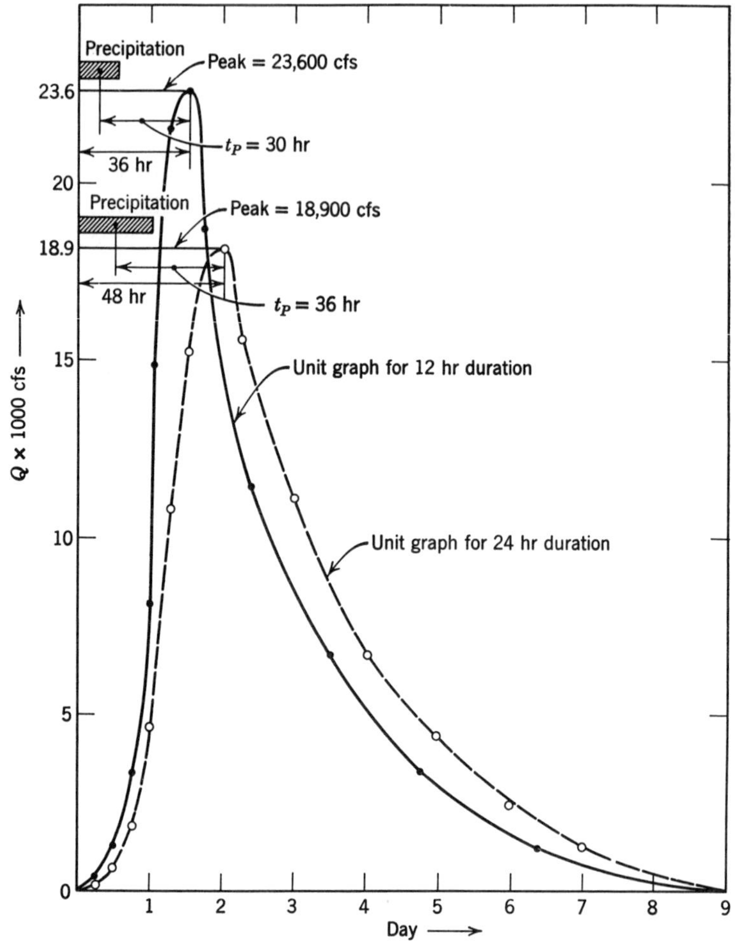

Fig. 2-52 Unit graphs for 12- and 24-hr duration, constructed with the help of the S-curve and unit graphs for storms of 18-hr duration. South Branch, Potomac River, Springfield, West Virginia.

from outlet to divide in miles, and L_c is the distance from the outlet to a point on the stream nearest the centroid of the basin. The coefficient C_t varies from 1.8 to 2.2, with lower values associated with basins of steeper slopes.

For the standard duration of rain t_R, Snyder proposed

$$t_R = \frac{t_P}{5.5} \tag{2.33}$$

For rains of this duration he found that the unit-hydrograph peak q_P was given by

$$q_P = \frac{645.6 C_P A}{t_P} \tag{2.34}$$

in which A is the drainage area in square miles, C_P is a coefficient ranging from 0.56 to 0.69, and 645.6 is the conversion factor derived in Eq. 2.31 to give q_P in cubic feet per second. For the time base T (in days) of the unit graph, Snyder adopted the expression

$$T = 3 + 3\frac{t_P}{24} \tag{2.35}$$

Equations 2.32 to 2.35 suffice to construct the unit graph for a storm of duration t_R. For any other duration t_R', the basin lag is defined as

$$t_P' = t_P + \frac{t_R' - t_R}{4} \tag{2.36}$$

and this new lag is used in Eqs. 2.34 and 2.35. Linsley [31] found that the coefficients C_t and C_P varied considerably when applied to drainage basins in the West.

EXAMPLE

Snyder's method has been applied to a small basin in the Appalachian Mountains, as shown in Fig. 2.53*a*, with $C_t = 2.0$ and $C_P = 0.6$, for the construction of unit graphs for 6-hr and 2-hr storm durations.

RESULTS

$A = 29$ sq mi
$L = 9$ mi. $L_c = 5.2$ mi
$t_P = 6.32$ hr (from Eq. 2.32)
$t_R = 1.15$ hr (from Eq. 2.33)
$t_P' = 7.35$ hr (from Eq. 2.36 and $t_R' = 6$ hr)
$= 6.53$ hr (from Eq. 2.36 and $t_R' = 2$ hr)
$T = 3.94$ days for $t_R' = 6$ hr
$= 3.82$ days for $t'_R = 2$ hr

Time to peak from beginning of rising limb $= (t_R'/2) + t_P' = 3 + 7.53 =$ 10.53 hr for storm of 6-hr duration and $1 + 6.53 = 7.53$ hr for storm of 2-hr duration.

$q_P = 1{,}480$ cfs (from Eq. 2.34) and $t_P' = 7.35$ hr, i.e., $t_R' = 6$ hr
$= 1{,}710$ cfs (from Eq. 2.34) and $t_P' = 6.53$ hr, i.e., $t_R' = 2$ hr

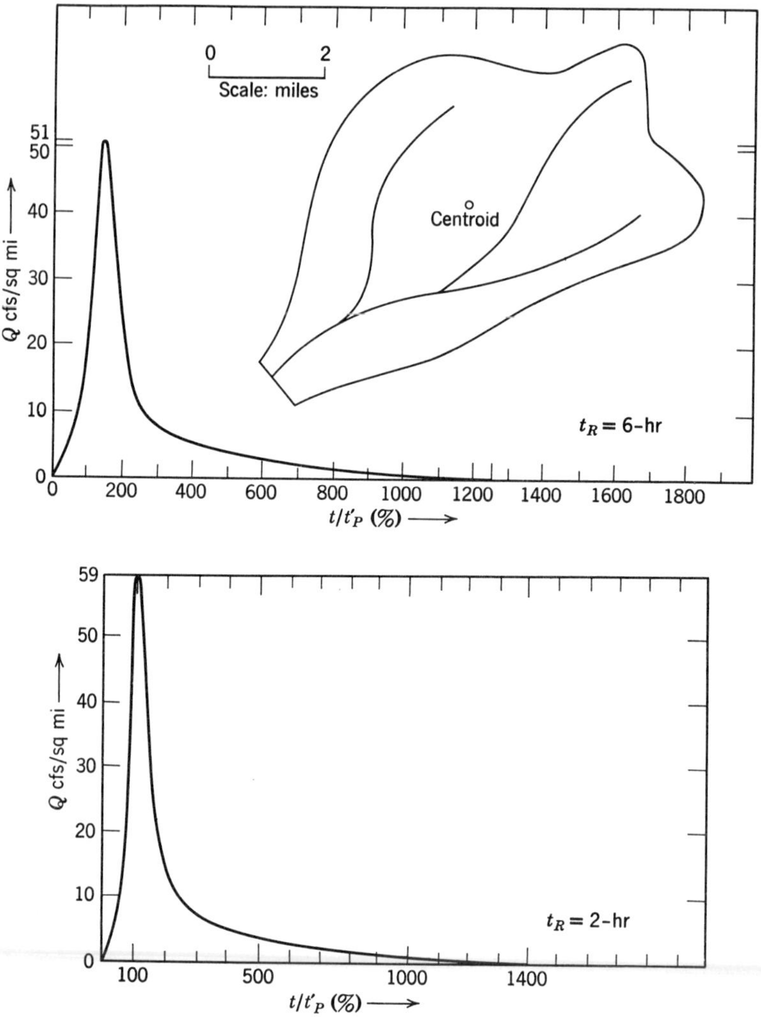

Fig. 2-53*a* Synthetic unit graphs using Snyder's method. Small basin in Appalachian Mountains (A_d = 29 sq mi). Replotted hydrographs for 6- and 2-hr duration.

The unit graphs obtained with these data have been replotted in Fig. 2.53*a* with a flow scale in cubic feet per second per square mile and a time scale in percent of lag. Subsequently these unit graphs were used to construct the total hydrograph for a storm with a given runoff distribution in time as indicated in Fig. 2.53*b*.

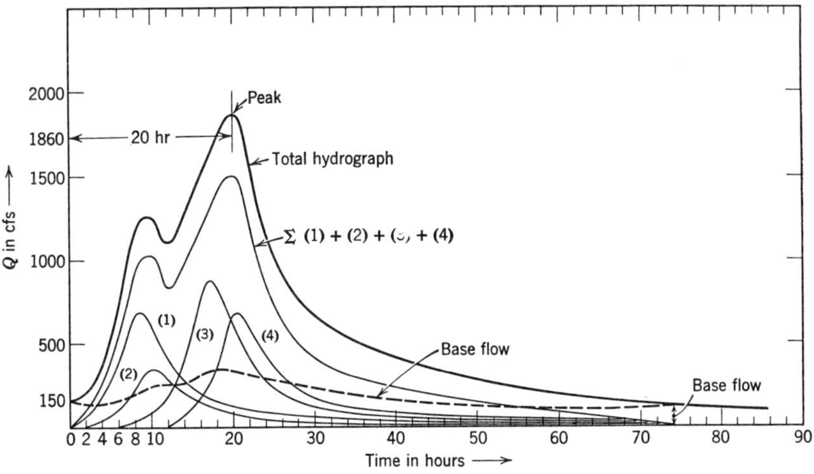

Fig. 2-53*b* Hydrograph constructed for given runoff. Small basin in Appalachian Mountains. Assume initial flow is 150 cfs and base flow receding with $K_r = 0.88$. Use unit graphs of Fig. 2-53*a*.

Time (hr)	Runoff (in.)
0–2	0.40
2–4	0.20
4–6	0
6–12	0.60
12–14	0.40

2.17 *Streamflow Routing* [21]

The movement of a flood wave through a stream channel is a highly complicated problem, not only because the flow varies with time as the wave progresses downstream but also because of the varying nature of the channel properties and the lateral inflow into the channel. The differential equations for unsteady flow in uniform channels were first solved by Massau [32], who graphically integrated the equations of characteristics. In 1964 Massau's method is still the most convenient way of solving the differential equations of flow. For engineering purposes, however, the differential equations may be replaced by difference equations, made simpler by the assumption that the flow in the channel is changing gradually with time and that the channel storage may be expressed simply in terms of the inflow and (or) outflow of the channel reach.

Channel storage has a damping effect on a flood wave travelling down a channel, attenuating the peak and therefore stretching the time base of the flood hydrograph. The purpose of any flood routing technique is to obtain a fair estimate of the magnitude of the damping effect.

In principle the reach of a river or a reservoir may be represented as in Fig. 2.54 in which I and O designate the inlet and the outlet flow rates. Flows at these points are considered in a series of successive time intervals, each of constant duration Δt, the so called routing period. At the beginning of each time interval the subscript 1 is assigned to both I and O; at the end of the routing period the subscript 2 is assigned to inflow and outflow. The routing period $\Delta t = t_2 - t_1$ is chosen to be sufficiently small so that

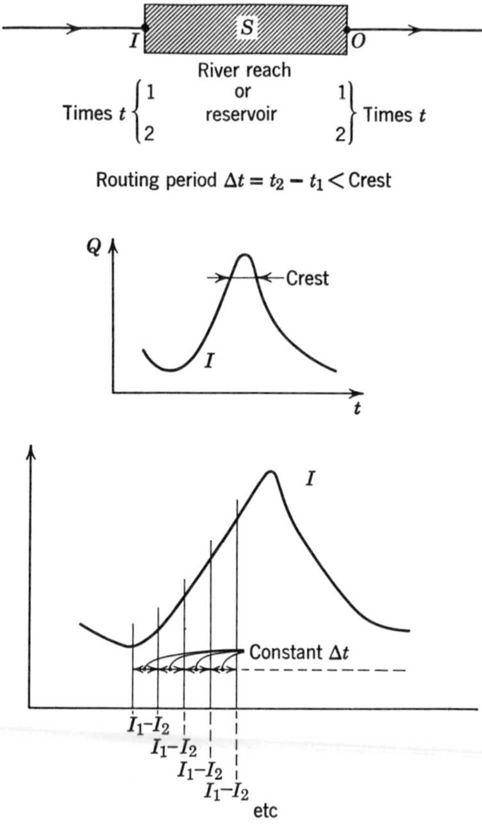

Fig. 2-54 Block diagram for flow routing.

no significant part of the crest of the inflow hydrograph (see Fig. 2.54) may be omitted in the course of the routing process. The river or reservoir storage is designated by the symbol S.

The known quantities at the start of the routing process are as follows:

I_1 and I_2 (Δt later) because the entire inflow hydrograph is given
O_1, the outflow at time t_1
S_1, the storage at time t_1

The unknown quantities in each step Δt of the routing process are two-fold:

O_2, the outflow at time t_2

S_2, the storage at time t_2

Although the chief concern of the hydrologist lies with O_2, he needs two equations to solve for O_2 and S_2 simultaneously. The first equation is one of continuity or conservation of mass for the water in the reach during the routing period Δt:

$$I_{\text{ave}} - O_{\text{ave}} = \frac{\Delta S}{\Delta t} \tag{2.37}$$

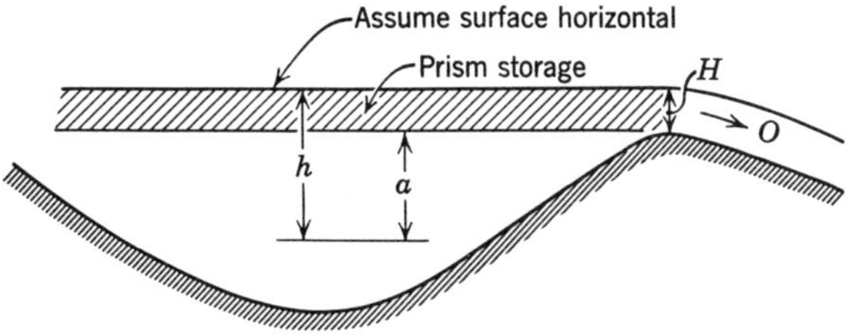

Fig. 2-55 Flood routing through a large reservoir. Given $O = f_1(H)$ and $S = f_2(h)$. But since $h = a + H$ it is possible to find $O = f_3(S)$. Value of a is irrelevant (may be zero).

For sufficiently small Δt, the average values of I and O may be approximately expressed as

$$I_{\text{ave}} = \frac{I_1 + I_2}{2}, \qquad O_{\text{ave}} = \frac{O_1 + O_2}{2}$$

and Eq. 2.37 may be written as

$$\frac{I_1 + I_2}{2} - \frac{O_1 + O_2}{2} = \frac{S_2 - S_1}{\Delta t} \tag{2.38}$$

The second relation that is needed may implicitly be written as

$$O_2 = f\,(S_2) \tag{2.39}$$

and is easy to be put in explicit form only when a flood wave is routed through a large reservoir.

ROUTING THROUGH A LARGE RESERVOIR

For a large reservoir, the simplifying assumption of horizontal pool level makes it possible to establish a simple relationship between the active storage S (Fig. 2.55) and the outflow O evacuated via the spillway. The outflow O is known as a function of the water level above the spillway crest, say

$$O = CLH^{3/2} \tag{2.40}$$

in which C is a weir coefficient, L is the width of the weir, and H is the head on the weir. Also, the active storage S may be computed as a function of H, say $S = \varphi(H)$, by summing up the products of areas relieved from a contour map with increments ΔH. By elimination of H between the latter equation and Eq. 2.40, we obtain an equation having the same form as Eq. 2.39.

EXAMPLE

A spillway of the ogee type of width $L = 200$ ft is planned to protect a dam from being overtopped by a peak flood estimated at approximately 14,000 cfs and occurring on the average once every 1,000 yr. The weir coefficient C is estimated to be 3.6 and the area-elevation relation for head H above the crest of the weir is as follows:

H (ft)	Area (acres)
0	1,400
5	1,420
10	1,450

Assuming (under the worst conditions) that the water surface of the reservoir is at the spillway crest when the flood hits the reservoir, what is the maximum depth of flow over the spillway at the time the flood occurs? This depth of flow has to be subtracted from the maximum permissible reservoir level in order to determine the height of the spillway crest. Large depths of flow may require deep excavations for the spillway, and flood protection may become very expensive [3].

To solve this problem, a time base T of three days is assumed for the inflow hydrograph I and for the shape of I, Commons' [33] dimensionless hydrograph (Fig. 2.56) is adopted. A routing period of 2 hr is used for the calculations.

Equation (2.38) is rewritten as

$$(I_1 + I_2) + \left(\frac{2S_1}{\Delta t} - O_1\right) = \frac{2S_2}{\Delta t} + O_2 \tag{2.41}$$

[3] A similar problem was posed to the author in 1956 when he was employed as an hydraulic engineer with SOFINA (Brussels, Belgium) in the redesign of the existing spillway for the protection of the earthen dam below the Necaxa Reservoir, Mexico. Actually the most difficult problem was to forecast a design flood. Franklin F. Snyder (reference 30) was consulted on this project.

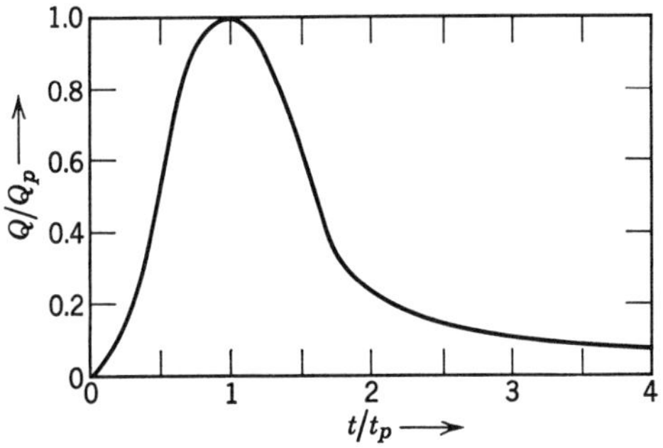

Fig. 2-56 Commons' dimensionless unit hydrograph. Q_p = Peak discharge; t_p = Time peak discharge occurs.

At the beginning of each successive calculation over the routing period, all terms on the left-hand side of Eq. 2.41 are known and therefore the right-hand side is known. Indeed, O is computed in cubic feet per second for different values of H, from Eq. 2.40, in which L and H are expressed in feet and plotted in Fig. 2.57. Graphs of $2S/\Delta t$, $2S/\Delta t \pm O$ are constructed graphically in Fig. 2.57 with the help of Table 2.10. This table is based on increments $\Delta H = 2$ ft.

$$\Delta t = 2\text{ hr} = \frac{D(ay)}{12} \quad \text{and} \quad 1\text{SFD} = 2\text{ AF},$$

so that $2S/\Delta t = 12(A_1 + A_2)$, expressed in cubic feet per second if A_1 and A_2 are the areas in acres, corresponding to heads H_1 and H_2.

The calculations are shown step by step in Table 2.11. After the left-hand side of Eq. 2.41 is computed at the beginning of a step, the total value of the right-hand side is known. A specific calculation proceeds from there as follows. Consider the value $t_2 = 8$ hr in Table 2.11. The

Table 2.10 Relationship between $\frac{2S}{\Delta t}$ and H

H (in ft)	0	2	4	6	8	10
0 (in 1,000 cfs)	0	2.05	5.75	10.6	16.3	22.9
A (in 1,000 acres)	1.4	1.408	1.416	1.426	1.438	1.450
$A_1 + A_2$		2.808	2.824	2.842	2.864	2.888
$A_1 + A_2$ (cumulative)		2.808	5.632	8.474	11.338	14.226
$\frac{2S}{\Delta t}$ (in 1,000 cfs)		33.7	67.5	101.5	136.2	171.0

Table 2.11 Flood Routing Calculations for a Large Reservoir

t_1 (hr)	$t_2 = t_1 + \Delta t$ (hr)	I_1 (in 1,000 cfs)	I_2 (in 1,000 cfs)	$I_1 + I_2$ (in 1,000 cfs)	$\frac{2S_1}{\Delta t} - O_1$ (in 1,000 cfs)	$\frac{2S_2}{\Delta t} + O_2$ (in 1,000 cfs)	H_1 (ft)	H_2 (ft)	O_2 (in 1,000 cfs)
0	2	0	1.69	1.69	0	1.60	0	0.08	0.05
2	4	1.69	4.74	6.43	1.0	7.43	0.08	0.40	0.30
4	6	4.74	8.82	1.56	6.0	19.56	0.40	1.05	0.90
6	8	8.82	12.65	21.47	15.0	36.47	1.05	1.95	2.00
8	10	12.65	13.90	26.55	29.0	55.55	1.95	2.90	3.40
10	12	13.90	13.20	27.10	44.0	71.10	2.90	3.85	5.40
12	14	13.20	10.50	23.70	58.0	81.70	3.85	4.40	6.70
14	16	10.50	7.22	17.72	66.0	83.72	4.40	4.50	7.00
16	18	7.22	4.62	11.84	68.0	79.84	4.50	4.35	6.60
18	20	4.62	3.39	8.08	65.0	73.01	4.35	3.95	5.60
20	22	3.39	2.82	6.21	60.0	66.21	3.95	3.60	4.50
22	24	2.82	2.38	5.20					
24	26	2.38	2.15	4.53					
26	28	2.15	1.92	4.07					
28	30	1.92	1.81	3.73					
30	32	1.81	1.69	3.50					
32	34	1.69	1.58	3.27					
34	36	1.58	1.47	3.05					
36	38	1.47							

computed value of $(2S_2/\Delta t) + O_2$ is 36.47×10^3 cfs. Figure 2.57 gives for this value a corresponding $H_2 = 1.95$ ft, and Fig. 2.57 inset gives a value $O_2 = 2.00 \times 10^3$ cfs. In the next step, H_1 is assigned the value 1.95 ft, and the corresponding value of $(2S_1/\Delta t) - O_1$ follows from Fig. 2.57 as

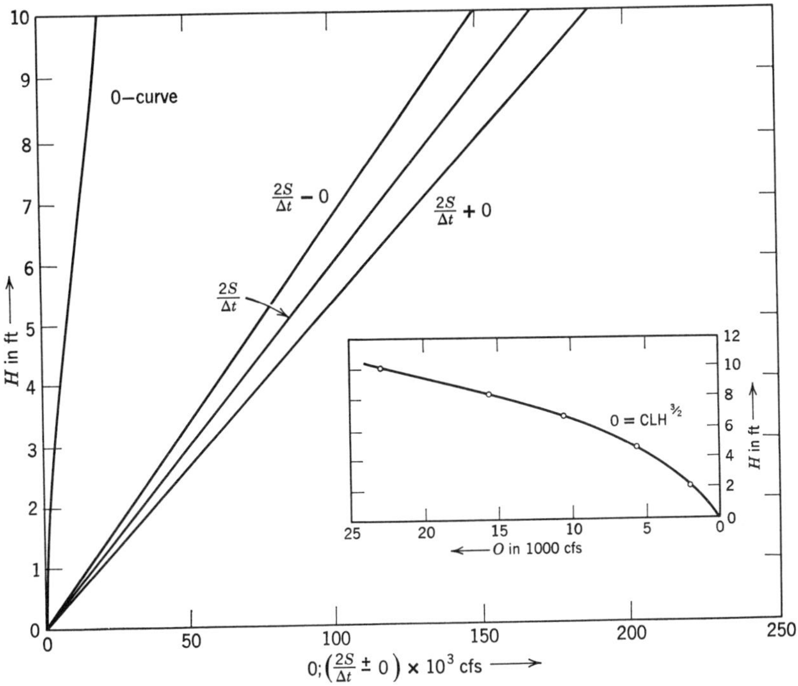

Fig. 2-57 Graphical construction of $\dfrac{2S}{\Delta t} \pm 0$. Inset is the discharge curve for spillway. $O = CLH^{3/2}$.

29.0×10^3 cfs. To this the values of I_1 and I_2 for $t_2 = 10$ hr are added, rendering the new value of $(2S_2/\Delta t) + O_2$, etc. The calculations are continued until the peak outflow is reached, and the results are plotted in Fig. 2.58. They show that there is a lag of $5\frac{1}{2}$ hr between the peak of inflow and that of outflow and a flood peak attentuation $(I - O)$ of 7,000 cfs. It should be noted that I should pass through the peak of O, because S is maximum for $I = O$, and also for maximum H of the reservoir, and therefore for maximum O. The maximum value of H corresponds to $O_{\max} = 7{,}000$ cfs and is 4.50 ft (Fig. 2.57 inset). The storage S is indicated in Fig. 2.58 by the shaded difference between I and O.

When the reservoir is gated, this method remains applicable; no new

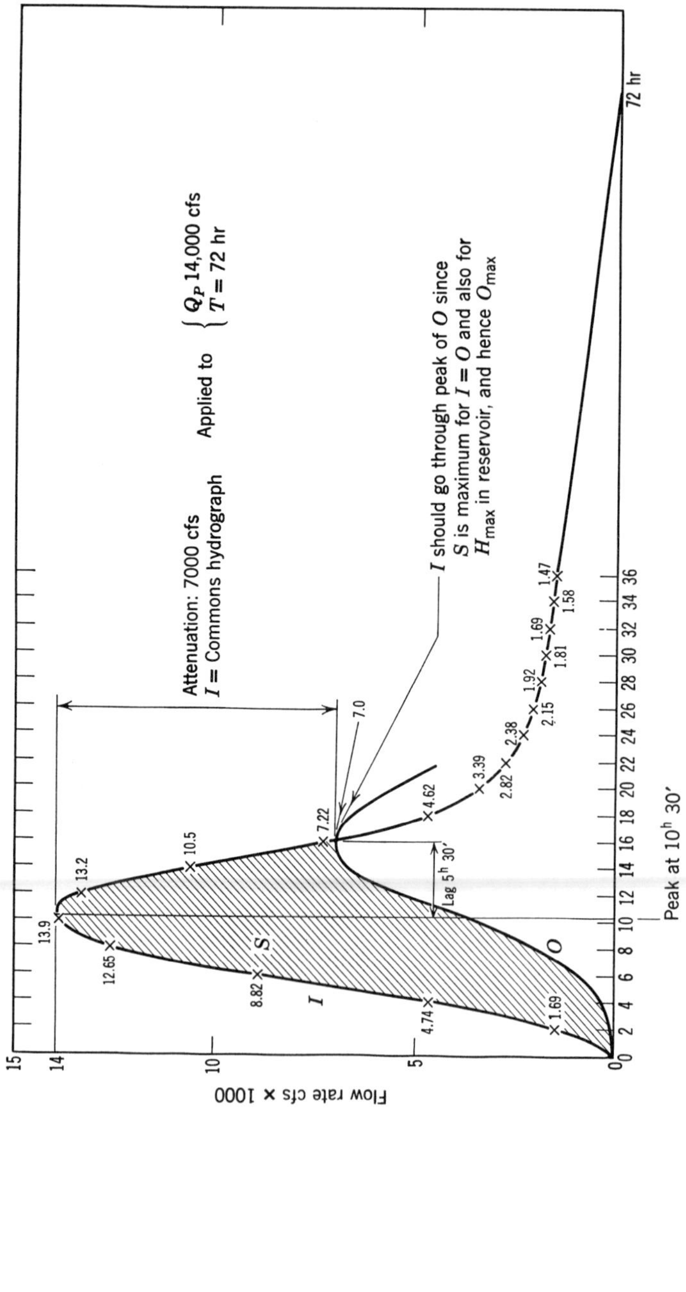

Fig. 2-58 Flood routing, I and O curves.

principles are involved, but Eq. 2.38 is modified and becomes

$$I_{\text{ave}} - O_{\text{ave}} - O_c = \frac{\Delta S}{\Delta t} \qquad (2.42)$$

in which O_c is the regulated outflow and the other symbols are as defined before.

ROUTING IN RIVER CHANNELS

The active storage in the reservoir routing problem consists only of prism storage because of the assumption of horizontal pool level. In a river reach, however, during the passage of a flood wave, beside the prism storage (beneath a line parallel to the river bottom) there is also wedge storage (overlying the prism storage and bounded by the water surface); therefore, there is no simple relationship between storage and outflow alone, as is the case for the reservoir. The wedge storage is positive during the rise of the flood wave, and negative during the recession of the flood wave. The effect of wedge storage is usually accounted for by including inflow as well as outflow in the storage equation. McCarthy [33] proposed the equation

$$S = K[xI + (1 - x)O] \qquad (2.43)$$

in which the constant x is a weighting factor expressing the relative influence of I and O, and K, known as the storage factor, has the dimensions of time and expresses a storage to discharge ratio. The constant x may be determined from the inflow hydrograph I_A and the outflow hydrograph O_B of the reach (Fig. 2.59). Indeed, S attains a maximum in time when I_A and O_B intersect each other in point C. At this point $dS/dt = 0$, and from Eq. 2.43, by differentiation with respect to time

$$\frac{1}{K}\frac{dS}{dt} = x\frac{dI}{dt} + (1 - x)\frac{dO}{dt} \qquad (2.44)$$

Hence

$$x\left(\frac{dI}{dt}\right)_C = -(1 - x)\left(\frac{dO}{dt}\right)_C \qquad (2.45)$$

and $\left(\frac{dI}{dt}\right)_C$ and $\left(\frac{dO}{dt}\right)_C$ may be determined from the tangents to the hydrographs in point C. Equation 2.45 will then yield the value of x. In

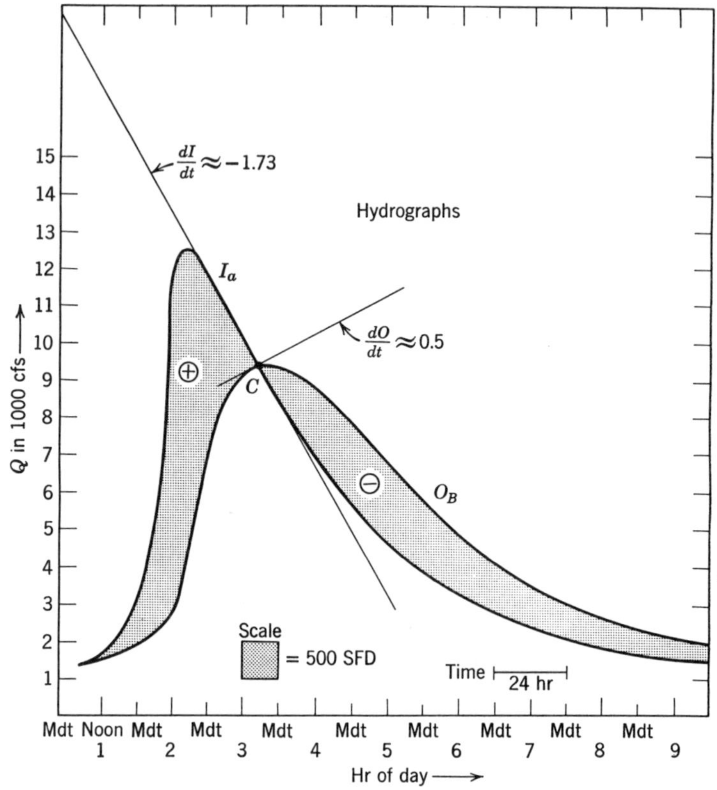

Fig. 2-59 Flow routing through a river reach *A-B*.

Storage as a Function of Time

Time	Number of Squares Increment	Σ Squares	*S* Storage *SFD*
1N	1	1	20
Mdt	13	14	280
2N	73	87	1740
Mdt	189	276	5520
3 Noon	70	346	6920
Peak 6 pm	6	352	7040
Mdt	−6	346	6920
4N	−36	310	6200
Mdt	−51	259	5180
5N	−51	208	4160
Mdt	−46	162	3240
6N	−42	120	2400
Mdt	−36	84	1680
7N	−27	57	1140
Mdt	−21	36	720
8N	−14	22	440
Mdt	−11	11	220
9N	−10	1	20

Muskingum x

$x(-1.73) = -(1 - x)0.5$

$x(1.73 + 0.5) = +0.5$

$x = 0.224$

Fig. 2.59, for example, where the hydrographs of a certain flood at the entrance A and exit B of a river reach are given, the value of x is 0.224. In Fig. 2.59, the storage S has been computed as a function of time by counting the squares between I_A and O_B for constant time increments

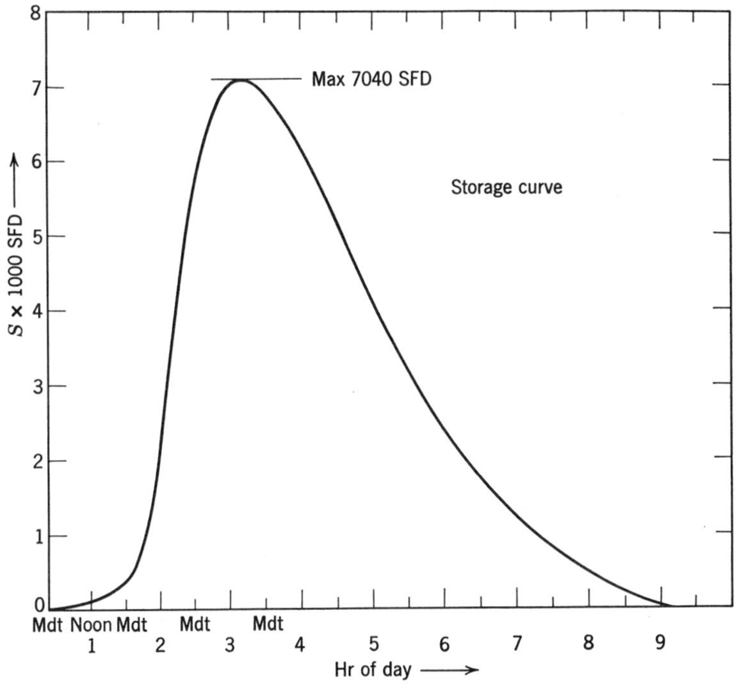

Fig. 2-60 Storage curve as a function of time for a flood moving down a river reach.

$\Delta t = 2.4$ hr and has been tabulated (see Fig. 2.59). The results are graphically represented in Figs. 2.60 and 2.61, and the rising and falling stages are indicated on the storage loop. To determine K, values of

$$[xI + (1 - x)O]$$

are plotted versus S for $x = 0.224$ in Fig. 2.62. They fit closely to a straight line and K is determined from the slope of that line.

K and x are called the Muskingum routing coefficients for the reach AB. When, with K and x determined, S is inserted from Eq. 2.43 in Eq. 2.38, the result may be expressed as

$$O_2 = c_0 I_2 + c_1 I_1 + c_2 O_1 \tag{2.46}$$

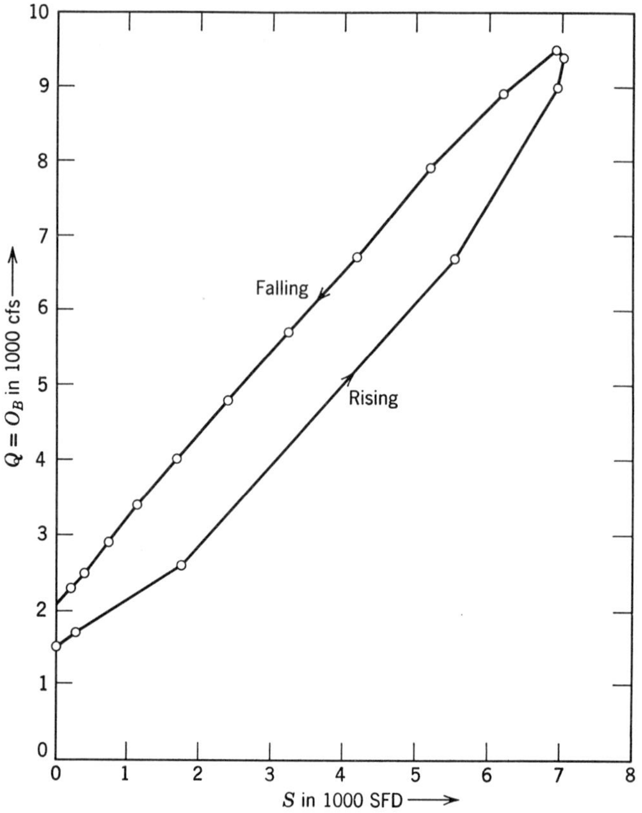

Fig. 2-61 Storage as a function of outflow at the exit station of the river reach.

in which

$$
\begin{aligned}
c_0 &= -\frac{Kx - 0.5\Delta t}{K - Kx + 0.5\Delta t} \\
c_1 &= \frac{Kx + 0.5\Delta t}{K - Kx + 0.5\Delta t} \\
c_2 &= \frac{K - Kx - 0.5\Delta t}{K - Kx + 0.5\Delta t}
\end{aligned}
\tag{2.47}
$$

The sum of the coefficients $c_0 + c_1 + c_2 = 1$.

Equation 2.46 is the Muskingum flood routing formula. With K and x determined, c_0, c_1, and c_2 are computed from Eq. 2.47. For a given hydrograph I and one starting value for O_1, the step-by-step computation of O_2 from Eq. 2.46 is easily effected.

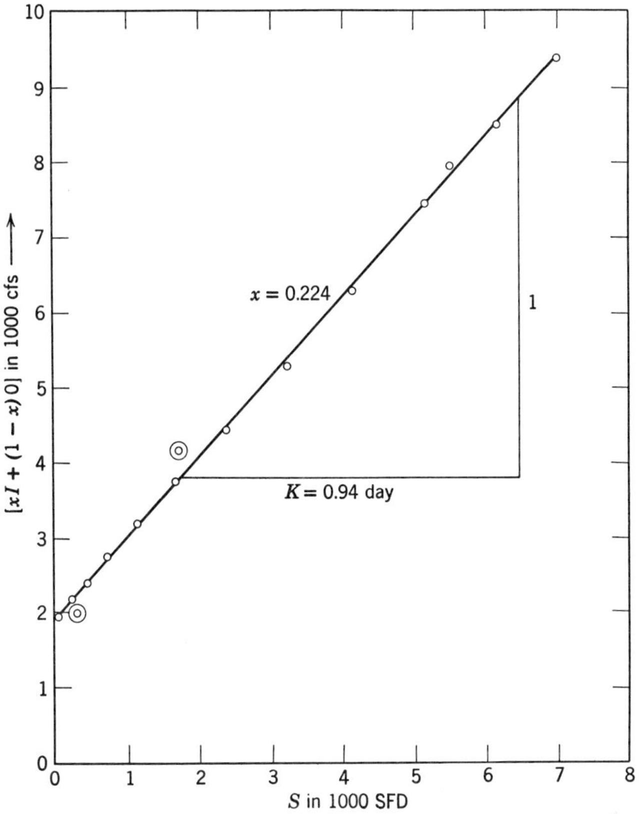

Fig. 2-62 Graph of $[xI + (1 - x)0]$ versus S for a trial value of $x = 0.224$ (see Fig. 2-59).

The problem of local or lateral inflow between the inlet and outlet of a river reach may be solved in a simple way only when the lateral inflow is concentrated near the inlet or outlet. In the first case, the lateral inflow is added to I, and the total flow is routed through the reach. In the latter case, the flood is routed and then the lateral inflow is added to O.

2.18 *Frequency Analysis of Runoff Data*

The designer of a hydraulic structure has to apply a safety factor to the normal capacity of the structure to allow for a maximum flood. This safety factor is somewhat similar to those used by the structural engineer in the design of a structure under live load. However, unlike the safety factors for structural design which are prescribed in building codes, and which usually have a narrow range of variation, the margin of safety left in the design of a hydraulic structure varies widely and is often a matter of speculation. In structures where there is no danger of loss of lives, the design for the maximum flowrate is guided by principles of engineering economy. It is possible to estimate the lifetime of the structure and the damage that would result if the flow capacity of the structure were exceeded once, twice, or any number of times during this lifetime. It is also possible to compute the incremental benefits that would result from the incremental costs required for a larger structure, and finally to choose the structure which has the maximum ratio of incremental benefits over incremental costs. On the other hand, when it comes to the design of the spillway for a dam backing up a huge reservoir and protecting a populated area, no flood should exceed the capacity of the spillway during the lifetime of the structure or during a much longer period. The latter period may be the average return period of the 1,000-yr flood or even that of the possible maximum flood, determined by the hydrometeorologist through maximization of storms by moisture adjustment and transposition of observed storms [34]. Moisture adjustment of a storm involves the estimation of the increased precipitation that could be expected if maximum atmospheric moisture were available. Transposition involves the adjustment of the precipitation of storms observed outside the problem area for differences in meteorological and topographical effects between the sites of precipitation and the problem area.

Sometimes the flood producing effectiveness of storms is increased by a redistribution of the precipitation in time. However, according to Paulhus [34] "it is obvious that maximization by always adjusting storm amounts upward and selecting the most critical chronological sequences can eventually lead to stream flow values so high that they would be ridiculous. Since transposition and moisture adjustment produce pronounced maximization for most basins, it is considered that they yield results that are high enough for practical design purposes without further maximization except in unusual cases."

It is evident from this that when the hydraulic engineer consults a hydrometeorologist, the engineer will have to assume his responsibilities and must accept a calculated risk.

The probable maximum precipitation is derived [34] by taking the results of depth-area-duration analyses (see section 2.6) for major storms that have or could have occurred in the drainage basin, adjusting them for maximum moisture content, and taking the envelope of the adjusted values for all storms to obtain the depth-area-duration curves of probable maximum precipitation.

THE N-YEAR EVENT

The hydrologist is interested in the N-year event, the event that may be expected to occur, on the average of once every N years; in other words, it is the event that has a return period T_r or recurrence interval of N years, N meaning the average number of years within which the specified event will be equaled or exceeded. In the case of floods, the data needed to determine the N-year event is the series of annual floods, i.e., the maximum flood peak of each year, whereby the second- and the lower-order events of each year are ignored, even if they are greater than the peaks of other years. Such a series is a true distribution series to which a distribution function may be fitted, and is therefore acceptable for rigorous analysis. In some empirical methods, as explained below, the use of all floods above a selected datum makes it possible to incorporate floods which are only second or third highest in a given year but which may be significantly higher than the peaks of other years. These floods make up a partial duration series which is not a true distribution series because the flood is not defined in terms of its occurrence in time, but according to its magnitude. When this series is used, it should be established that two consecutive events in any one year are independent. Therefore only one peak per complex storm is listed.

Both annual and partial series may be applied in the following computations of the return period T_r.

(*a*) "California" method [35]

$$T_r = \frac{t}{m'} \tag{2.48}$$

in which t is the number of years of record, m' is the rank of an item in a series arranged in decreasing order of magnitude ($m' = 1$ for highest magnitude).

(*b*) Hazen's formula [36]

$$T_r = \frac{2t}{2m' - 1} \tag{2.49}$$

(*c*) Gumbel's formula [37]

$$T_r = \frac{t}{m' + c - 1} \tag{2.50}$$

in which c is a correction given as a function of $(n + 1 - m')/n$ in Fig. 2.63, n being the number of items in a series from t years of record. The return periods computed according to these formulas may be plotted versus the magnitude of the respective events and a curve may be fitted by eye to the data. For short return periods, $T_r \leqslant t/5$ the results of Eqs. 2.48,

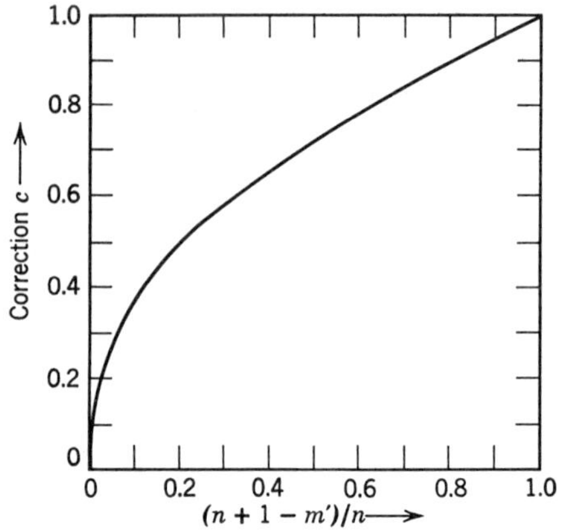

Fig. 2-63 Correction c for computation of T_r based on distribution of largest values. (After Gumbel.)

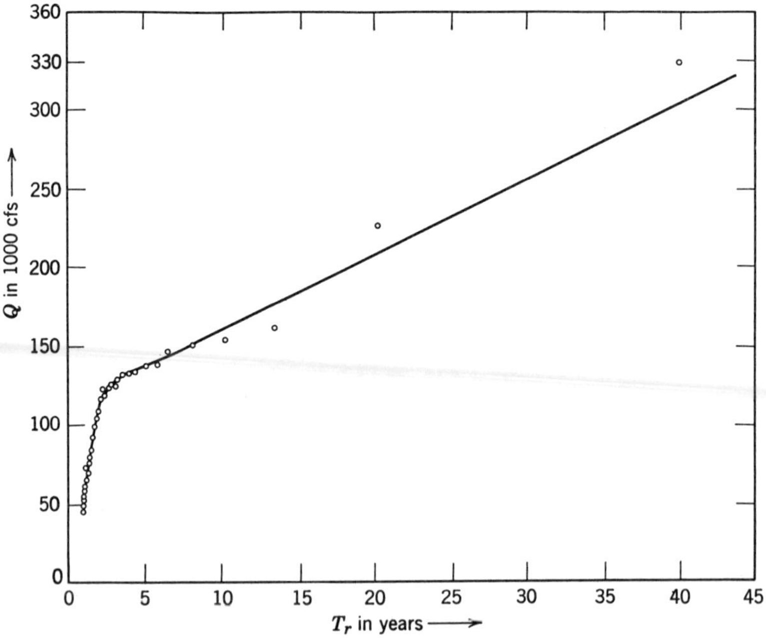

Fig. 2-64 Annual floods of Delaware River at Trenton, N.J. Frequency analysis. Period: 1921–1960 ($t = 40$, $n = 40$).

2.49, and 2.50 are in the same range, and it is easy to fit a curve to the plotted points, as may be seen from Table 2.12 and Fig. 2.64, in which the annual floods of the Delaware River at Trenton, N.J. are examined. However, the higher events give a significant difference in the return period depending on the formula used, and it would be highly inaccurate to extrapolate from the curve of Fig. 2.64, the return period for a flood higher than any of those observed.

To make extrapolation more reliable, theoretical frequency distributions using the entire data series have been proposed. Gumbel's theoretical method [38] is briefly outlined here.

Let X be an exponentially distributed variable of which the extreme values are observed in m samples of equal size s; the theory of extreme values states that as m and s approach infinity the cumulative probability P that any of the m extremes will be less than X is given by

$$P = e^{-e^{-y}} \tag{2.51}$$

in which e is the base of natural logarithms and y, designated as the reduced variate, is given by

$$y = a(X - X_f) \tag{2.52}$$

The mode of distribution X_f and the dispersion parameter a have been determined by Gumbel for a finite number m of samples as

$$X_f = \bar{X} - \sigma_x \frac{\bar{y}_m}{\sigma_m} \tag{2.53}$$

$$a = \frac{\sigma_m}{\sigma_x} \tag{2.54}$$

in which $\bar{X}$ is the arithmetic mean, σ_x the standard deviation, and $\bar{y}_m$ and σ_m are functions only of the sample size m (Table 2.13). Equation 2.51 is an expression of probability of nonoccurrence. The return period is computed from

$$T_r = \frac{1}{1 - P} \tag{2.55}$$

Powell [39] introduced the Gumbel Probability Paper (Fig. 2.65), which has the property that the flood-frequency curve may be approximated by a straight line. In designing this paper, it should be noticed that the reduced variate is a linear function of X, according to Eq. 2.52. Therefore a linear scale for y is plotted along the upper abscissa axis of the graph. A series of values of T_r is now assumed, starting with 1.01 and ending with the maximum T_r that seems reasonable for the problem. The corresponding values

Table 2.12 Annual Floods of the Delaware River at Trenton, N.J., Frequency Analysis. Period 1921–1960 ($t = 40$, $n = 40$)

Rank m'	Year	Discharge, 1,000 cfs	T_r years
1	1955	329	40
2	1936	227	20.20
3	1942	161.2	13.56
4	1925	154	10.09
5	1940	151.6	8.16
6	1933	147	6.74
7	1949	139.1	5.78
8	1953	139	5.06
9	1951	133.2	4.50
10	1956	133	4.04
11	1924	132	3.68
12	1935	129	3.41
13	1948	125.6	3.15
14	1938	125	2.92
15	1960	124	2.70
16	1927	123	2.53
17	1943	118.9	2.40
18	1928	116	2.25
19	1958	108	2.13
20	1921	108	2.03
21	1922	105	1.93
22	1939	99.5	1.86
23	1947	98.5	1.78
24	1929	84.8	1.69
25	1959	84.8	1.69
26	1946	82.3	1.58
27	1945	82.2	1.50
28	1952	82.2	1.50
29	1934	80.0	1.40
30	1950	79.8	1.35
31	1944	78.0	1.32
32	1957	77.5	1.27
33	1923	74.8	1.23
34	1937	74.2	1.21
35	1932	66.1	1.16
36	1941	56.8	1.14
37	1931	53.2	1.10
38	1926	48.1	1.07
39	1930	47.4	1.05
40	1954	46.3	1.02

Gumbel's formula

$$T_r = \frac{t}{m' + c - 1}$$

Table 2.13 Expected Means and Standard Deviations of Reduced Extremes
(After Gumbel)

m	$\bar{y}_m$	σ_m	m	$\bar{y}_m$	σ_m
20	0.52	1.06	80	0.56	1.19
30	0.54	1.11	90	0.56	1.20
40	0.54	1.14	100	0.56	1.21
50	0.55	1.16	150	0.56	1.23
60	0.55	1.17	200	0.57	1.24
70	0.55	1.19	∞	0.57	1.28

of P and y are computed from Eqs. 2.55 and 2.51, and the values of y are marked on the upper abscissa axis. The positions of the corresponding values of T_r on the lower abscissa axis follow immediately. Figure 2.65 gives the flood frequency curve for the Delaware River at Trenton, New Jersey (40 yr period, 1921–1960), based on Gumbels theoretical method. The return period for the 1955 flood of 329,000 cfs (the highest on record) is computed as 232 yr. The return period for a flood of 360,000 cfs is 469 yr, and the 1,000-yr flood is about 400,000 cfs.

In another example, the annual peak flows of the Raritan River at the Manville and Bound Brook, New Jersey, stations, located respectively upstream and downstream from the confluence of the tributary Millstone River, are analyzed together with the annual peak flows of the Millstone River at Blackwells Mills, New Jersey, for a 10-yr period of record (1945–1954). The analysis is based on:

(a) The record of the Bound Brook station only.

(b) The records at Blackwells Mills and Manville, assuming complete dependence. Events of equal probability are assumed to occur simultaneously and the total flow below the confluence is the sum of the flows in the two streams having the same return period.

(c) The records at Blackwells Mills and Manville, assuming complete independence. The probability of any two flows occurring simultaneously is the product of the probability that they will occur independently.

In the analysis based on (*a*) a peak flow of 35,000 cfs or greater might occur at Bound Brook once every 29.4 yr, and in the analyses based on (b) and (c) the return periods are respectively 9 yr (complete dependence) and 90 yr (complete independence) for the same peak flow of 35,000 cfs (Fig. 2.66).

THEORETICAL DISTRIBUTION OF THE RETURN PERIOD

Thomas [40] studied the theoretical distribution of the return period, based on the probability J of the occurrence of an event in any N-year

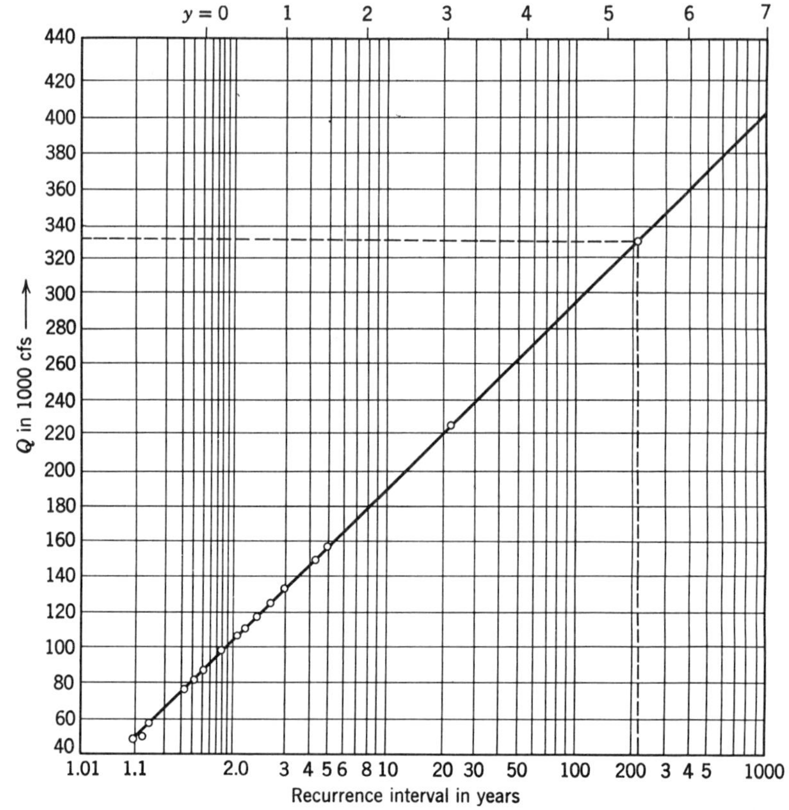

Fig. 2-65 Annual floods of Delaware River at Trenton, N.J., analyzed after Gumbel's theoretical method during the period 1921–1960. T_r for 1955 flood of 329,000 cfs is computed to be 232 yr; T_r for a flood of 360,000 cfs is computed to be 469 yr; the 1000-yr flood is computed to be about 400,000 cfs. (Not all data are plotted.)

period. It is not certain that the *N*-year event will occur every *N* years, and there is always a probability that the *N*-year event will reoccur after a time lapse shorter than *N* years. The likelihood of the reoccurrence of the *N*-year event decreases as the possible interval of reoccurrence becomes smaller. This is clarified by Table 2.14, based on the equation

$$J = 1 - P^N \tag{2.56}$$

in which J is as defined above, P is the probability of nonoccurrence of an event in any year, and P^N is the probability of nonoccurrence of that event in N years.

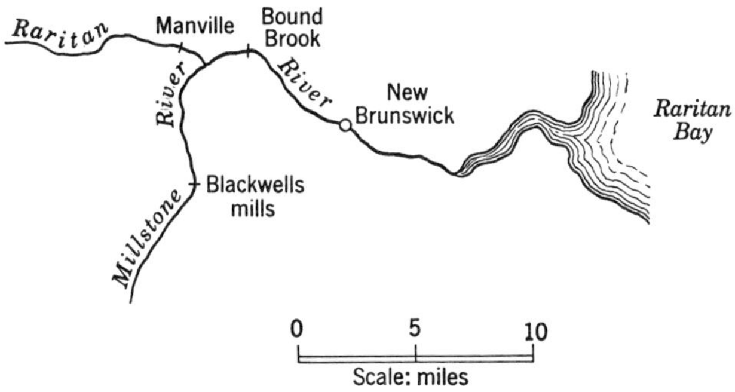

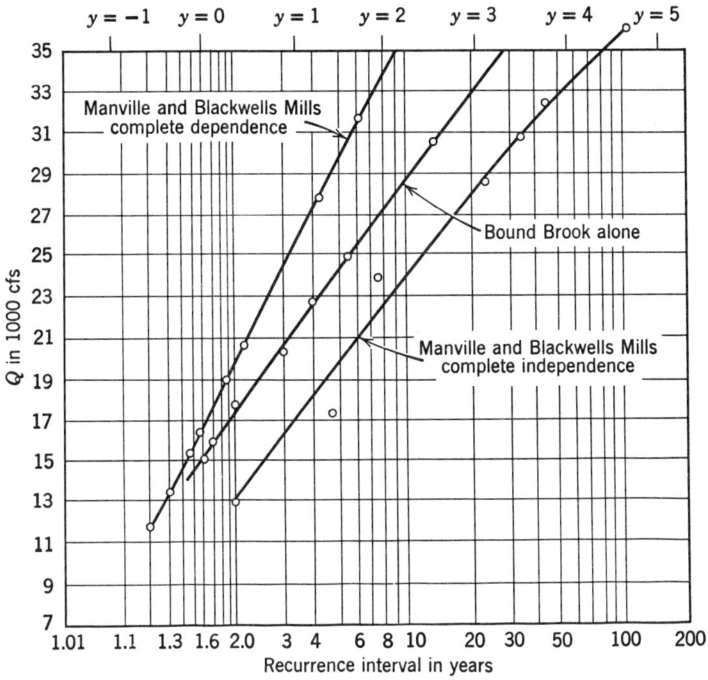

Fig. 2-66 Frequency analysis for floods of Raritan River at Bound Brook, N.J., during a 10-yr period, 1945–1954.

Table 2.14 shows for example that there is a 99% probability that the 1,000-yr flood will not reoccur after only 10 yr and that there is only a 1% probability that any two floods of the magnitude of the 1,000-yr flood will actually be separated by 4,620 yr.

Table 2.14 Theoretical Distribution of the Return Period
(After Thomas) [40]

Average Return Period, $\bar{T}_r$	Actual Return Period, T_r, Exceeded Various Percentages of the Time						
	1%	5%	25%	50%	75%	95%	99%
2	8	5	3	1	0	0	0
5	22	14	7	3	1	0	0
10	45	28	14	7	3	0	0
30	137	89	42	21	8	2	0
100	459	300	139	69	29	5	1
1,000	4,620	3,001	1,400	693	288	51	10
10,000	46,200	30,001	14,000	6,932	2,880	513	100

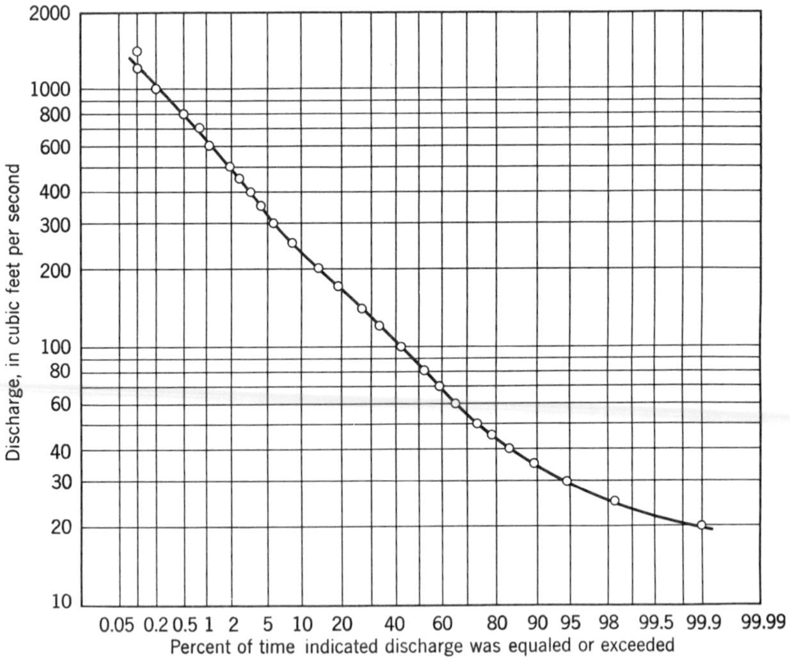

Fig. 2-67 Duration curve of daily flows, South Branch Raritan River near High Bridge, N.J., 1918–1958, on logarithmic-probability paper. (Courtesy of State of New Jersey, Division of Water Policy and Supply.)

Thomas [40] also computed the average return periods for various levels of probability [Table 2.15], as a function of the number of years of record N and the rank from the top, m'. The table shows, for example, that there is about a 25% probability that the maximum event will reoccur on an average of every 15 yr if the record period is 20 yr, and that there is about a 99% probability that the maximum event will reoccur on an average of at least every 2,000 years for a 20-yr. period.

FLOW-DURATION CURVES

A flow-duration curve (Fig. 2.67) is a cumulative frequency curve that shows the percentage of time that specified discharges have been equaled or exceeded. It uses the full series of data, i.e., all daily flows of record, which are of necessity interrelated and therefore not suitable for flood forecasting purposes (data must be independent). The flow-duration curve is used in determining the firm power, or primary power, on which a hydroelectric plant may depend for production at all times or for a given per cent, say 95, of the time. The flow-duration curve is also useful in comparing one stream with another.

RAINFALL INTENSITY-DURATION-FREQUENCY CURVES

The U.S. Weather Bureau [41], in making a frequency analysis by the method of extreme values after Gumbel, has determined a series of curves (Fig. 2.68) giving the rainfall intensity as a function of duration for a parametric value of the return period T_r from 2 yr to 100 yr. De Wiest and Waters [42] used these curves to evaluate a design flood with a return period of 150 yr determined by the U.S. Army Corps of Engineers for a proposed dam-site on the San Francisquito Creek, near Palo Alto, California (Fig. 2.69). San Francisquito Creek stream flow records to date (1955) covered only 17 yr. Any extrapolation of this small amount of data to 150 yr would have resulted in a very poor basis for engineering calculations. However, rainfall records were available for the last 53 yr. Using the rainfall data, the Corps of Engineers constructed curves of maximum precipitation over a time period (i.e., 4 hr, 8 hr, or 24 hr) versus return periods for each of the major storms. With the 53-yr record, a fairly accurate estimate of the rainfall with a return period of 150 yr could be made. As an alternative, use of the set of curves of Fig. 2.68 greatly reduced the work necessary to complete the curve of Fig. 2.70, showing the maximum 24-hr precipitation of Palo Alto versus its return period. Values for San Francisco and San Jose were averaged as the Weather Bureau publication does not include data for Palo Alto. By extrapolation from this curve, the maximum 24-hr precipitation with a return period of 150 yr was determined to be 4.93 in. Next, from data

Table 2.15 Average Return Periods for Various Levels of Probability
(After Thomas) [40]

Rank from Top, m'	No. of Years of Record, t	Probability 0.01	0.25	0.50	0.75	0.99
1	2	1.11	2.00	3.41	7.46	200
	3	1.28	2.70	4.85	10.9	299
	5	1.66	4.13	7.73	17.9	498
	10	2.71	7.73	14.9	35.3	996
	15	3.78	11.3	22.1	52.6	1490
	20	4.86	14.9	29.4	70.0	1990
	30	7.03	22.1	43.8	104	2980
	60	13.5	43.8	87.0	209	5970
2	3	1.06	1.48	2.00	3.06	17.0
	4	1.16	1.84	2.59	4.12	23.8
	6	1.42	2.57	3.78	6.20	37.4
	11	2.13	4.41	6.76	11.4	71.1
	16	2.87	6.27	9.74	16.6	105
	21	3.61	8.12	12.7	21.8	138
	31	5.11	11.8	18.7	32.2	206
	61	9.62	23.0	36.6	63.4	408
3	4	1.05	1.32	1.63	2.19	7.10
	5	1.12	1.56	2.00	2.78	9.47
	7	1.31	2.06	2.75	3.95	14.1
	12	1.86	3.32	4.62	6.86	25.6
	17	2.44	4.59	6.48	9.76	37.2
	22	3.03	5.86	8.35	12.6	48.6
	32	4.21	8.41	12.1	18.4	71.6
	62	7.76	16.1	23.3	35.8	140
4	5	1.03	1.24	1.46	1.83	4.50
	6	1.09	1.42	1.73	2.24	5.78
	8	1.25	1.80	2.27	3.04	8.26
	13	1.70	2.77	3.63	5.02	14.4
	18	2.18	3.74	5.00	7.00	20.5
	23	2.67	4.72	6.36	8.98	26.6
	33	3.66	6.67	9.07	12.9	38.7
	63	6.63	12.5	17.2	24.8	75.2
5	6	1.03	1.19	1.36	1.64	3.40
	7	1.08	1.34	1.57	1.95	4.23
	9	1.21	1.64	2.00	2.55	5.85
	14	1.59	2.43	3.07	4.05	9.81
	19	2.01	3.22	4.14	5.54	13.7
	24	2.43	4.02	5.21	7.02	17.7
	34	3.28	5.60	7.35	9.99	25.5
	64	5.86	10.4	13.8	18.9	49.0

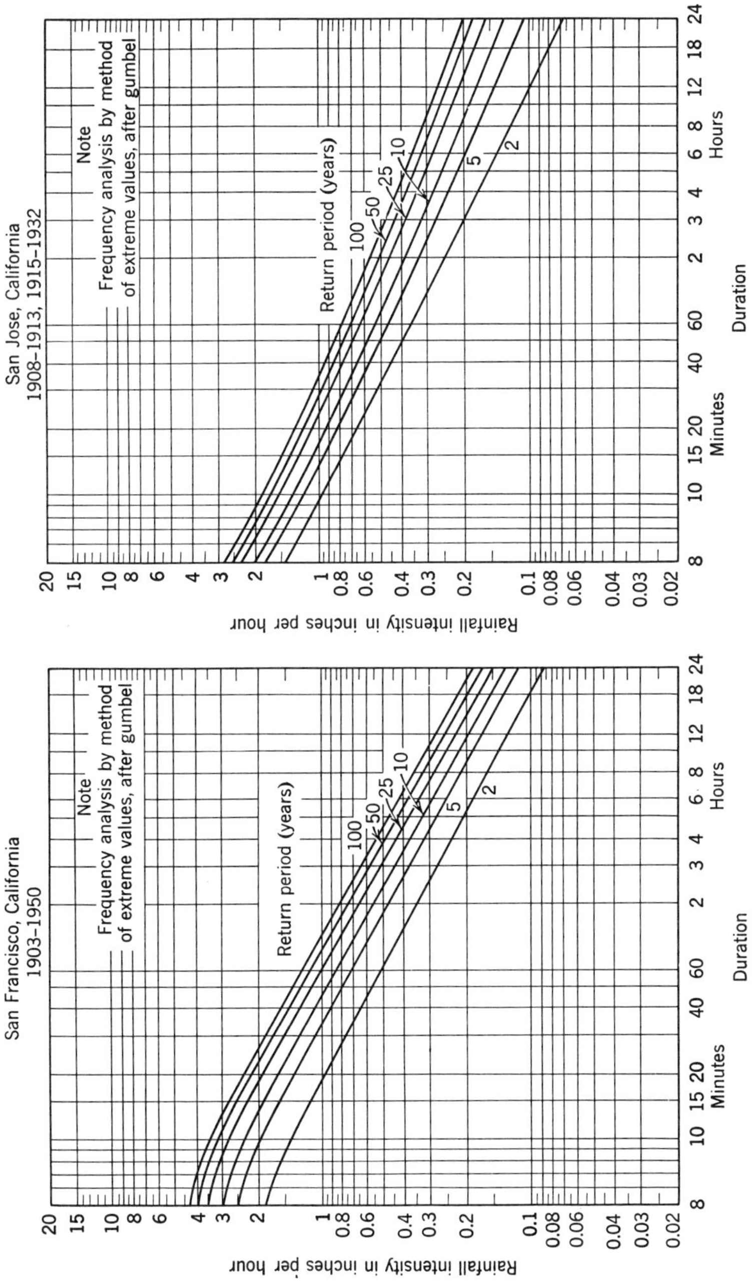

Fig. 2-68 Rainfall intensity-duration-frequency curves. (After *U.S. Weather Bureau Special Report* 25.)

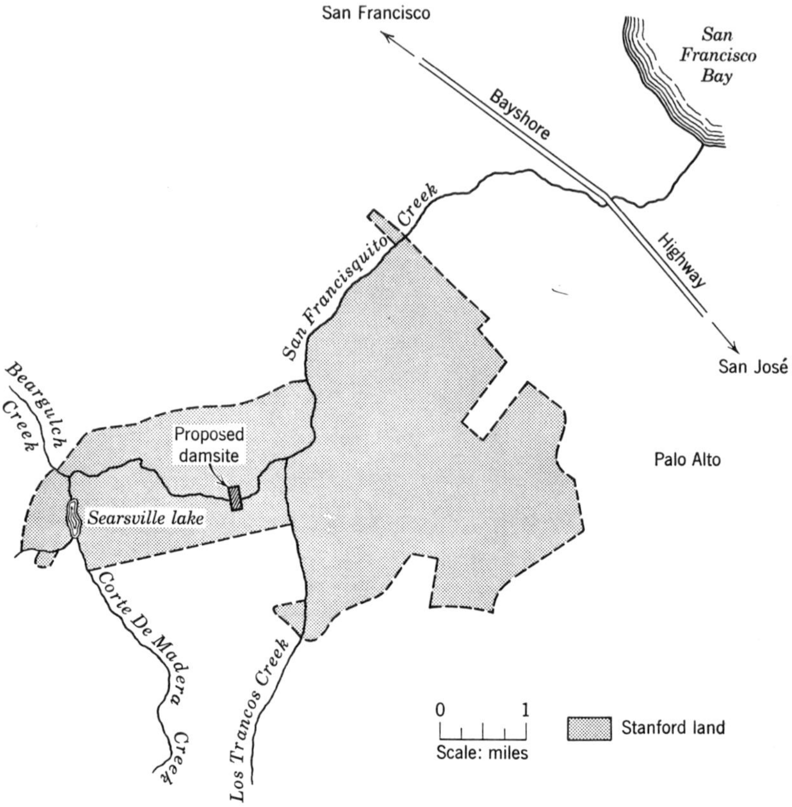

Fig. 2-69 Proposed damsite on San Francisquito Creek, near Palo Alto, Calif.

given by the *USGS Water Supply Papers*, the ten largest floods on record were plotted versus the corresponding rainfall over a 24-hr period (Fig. 2.71). Although a considerable amount of scatter was found for large flows, the curve was completed and the peak runoff for the 150-yr rainfall was found to be 8,200 cfs, in close agreement with the estimate of 8,400 cfs set forth by the Corps of Engineers. The method explained above represents a good way for engineers to extend a short period of creek runoff for a long period of time. The method, however, presupposes that a storm with a return period of 150 yr will result in a flood with a similar return period. This, of course, does not necessarily happen because of the effect of varying antecedent conditions and other factors.

2.19 *Reservoirs*

Reservoirs may be constructed on streams for purposes such as irrigation, production of water power, municipal water supply, flood protection, and

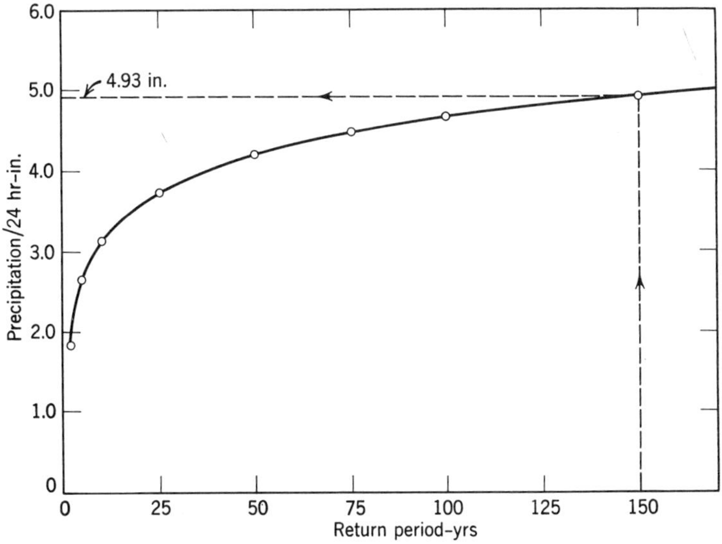

Fig. 2-70 Maximum 24-hr precipitation versus return period, Palo Alto, Calif.

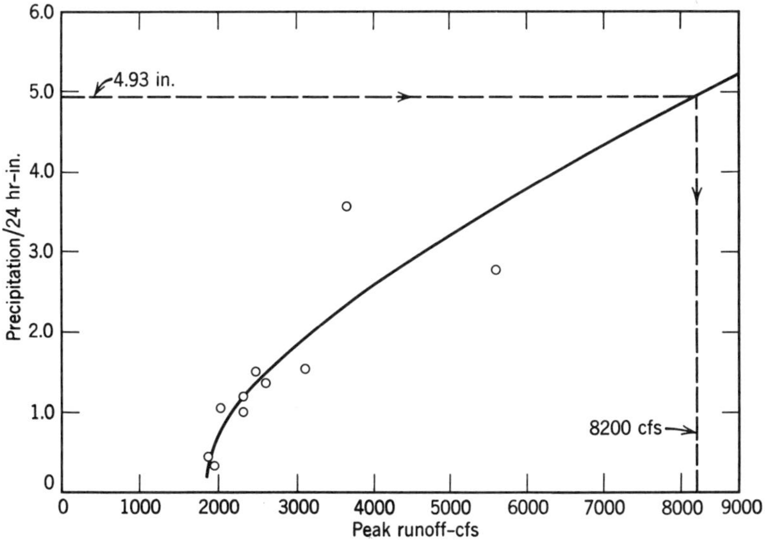

Fig. 2-71 Peak runoff in San Francisquito Creek versus Palo Alto, Calif. precipitation.

flow regulation. Any one reservoir may be used for several of these purposes at the same time. The capacity of the reservoir, in the case of flow regulation, for example, will depend on whether equalization of flow is sought on a short term basis, say one year (seasonal storage), or over a period of years (hold-over storage) [43]. In the first case the necessary storage is relatively small, although if the purpose is one of low flow augmentation (waste disposal), fresh water volumes as large as four times the amount of treated waste may be required. In the latter case, the hold-over capacity is large, sometimes as much as double the average annual runoff. A multiple-purpose reservoir may be constructed partly for seasonal storage (flood control) and partly for hold-over storage (conservation storage). In the case of the proposed San Francisquito Dam (Fig. 2.69), which would back up a reservoir of 13,100 AF, 71% of the storage capacity would be used exclusively for flood control and 29% of the capacity would be used for water conservation. Sometimes storage for water power is a by-product of conservation storage but usually it is seasonal storage.

DESIGN OF A STORAGE RESERVOIR

Designing a reservoir or determining its safe yield involves the solution of two major problems. One of these is finding the sequence of years with the lowest annual flows during a reasonable period of time, generally from 50 to 100 yrs. The other is the construction of mass curves of supply and demand for the sequence of dry years, normally on a monthly basis. Often stream flow records are available only for a short period, around 10 yr, and a synthetic record of sufficient duration must be built. One method of extending the record consists of comparing the runoff of the stream which is to be developed with that of nearby streams having older runoff records, or with a combination of the flow in these streams in order to obtain a correlation between the records of the stream that is to be developed and the records of the other nearby streams. Another method of extrapolation is to compare the runoff in the stream with the precipitation recorded at nearby stations or with any suitable combination of these records. All these correlations are made on an annual basis because they are used in determining the critical dry period which may span two years. It is always fortunate when this critical period falls within the actual period of available record, and not within some other part of the synthetic record.

The preliminary design of the reservoir is most conveniently made with the help of a mass curve (Fig. 2.72) in which the cumulated monthly inflow volumes, corrected for average losses, are plotted versus the time in months of the critical period. A mass curve makes it possible to

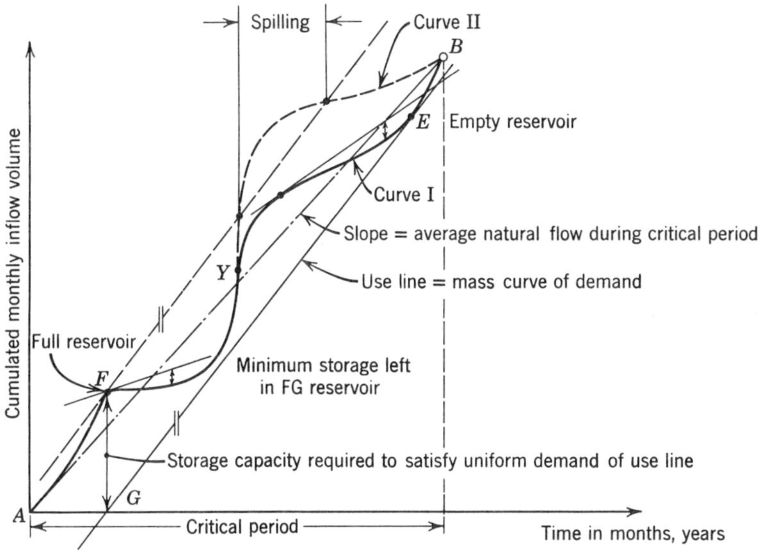

Fig. 2-72 Use of mass-curve in preliminary design of reservoir.

compute the necessary storage as a first approximation, as all corrections, such as those due to evaporation and seepage loss, are made only for average reservoir area and pool level. A detailed study of monthly flow losses must be based on a study of reservoir operation. Actually a simulation is made of the reservoir operation which would have taken place during the critical period if specific operational rules were followed.

Sometimes it is necessary to extend the monthly flow records by precipitation-runoff correlations. If the gage station is not at the damsite, its record must be adjusted for any lateral inflow. Negative flows in the mass curve are possible after correction of the available flow at the head of the proposed reservoir for seepage loss, evaporation loss, differential in runoff after and before the construction of the reservoir (see Section 2.21) and minimum flow requirements downstream.

The mass curve of demand, also called use line, is drawn tangent to the mass curve of supply (Fig. 2.72) in its lowest point *E*, and the maximum deviation *FG* between the two curves determines the required storage which will satisfy the uniform demand in this specific case. The use line, of course, may assume a curvilinear shape for a demand varying with time. Two possible mass curves of supply are represented in Fig. 2.72, namely I and II, having a common part *AY*. A line is drawn parallel to the use line through the point *F* where the reservoir is full. It shows that the reservoir will spill in the case of curve II, and considering the fact that the slope of the line AB is indicative of the average natural flow during the critical period and therefore as small as possible, it becomes obvious that beyond the critical period water will be abundant most of the time and large quantities of water may have to be spilled. Emphasis therefore should be put on the fact that a storage reservoir is constructed to meet water requirements in case of severe drought.

The safe yield of a reservoir is determined as the maximum possible uniform flow. To determine the safe yield for a given storage, tangents are drawn to the mass curve of supply so that the maximum deviation between the tangents and the curve is equal to the given storage. The tangent with the minimum slope determines the safe yield. Sometimes the costs of building a reservoir for a desirable safe yield prove to be excessive, and it is necessary to settle for a reasonable yield value, in which occasional deficiencies in water supply are permissible. This is the case for many irrigation projects.

In the design of a reservoir, allowance should be made for a progressive growth of water demand and for a loss of storage capacity due to sedimentation. The effects of sedimentation not only reduce the available capacity of the reservoir but also increase the losses due to evapotranspiration.

CAPACITY LOSS DUE TO SEDIMENTATION

The rate of sediment deposition in a reservoir frequently determines the useful life of the project. However, it is difficult to predict this rate with accuracy, as observations have shown that the rate of sediment accumulation decreases gradually. It is also important to locate the major deposition areas [44], especially when the reservoir satisfies the double purpose of flood control and water conservation and when the upper part of the storage is reserved for flood control. It is found that considerable sedimentation occurs at the entrance of the reservoir at shallow depths and absorbs a large portion of the storage allocated for flood control.

To predict the anticipated loss of storage in a reservoir, it is necessary to know the rate of sediment inflow. In a simple computation as is shown in Table 2.16, the average annual sediment inflow is used. This average inflow is often obtained by means of a synthetic record, as very few records of sediment sampling exist. In most instances, as is also the case for streamflow measurements, a station will be established on the stream when there is a reasonable chance that the stream will be developed for water conservation, flood protection, or hydro-power. A few years will elapse between the preliminary study and the erection of the hydraulic structures and it is this period of time that will make up the record which has to be extrapolated. The transported sediment is usually measured in

Table 2.16 Time Required to Fill Lake Mead to 80% *Capacity*

Capacity in 1,000 AF	Capacity-Inflow Ratio, C/I	Trap Efficiency in %		Annual Sediment Trapped in 1,000 AF	Increment Reservoir Volume in 1,000 AF	Years to Fill
		At Indicated C/I	Average for C/I Increment			
27,930	2.15	98				
			98	194	4,930	25.5
23,000	1.78	98				
			97.5	193	5,750	30
17,250	1.33	97				
			97	192	5,750	30
11,500	0.88	97				
			96.5	191	5,900	31
5,600	0.43	96				
(20% of initial capacity)					Total	116.5 years

parts per million by weight and the total weight of transported solids must be converted in volume, after settlement in the reservoir, in order to compute the reduced capacity. The problem here is to estimate the specific weight of the deposited materials, which increases with depth as the materials become more compacted. Bondurant [44] estimates that specific weights of silt, sand, and mixtures thereof may vary from 70 to 100 lb per cu ft. Clay sediments may be deposited in a number of ways; disperse particles tend to settle very slowly, flocculated clays, on the other hand, usually settle rapidly. Clay may also be found in a thixotropic state (i.e., solid when at rest, but fluid when under mechanical stress). In this state, clay particles usually settle rapidly, but are susceptible to being moved along the floor of the reservoir in density flows. Bondurant [44] cites for the initial dry weight of clay deposits in a flocculated or thixotropic state in Lake Mead a value of about 7 lb per cu ft at the surface to about 30 lb per cu ft at a depth of 50 ft. No values for the specific weight of the thixotropic clay are available below 50 ft, although the deposits extend to a depth of more than 150 ft. Bondurant expresses the belief that this material at Lake Mead would be susceptible to discharge from the reservoir if outlets were available at the proper elevation. His opinion seems correct in view of results obtained by French engineers [45] with dams in Algeria. The extremely fine sediments, consisting of silt and clay for 85 to 99% of the total weight, were successfully evacuated via horizontal holes drilled in the lower part of the dam while flowing at a specific gravity of 1.4 to 1.5.

Brune [46] made a study of the trap efficiency of reservoirs as a function of the reservoir capacity-annual inflow ratio, in which the trap efficiency is the proportion of inflowing sediments retained in a reservoir. He analyzed data for some forty reservoirs, leading to a median curve for normal ponded reservoirs as given in Fig. 2.73. This curve is useful in determining the life expectancy of a reservoir. Table 2.16 gives the computations of the time required to fill Lake Mead to 80% of its capacity. The capacity and the average annual inflows as given by Maddock [43] are respectively 27,930,000 AF and 12,987,000 AF; the average sediment inflow is estimated at 198,000 AF. If 10% of the sediments were vented

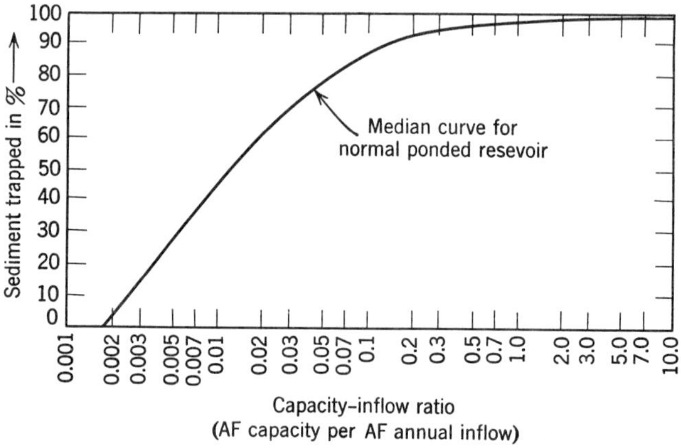

Fig. 2-73 Reservoir trap efficiency as related to capacity-inflow ratio. (After Brune.)

annually, it would take 128 yr to fill the reservoir to an 80% capacity. These computations have assumed an unfavorable constant rate of sediment deposition in the reservoir, and it is realistic to expect a longer lifetime than the one resulting from these elementary calculations.

◆ Applications to Hydrologic Problems in New Jersey

2.20 *Development of the Pennsauken Aquifer near Princeton, N.J.* [47]

When a river cuts through unconsolidated sedimentary deposits, there is a constant exchange of water between the voids of these deposits and the river itself. The water bearing strata or aquifers adjacent to the river may then be developed for ground water. The interrelation between surface water and ground water and the resulting flow pattern of water on its subterranean travel depend largely on the stage of rivers intersecting the ground-water basin and on the hydraulic conductivity of the formations, and the pattern may be altered by man-made hydraulic structures established on the rivers.

Both water quality and quantity are of vital importance in densely populated areas, often the seat of heavy industrialization. Knowledge of the influence of surface water upon the flow pattern of a ground-water basin is a prerequisite for the rational use of water and for pollution control. Indeed, pollutants are too frequently discharged in surface streams and subsequently penetrate into the ground before they can be ejected into the ocean. In the study of the Pennsauken aquifer, quantity is the main objective and the safe yield of the source of water supply must be determined. For relatively large systems at least, this involves a steady state analysis (Definition see section 3.2) in which average values of precipitation, evaporation, surface runoff and percolation on either annual or monthly bases are used. This approach belongs to the domain of the consulting engineer [47]; it is shown mathematically that terms accounting for evaporation and recharge and which are responsible for the build-up of a ground-water mound are not significant in the particular case of steady flow in strata underlain by a vast body of water at constant head. However, for smaller systems, if the available water is severely depleted by an unfavorable coincidence of prolonged drought and excessive draft during a critical period shorter than that of the steady state analysis, and if the resistance to flow delays the response to the demand beyond tolerable limits, an unsteady state analysis (Definition see section 3.2) for time-dependent input data and output requirements may become imperative before a decision as to the feasibility of carrying or not carrying out a certain project can be made. Such a study cannot be made without an adequate knowledge of the surface hydrology of the basin.

The Pennsauken deposits are located in the upper Millstone River watershed (Fig. 2.74), part of a lowland area stretching from the Raritan Estuary to the Delaware River below Trenton, N.J. This area is covered with a formation composed chiefly of coarse sand, a subordinate amount of gravel, and a slight admixture of material of a clayey nature. A master stream once drained this lowland into the valley of the lower Delaware River. For a long period, this master stream and its tributaries eroded the Cretaceous formations of the valley which were underlain by a Magothy-Raritan base. The easily erosible formations provided much debris that was transported by the rivers and progressively deposited in the lower stream reaches. Finer sediments were carried farther downstream while the heavier sands and coarse gravel gradually filled the deep main trough created by the master stream and extending down to the basal Margothy-Raritan formations. The aggradation by deposition of the main valley was done largely by reworked Cretaceous sediments called Pennsauken, and the valley itself is often referred to as Pre-Pennsauken. The wide cross-state lowland that was shaped in this manner between the

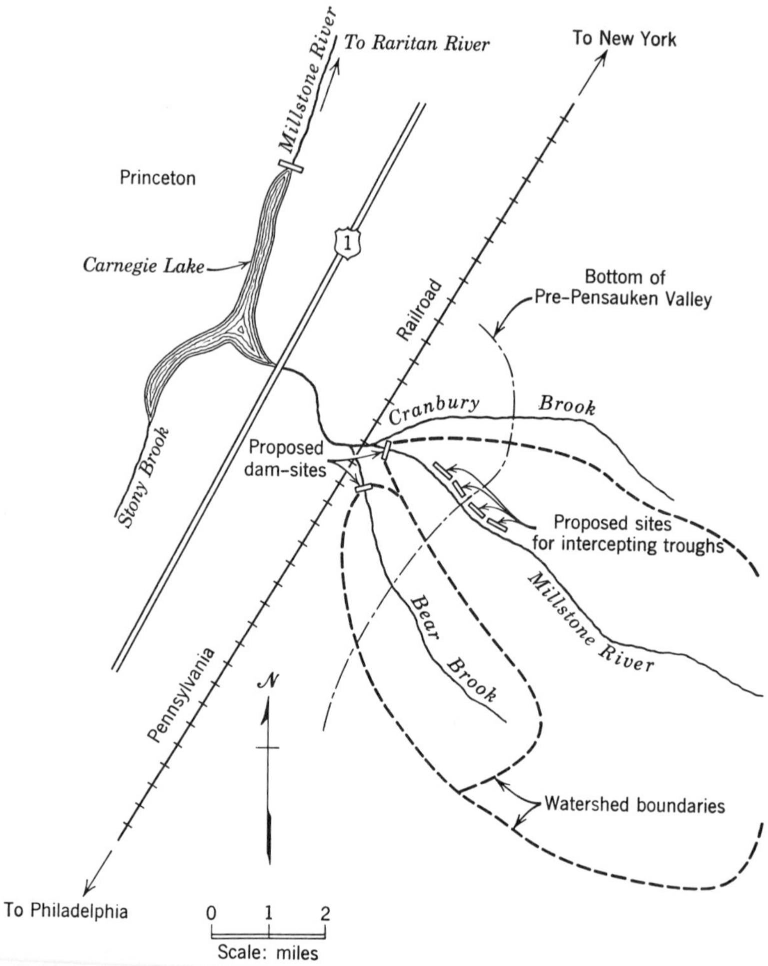

Fig. 2-74 Geographical location of upper Millstone River, N.J. watershed.

New Jersey Turnpike and Princeton's Lake Carnegie rarely rises above an elevation of 120 ft above sea level, with the 80 ft contour closely paralleling the reaches of the Millstone River and of Bear Brook, as shown in Fig. 2.74. Below an elevation of 150 ft, practically all of the watershed is covered by very porous Pennsauken deposits. Surface permeability is so high that even the heaviest rainstorms leave few traces of direct runoff, a feature also substantiated by the absence of gullies and water washes even in freshly plowed fields. Therefore, it is not unreasonable to assume that approximately 50% of the mean annual precipitation (around 45 in.), percolates to the ground-water basin. Considering that the average runoff per square mile from that part of the Millstone watershed which is underlain by Pennsauken deposits is much larger than that of the remainder of the watershed, mainly drained by surface runoff, it is concluded that the ground-water reservoir extends far beyond the surface divides delineating the watershed. Flow from that ground-water reservoir sustains the base flow of the Millstone River.

The late Homer Sanford [47] used streamflow records of the Neshanic River at Reaville (Hunterdon County, northwest of Princeton) which drains a basin with characteristics similar to the drainage area of the lower Millstone River, to estimate the flow of the latter at approximately 141 mgd (million gallons per day). Subtracting this quantity from the total flow of the Millstone River, Sanford found an average discharge of 98 mgd for the Millstone above Lake Carnegie and, weighted on an areal basis, a runoff of 56 mgd for the Millstone-Bear Brook watershed (Fig. 2.74).

Most of the recharge to the ground-water table occurs during the rainy season of early November to late May. With the resulting build-up of the ground-water level, effluent discharge into the rivers increases. This increase ultimately becomes a net loss of fresh water to the ocean. Sanford considered the major part of the 56 mgd runoff to be a waste and proposed to reduce this loss by the construction of two low earth dams at the sites indicated in Fig. 2.74. Water behind these dams would reach an elevation of 70 ft and the dam crests would be approximately 10 ft higher. The increased water level in the rivers would automatically decrease the hydraulic gradient from the ground-water reservoir to the rivers and hence limit the effluent discharge. Simultaneously, the ground water would be pumped to make available more storage space for retention of a higher percentage of the winter-spring rainfall, or for quick recharge from flash summer runoff. The project therefore is essentially one of bank storage [53], aiming at the prevention of fresh water losses to the ocean of the order of 56 mgd. The project is typical of the Atlantic Coastal Plains of the United States where vast quantities of fresh water are lost to the ocean and where plans for conservation of this precious mineral are still waiting to be made. It is in the densely populated and highly industrial area between Washington and New York that water will become scarce in the not so distant future.

For the withdrawal of water from the ground Sanford proposed the hydraulic dredging of deep intercepting troughs, parallel with the Millstone River, to a depth of 50 ft below pool level behind the dams. Sanford suggested four possible sites, two on each side of the bottom of the pre-Pennsauken Valley (Fig. 2.74). To obtain an estimate of the yield of the aquifer into the troughs, he prescribed a long-time pumping test from a pilot trench excavated over a length of 1,000 ft and with a depth of 15 ft at the site of the proposed permanent troughs. Sanford assumed that a trench of these dimensions would support pumpage at a rate of not less than 5 mgd. Indeed, flow tests made by the Surface Water Branch of the USGS, Trenton, N.J., have indicated stream discharge upstream from the pilot test sites that is materially in excess of the suggested 5 mgd rate of pumpage.

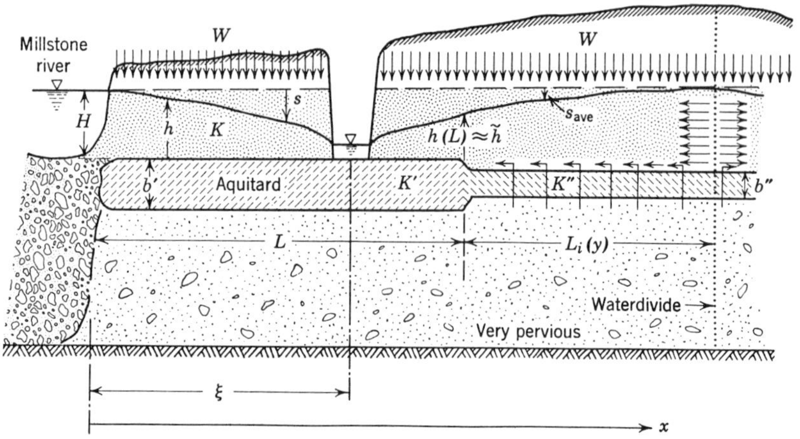

Fig. 2-75 Mathematical model of idealized field conditions for flow into trench on upper Millstone River, N.J.

De Wiest [47] has tested Sanford's hypothesis for steady flow with the help of a mathematical model (Fig. 2.75) and has also examined alternative solutions involving well fields. At the time (March 1964) of the writing of this textbook the author is engaged in the analytic investigation of the dynamic storage (unsteady state analysis) of the project, as well as in an R-C electric analog model study of same.

2.20 *Preliminary Study to Determine the 1,000-Year Design Flood for the Spillway of Spruce Run Lake Reservoir* (N.J.)

The construction of Spruce Run Lake Reservoir by the erection of a dam at the confluence of the Mulhockaway and the Spruce Run, tributaries of the South Branch of the Raritan River (Fig. 2.76), was authorized by the New Jersey Water Supply Law of 1958 to provide storage facilities to meet the immediate and near-future water demands of the northeastern metropolitan and Raritan Valley areas [48]. Its name is often linked to that of Round Valley, an off-channel reservoir, the construction of which

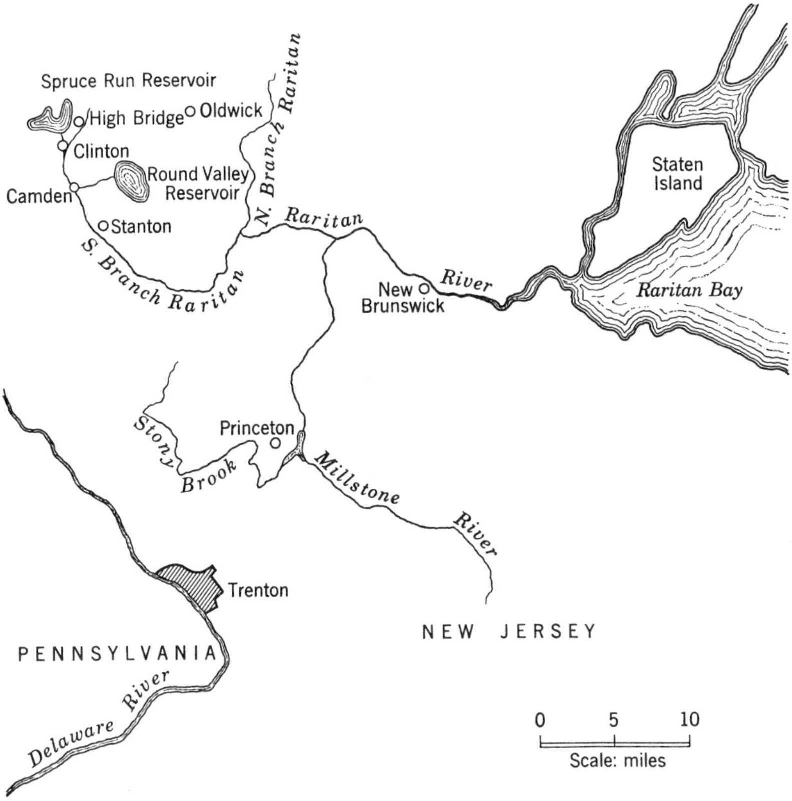

Fig. 2-76 Geographical location of Spruce Run, N.J.

was authorized by the same 1958 law. Spruce Run Lake Reservoir is expected to provide 11 billion gallons multiple-purpose storage for various streamflow regulation uses. In March 1964 the project was in the stage of completion. The reservoir has been created by the construction of a rolled earth fill dam with an impervious core across the valley of Spruce Run at its confluence with Mulhockaway Creek. An overflow spillway channel is constructed in bedrock in a saddle at the south end of the dam and is designed to discharge 2.5 times the estimated maximum flood on Spruce Run and Mulhockaway Creek. For lack of observed streamflow records on either Spruce Run or Mulhockaway Creek, the natural runoff from these streams has been assumed to equal the average of the observed streamflows recorded on the South Branch of the Raritan River at the High Bridge and Stanton gaging stations. The 1,000-yr design flood as given by reference 48 is 24,000 cfs.

De Wiest [49] has made a preliminary study of the aforementioned 1,000-yr design flood using techniques described in this chapter which may be summarized as follows:

(*a*) *Available hydrologic data*

Stage-hydrograph records of Spruce Run at Clinton, drainage area 41.3 sq mi, 4-yr period 1959–1962.

Daily precipitation and API at Oldwick for 1957–1962.

Daily precipitation at Clinton for a 15-yr period 1943–1957.

Hourly precipitation at Trenton for all of the 20-yr period 1943–1962 for which streamflow records should be synthesized.

(*b*) *Design method*

Fourteen storms were selected from the 1959–1962 period, and it was hoped to develop a coaxial relation between total precipitation and runoff for given API and storm durations of the same type as shown in Fig. 2.47. However no storm of extremely heavy precipitation was observed during the 1959–1962 period, and when the storm runoff was separated from the base flow and plotted versus the precipitation at nearby Oldwick, the result fitted closely to a straight line (Fig. 2.45) relatively independent of API conditions and with a runoff coefficient $k = 0.8$. This high value of k is entirely compatible with the physiography of the watershed and its location in the foothills of the Kittatinny Mountains. Instead of a correlation like that of Fig. 2.47 which could not be established, the curves of Fig. 2.77 were derived in the following way. The storm runoff for the fourteen selected floods of the 1959–1962 record of Spruce Run at Clinton was plotted versus the API at Oldwick, and the amount of precipitation observed at Oldwick during the storms was noted next to each point on the graph. Next the maximum amounts of annual precipitation observed in Clinton were determined for the period 1943–1957 and for Oldwick during 1957–1958, from the Climatological Data for New Jersey compiled by the U.S. Weather Bureau. The API's were computed for each case from the same Climatological Data and the storm runoff was determined from Fig. 2.45, taking into consideration the fact that in all but two cases of maximum annual floods (July 1945 and August 1955) the antecedent precipitation index was moderate. The combined data for the periods 1943–1957 (Clinton precipitation) and 1957–1962 (Oldwick precipitation) were plotted in Fig. 2.77. The increment in runoff between Figs. 2.45 and 2.77 was 1.8 to 2.1 in. for the July 1945 storm (API 3.93) and 3.30 to 4.0 in. for the August 19, 1955 storm (API 7.15).

The duration of the storms in Clinton and Oldwick were estimated from the hourly precipitation of concurrent storms at Trenton. For simplicity's sake these durations were averaged at 8 hr and 14 hr, seemingly typical

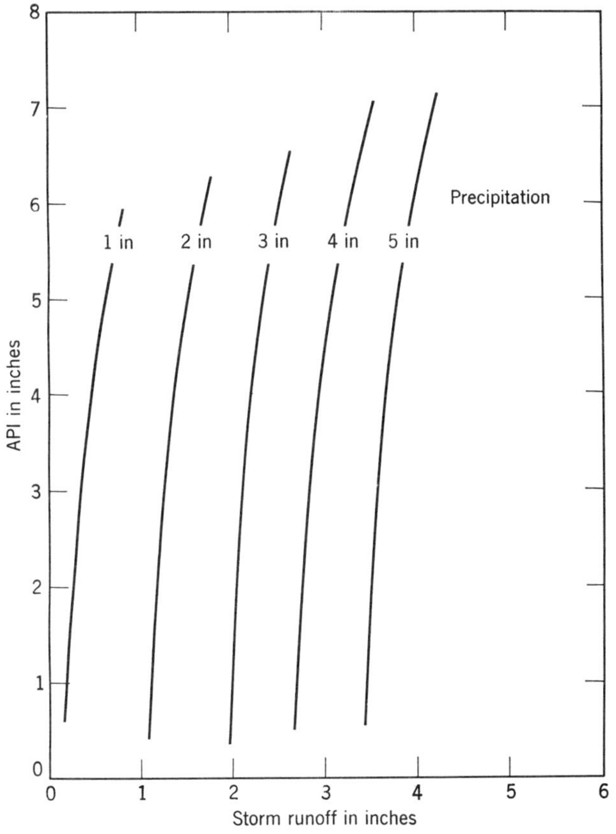

Fig. 2-77 Runoff-precipitation relationship depending on API for Spruce Run, N.J.

periods for most of the storms. Unit hydrographs for 8 and 14 hr durations were constructed from the observed floods of Spruce Run during the 1959–1962 period, and the peak flow for every storm of the 1943–1958 period was determined. Thus a sequence of 20 years of maximum floods was obtained and Gumbel's theoretical method was applied. The 1,000-year flood was determined to be 16,600 cfs as compared to the 24,000 cfs used by the State of New Jersey. The 1955 flood has a recurrence period of 28 yr as compared to a recurrence period of 232 yr for the 1955 flood of the Delaware River in Trenton, based on a 40-yr record (Fig. 2.65). However, the recurrence period for the 1955 flood of the Delaware River in Trenton based on the 1943–1962 record is found to be only 99 yr. The fact that the 1955 Spruce Run flood determined with the data available at the present has a return period of only 28 yr as compared to a reasonably expected value of 100 yr (for the 20-yr period of record) would indicate that 16,600 cfs is a low value for the estimated 1,000-yr flood and that the actual 1,000-yr design flood of 24,000 cfs is realistic and not overpessimistic.

REFERENCES

1. Jones, P. B., G. D. Walker et al., "The Development of the Science of Hydrology," Texas Water Commission, *Circular* 63–03, April 1963.
2. Batisse, M., "Le programme de recherches a long-terme de l'UNESCO dans le domaine de l'hydrologie scientifique," *Journal of Hydrology*, Vol. 1, pp. 166–176 (1963).
3. Leet, L. D. and S. Judson, "Running Water," *Physical Geology*, Chapter 10, pp. 159–197, Prentice-Hall, 1958.
4. Lobeck, A. K., "Streams in General," *Geomorphology, an Introduction to the Study of Landscapes*, Chapter V, pp. 155–183, McGraw-Hill, 1939.
5. Horton, R. E., "Discussion of Paper, Flood Flow Characteristics, by C. S. Jarvis," *ASCE Transactions*, Vol. 89, pp. 1081–1086 (1926).
6. Linsley, R. K., M. A. Kohler, and J. L. Paulhus, *Hydrology for Engineers*, Chapter 3, McGraw-Hill, New York, 1958.
7. Thiessen, A. H., "Precipitation for Large Areas," *Monthly Weather Review*, Vol. 39, pp. 1082–1084 (July 1911).
8. Horton, R. E., "Discussion on Distribution of Intense Rainfall," ASCE, *Transactions*, Vol. 87, 578–585 (1924).
9. Corps of Engineers, U.S. Army, "Storm Rainfall in the United States Depth-Area-Duration Data," War Department, 1945.
10. Miller, D. W., J. J. Geraghty, and R. S. Collins, *Water Atlas of the United States*, published by Water Information Center, Port Washington, Long Island, New York, 1962 (40 pp.).
11. Krick, I. P. and R. Fleming, *Sun, Sea and Sky*, Chapter 1, pp. 15–39, J. B. Lippincott Co., Phildelphia and New York, 1954.
12. Linsley, R. K., M. A. Kohler, and J. L. Paulhus, *Applied Hydrology*, McGraw-Hill, New York, 1949 (689 pp.).
13. Leopold, L. B. and W. B. Langbein, *A Primer on Water*, U.S. Geological Survey, 1960 (50 pp.).
14. Polubarinova-Kochina, P. Ya, *Theory of Ground Water Movement*, p. 38, Princeton University Press, Princeton, 1962.
15. U.S. Weather Bureau, Washington, D.C., "Psychrometric Tables," *Circular* 235 (1941).
16. Meyer, A. F., *Evaporation from Lakes and Reservoirs*, Minnesota Resources Commission, St. Paul, Minn., 1942.
17. McEwen, G. F., *Results of Evaporation Studies*, Scripps Institute of Oceanography, *Technical Series*, Vol. 2, pp. 401–415 (1930).
18. Anderson, E. R., L. J. Anderson, and J. J. Marciano, "A Review of Evaporation Theory and Development of Instrumentation," *U.S. Navy Electronics Laboratory Report* 159, pp. 3–37 (February 1950).
19. Baker, D. and H. Conkling, *Water Supply and Utilization*, Wiley, New York, 1930 (495 pp.).

20. Horton, R. E., "The Role of Infiltration in the Hydrologic Cycle," *Transactions American Geophysical Union*, Vol. 14, pp. 446–460 (1933).
21. Gilcrest, B. R., "Flood Routing," in *Engineering Hydraulics*, Chapter X, pp. 635–710, H. Rouse (Ed.), Wiley, New York, 1950.
22. Stevens, J. C., "A Method of Estimating Stream Discharge from a Limited Number of Gagings, " *Engineering News*, pp. 52–53 (July 18, 1907).
23. Posey, C. J., "Gradually Varied Channel Flow," in *Engineering Hydraulics*, Chapter IX, pp. 589–634, H. Rouse (Ed.), Wiley, New York, 1950.
24. New Jersey (State of), Department of Conservation and Economic Development, "Surface Water Supply of New Jersey, Stream Flow Records, October 1955 to September 1960," *Special Report* 20, 1963 (425 pp.).
25. Barnes, B. S., "Discussion of Analysis of Runoff Characteristics by O. H. Meyer," *ASCE Transactions*, Vol. 105, p. 106 (1940).
26. Langbein, W. B., "Some Channel Storage and Unit Hydrograph Studies," *Transactions American Geophysical Union*, Vol. 21, pp. 620–627 (1940).
27. Sherman, L. K., "Streamflow from Rainfall by the Unit-Graph Method," *Engineering News Record*, Vol. 108, pp. 501–505 (1932).
28. Kohler, M. A. and R. K. Linsley, "Predicting the Runoff from Storm Rainfall," *U.S. Weather Bureau Research Paper* 34, 1951.
29. Hoel, P.G., *Introduction to Mathematical Statistics*, pp. 126–129, Wiley, New York, 1947.
30. Snyder, F. F., "Synthetic Unit Hydrographs," *Transactions American Geophysical Union*, Vol. 19, Part I, pp. 447–454 (1938).
31. Linsley, R. K., "Applications of Synthetic Unit Graphs in the Western Mountain States," *Transactions American Geophysical Union*, Vol. 24, Part 2, pp. 580–587 (1943).
32. Massau, J., "Mémoire sur l'intégration graphique des équations aux dérivées partielles" *Annales de l'Association des Ingénieurs Sortis des Ecoles Spéciales de Gand*, Vol. 23, p. 95 (1900).
33. McCarthy, G. T., "The Unit Hydrograph and Flood Routing," *Conference of North Atlantic Division, U.S. Corps of Engineers*, June 1938.
34. Paulhus, J. L. H. and C. S. Gilman, "Evaluation of Probable Maximum Precipitation," *Transactions American Geophysical Union*, Vol. 34, pp. 701–708 (October 1953).
35. California Department of Public Works, "Flow in California Streams," *Bulletin* 5 (1923).
36. Hazen, A., *Flood Flows*, Wiley, New York, 1930.
37. Gumbel, E. J., "On the Plotting of Flood Discharges," *Transactions American Geophysical Union*, Vol. 24, Part II, pp. 699–719 (1943).
38. Gumbel, E. J., "Statistical Theory of Extreme Values and Some Practical Applications," *U.S. National Bureau of Standards—Applied Mathematics, Series* 33 (February 1954).
39. Powell, R. W., "A Simple Method of Estimating Flood Frequency," *Civil Engineering*, Vol. 13, pp. 105–107 (1943). With discussions by W. E. Howland and E. J. Gumbel, respectively on pp. 185 and 438.
40. Thomas, H. A., Jr., *The Reliability of Hydrologic Predictions*, a paper presented at the meeting of the American Geophysical Union, Cambridge, Mass., September 1947. See also: "Frequency of Minor Floods," *Journal, Boston Society of Civil Engineers*, Vol. 35, pp. 425–442 (1948).
41. U.S. Weather Bureau, *Technical Paper* 25, *Rainfall Intensity-Duration-Frequency*

Curves for the United States, Alaska, the Hawaiian Islands and Puerto Rico, 1955.
42. De Wiest, R. J. M. and G. Watters, *An Investigation into Proposals for Flood Control on San Franciscquito Creek*, Student Report, Civil Engineering 262, Stanford University, Spring Quarter, 1958.
43. Maddock, Th., "Reservoir Problems with Respect to Sedimentation," *Proceedings of the Federal Inter-Agency Sediment Conference*, pp. 9–17, Denver, 1947.
44. Bondurant, D. C., "Reservoir sedimentation, barrages et bassins de retenue," *Colloquium* 14, pp. 185–195, Université de Liège, Belgium (1959).
45. Thévenin, J., "Etude de la sédimentation des barrages-réservoirs en Algérie et moyens mis en oeuvre pour préserver les capacités," *Colloquium* 14, pp. 227–264, Université de Liège, Belgium (1959).
46. Brune, G. M., "Trap Efficiency of Reservoirs," *Transactions American Geophysical Union*, Vol. 34, pp. 407–418 (June 1953).
47. De Wiest, R. J. M., "Replenishment of Aquifers Intersected by Streams," *ASCE Journal of the Hydraulics Division*, pp. 165–191 (November 1963).
48. New Jersey, State of, Department of Conservation and Economic Development, Division of Water Policy and Supply, *Special Report* 15, August 1958.
49. De Wiest, R. J. M., *Preliminary Study of the 1,000 year-Design Flood for the Spillway of Spruce Run Lake Reservoir*. Paper presented at the 10th Annual Meeting of the New Jersey Academy of Sciences, Seton Hall University, April 4, 1964.
50. Gray, D. M., "Synthetic Unit Hydrographs for Small Watersheds," *ASCE Journal of the Hydraulics Division*, pp. 33–54 (July 1961).
51. Eagleson, P. S., "Unit Hydrograph Characteristics for Sewered Areas," *ASCE Journal of the Hydraulics Division*, pp. 1–25 (March 1962).
52. Mesnier, G. N. and K. T. Iseri, "Selected Techniques in Water Resources Investigations," *USGS Water Supply Paper* 1669-Z, pp. 1–64 (1963).
53. Cooper, H. H. and M. I. Rorabaugh, "Ground-Water Movements and Bank Storage due to Flood Stages in Surface Streams," *USGS Water Supply Paper* 1536-J, pp. 343–366 (1963).

CHAPTER THREE

Ground-Water Flow

◆ Elementary Concepts and Definitions

3.1 *Introduction*

Ground-Water flow, although an important aspect of the science of geohydrology, is only a special case of the flow of fluids through porous media. Consequently many concepts of the flow of fluids in porous media have passed into the more restricted domain of ground-water flow without being adequately defined or redefined in the process. This puts an unnecessary burden upon the reader who has a limited background in fluid mechanics in particular and in continuum mechanics in general. Sometimes an elementary concept, such as moisture content of a porous medium, is used without specifying its volumetric or gravimetric nature. Soil physicists may prefer to use the volumetric and soil mechanicians the gravimetric. Other basic concepts have become so familiar to engineers and physicists that their definition is omitted as a rule. These concepts are briefly outlined in this chapter in order to facilitate the reading of this textbook.

3.2 *Major Subdivisions of Ground-Water Flow*

The subject of ground-water flow may be broken down into several subdivisions according to the dimensional character of the flow, the time dependency of the flow, the boundaries of the flow region or domain, and the properties of the medium and of the fluid.

DIMENSIONAL CHARACTER

All ground-water flow in nature (Fig. 3.1) is to a certain extent three-dimensional. This means that if it were possible to measure the velocity of a water particle percolating through the soil and to represent the

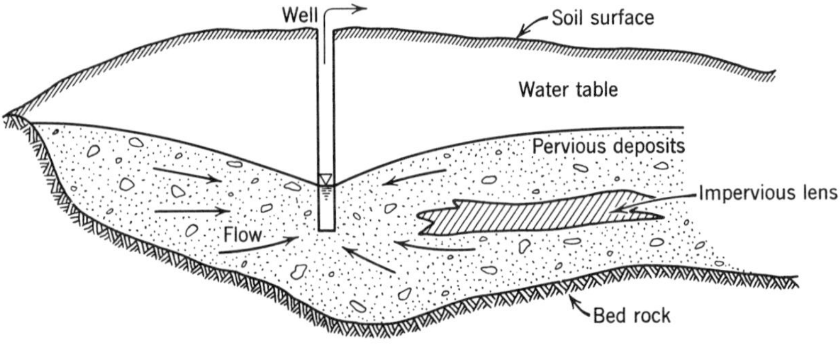

Fig. 3-1 Three-dimensional flow to a well.

velocity by a vector[1] of a length proportional to the magnitude of the velocity and parallel to the direction of the flow, the vectors at any number of points along the flow path would not be coplanar. It is equivalent to saying that the velocity vector in any point has components along three main axes, x, y, and z, of a cartesian coordinate system, or even more simply, that the velocity is a function of x, y, and z. The cross section through the axis of the well of Fig. 3.1 shows a flow picture that is different from that of any other cross section through the axis of the well. Temporarily, we may think of the true velocity of water in the pores of the soil, although as will be seen in section 4.2, it is not this true velocity which occurs in the equations of ground-water flow.

The difficulty in solving ground-water problems depends on the degree to which the flow is three-dimensional. It is practically impossible to solve

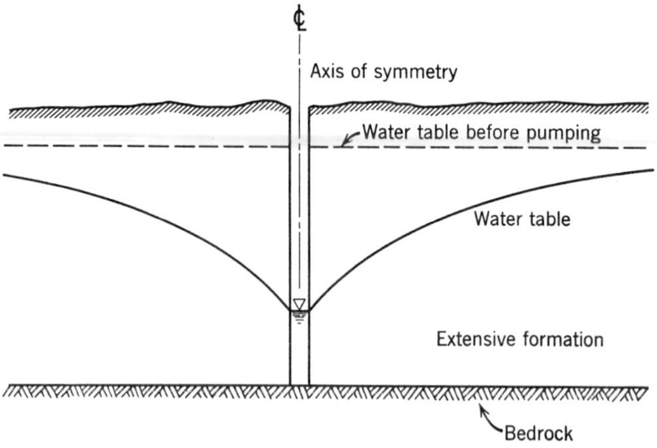

Fig. 3-2 Three-dimensional flow to a well with radial symmetry.

[1] A vector is an oriented line segment with given length and direction.

analytically a natural three-dimensional ground-water flow problem unless symmetry features (Fig. 3.2) of the problem make it possible to put the equations into two-dimensional form. Another particular kind of flow is that in which the aforementioned velocity vectors would be nearly coplanar, in which case it is justifiable to treat the problem as a two-dimensional one (Fig. 3.3). Fortunately, this approximation can often be made and the majority of all engineering problems are approached in this way. Even then, two-dimensional problems may be hard to solve because of difficult boundary conditions. The solution may be simplified further

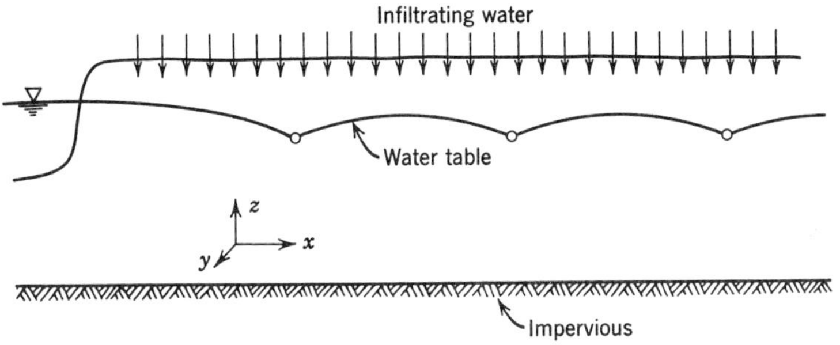

Fig. 3-3 Horizontal drains parallel to a very long rectilinear ditch (y-direction). The flow picture is independent of the y-direction when it is far away from the ends of the drains and ditch.

by reducing the dimensionality of the problem to one. It is evident that in this step-by-step reduction a significant error may be introduced and that it remains for the common sense and the engineering practice of the hydrologist to estimate that error.

In this textbook we illustrate techniques for the solution of problems made two-dimensional either by assumption or by special symmetry features.

TIME DEPENDENCY

Ground-water flow may be evaluated quantitatively by the knowledge of the velocity, pressure, density, temperature, and viscosity of water percolating through a geologic formation. These characteristics of water are commonly the unknown variables of the problem and may vary in place, from point to point in the formation, and in time (i.e., assume different values as time goes on). If the unknown or dependent variables are functions of the space variables or independent variables x, y, z only, the flow is steady. On the other hand if the unknowns are also functions of time, the flow is unsteady or time-dependent. Time dependency will make it more difficult to solve the problem, time playing to some extent the role of a fourth dimension. However, it cannot be said that time adds the same difficulty to a problem as does the addition of a dimension. In general it is easier[2] to solve a two-dimensional unsteady flow than a three-dimensional steady flow problem. Therefore, problems of the two-dimensional type will be solved in this textbook.

Steady flow is given special attention in Chapter 5. It may be conceived of as a limit case of unsteady flow, as time goes to infinity, or else as the average of unsteady flow over a given period.

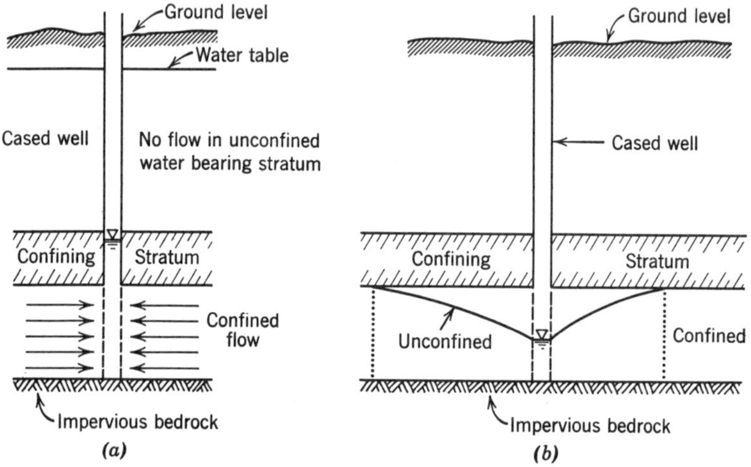

Fig. 3-4 (*a*) Confined flow to a well; (*b*) transition from confined (outer region) to unconfined flow condition (near well).

BOUNDARIES OF FLOW REGION OR DOMAIN

Ground-water flow is confined when the boundaries or bounding surfaces of the medium (i.e., the space made up by the water-filled pores) through which the water percolates are fixed in space for different states of flow. Ground-water flow is unconfined, on the other hand, when it possesses a free surface, the position of which varies with the state of flow. Unconfined flow is sometimes referred to as flow with a water table. Ground-water flow may pass from the confined type to the unconfined type and vice versa. This is illustrated in Fig. 3.4. In Fig. 3.4*a* a well is screened at its lower part and water is withdrawn from the confined lower stratum. Water is also ponded above the confining stratum. If the well were screened in that part, unconfined flow would take place there.

[2] This is because the partial differential equations of ground-water flow have higher derivatives with respect to the space variables than with respect to the time variable.

Figure 3.4*b* shows that heavy draft from the lower stratum may create a zone of unconfined flow adjacent to the well, while the exterior region remains confined. The unconfined flow may disappear for a smaller pumping rate.

PROPERTIES OF THE MEDIUM AND OF THE FLUID

A medium is called isotropic if its properties at any point are the same in all directions from that point. It is called anisotropic if, on the other hand, some properties are affected by the choice of direction in a point. The medium is of heterogeneous composition if its nature, properties or conditions of isotropy or anisotropy vary from point to point in the medium; it is homogeneous if its nature, properties and isotropic or anisotropic conditions are constant over the medium.

A fluid is homogeneous essentially when it is single phased. A mixture of completely miscible salt water and fresh water is treated as homogeneous, although the density of the mixture may vary from point to point. A dispersed mixture of oil and water, however, is heterogeneous. Oil and water may be treated as immiscible for some purposes, exhibiting a liquid-liquid interface on both sides of which the flow may be homogeneous.

The combination of properties of the medium and of the fluid in such characteristics as the hydraulic conductivity (see section 4.5) leads to heterogeneous flow if either medium or fluid are heterogeneous. This textbook is restricted to the description of homogeneous flow.

Another subdivision is that of saturated and unsaturated flow. Flow is saturated if the voids of the medium are completely filled with fluid in the phase of the main flow. The flow is unsaturated if this is not the case. Deep percolating ground-water flow is always saturated and of prime importance to the ground-water geologist and engineer prospecting for a source of water supply. Above the saturated medium there is a zone of unsaturation of great interest to the soil physicist and to the agronomist and occasionally to the civil engineer (e.g., when he deals with the drainage of road bases). This textbook concentrates on saturated flow.

3.3 *Geologic Formations in Ground-Water Flow—Nomenclature*

Aquifer: a geologic formation or stratum containing water in its voids or pores that may be removed economically and used as a source of water supply. Unconsolidated alluvial deposits of sand and gravel, consolidated sandstones are examples of water-bearing strata.

Aquiclude: a geologic formation so impervious that for all practical purposes it completely obstructs the flow of ground water (although it may be saturated with water itself), and completely confines other strata with which it alternates in deposition. A shale is an example.

Aquitard: a geologic formation of a rather impervious and semi-confining nature which transmits water at a very slow rate compared to the aquifer. Over a large area of contact, however, it may permit the passage of large amounts of water between adjacent aquifers which it separates from each other. Clay lenses interbedded with sands, if thin enough, may form aquitards.

3.4 *Rock and Soil Composition*

Rock and soil form the porous medium in which water is collected by water-bearing strata and through which it flows under influence of various forces as explained in Chapter 4. This porous medium has a solid matrix or skeleton, an assemblage of solid mineral grains separated and surrounded by voids, pores, or interstices which may be filled with water, gases, or organic matter. In this textbook the term rock applies to a cemented or consolidated porous medium such as sandstone or limestone, whereas soil refers to unconsolidated or loose, uncemented material, such as sand, gravel and clay.

POROSITY, VOID RATIO

A given volume V_0 of porous medium contains V_s of solids and V_v of voids. V_v of voids may be subdivided into V_w of water and V_g of gases and organic matter. The porosity n is defined as

$$n = V_v/V_0 \tag{3.1}$$

whereas the void ratio e is defined as

$$e = V_v/V_s \tag{3.2}$$

Porosity and void ratio are interrelated by the expression

$$e = n/(1 - n) \tag{3.3}$$

Void ratio is commonly used in soil mechanics and has the advantage of a larger numerical variation. Thus for a natural soil porosity range of 0.3 to 0.6, the corresponding void ratio range is 0.43 to 1.5. In ground-water flow, the concept of void ratio is almost never used and porosity is almost always used in the sense of effective porosity, as will be defined later (see section 3.7). Porosity of consolidated materials depends on the degree of cementation, the state of solution, and fracturing of the rock; porosity of unconsolidated materials depends on the packing of the grains, their shape, arrangement, and size distribution. Small grains will fit into the openings left between grains of large diameter, and thus a medium with nonuniform size distribution will have a smaller porosity than one in which the grains are well sorted. The effect of packing upon porosity may be evaluated numerically in the case of grains of uniform spherical shape. It has been shown by Graton and Fraser [1] that the loosest packing is that of a cubical array of spheres, for which $n = 0.476$, and the tightest packing is that of a rhombohedral array, with $n = 0.26$ (Fig. 3.5). Intermediate arrays, of course, have porosities lying between these limits. Figure 3.6 shows several types of rock interstices and the relation of rock texture to porosity. Porosity is a measure of the water-bearing capacity of a formation, and, as may be expected, also plays a role in the capability of a formation to transmit water. This capability is expressed by the fluid (hydraulic, in the case of water) conductivity of the formation which will be defined in section 4.5.

However, the relationship between porosity and hydraulic conductivity is not a simple one and other factors besides porosity affect the value of hydraulic conductivity. Thus, pore size, of the same order of magnitude as grain size, is far more important than porosity as regards the water-transmitting capability of a stratum. Sands with relatively large rounded or angular grains may have smaller porosity than clays which are composed of tiny platelike particles with a large specific surface causing high molecular forces between the clay and water particles. In spite of their smaller porosity sand formations are pervious and good aquifers while clay formations form aquicludes or at best aquitards. Methods of measuring porosity may be found in a number of textbooks [2, 3].

3.5 *Simple Medium Properties*

MOISTURE CONTENT

The gravimetric moisture content w as defined in soil mechanics is

$$w = W_w/W_s \tag{3.4}$$

in which W_w is the weight of the water and W_s is the weight of the solid particles of a given volume of porous medium. The volumetric moisture content c as defined in soil physics is a volume concentration of water in soil

$$c = \mathcal{V}_w/\mathcal{V}_0 \tag{3.5}$$

DEGREE OF SATURATION

It is defined as the ratio of the volume of water to the volume of pores

$$S = \mathcal{V}_w/\mathcal{V}_v \tag{3.6}$$

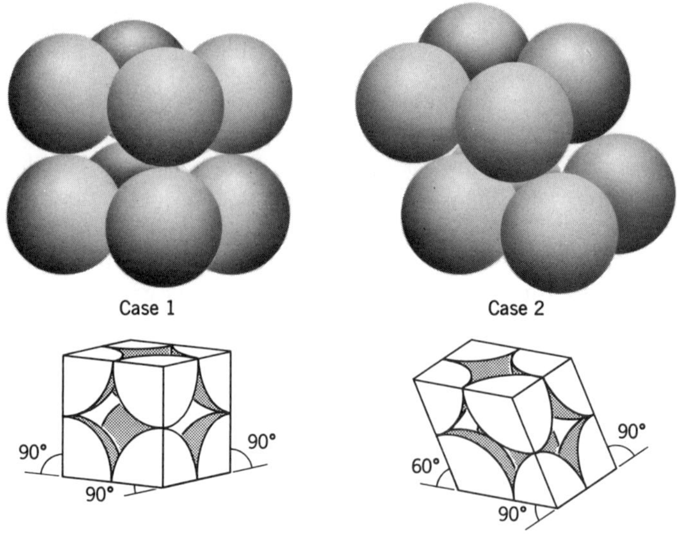

Fig. 3-5 Packing of spherical grains. Unit cells of cubic (Case 1) and rhombohedral (Case 2) packing. (After Graton and Fraser.)

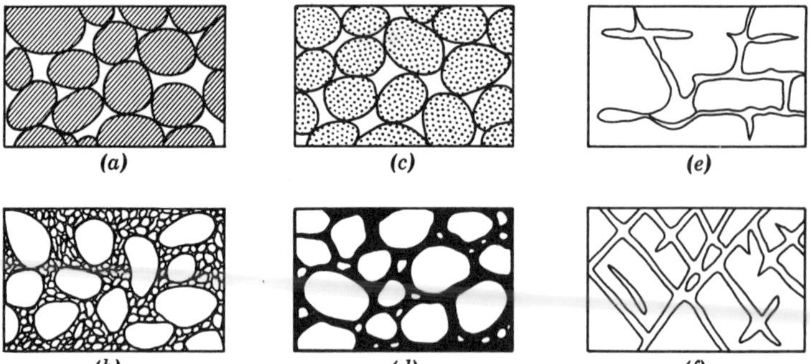

Fig. 3-6 Rock interstices and the relation of rock texture to porosity. Diagram showing several types of rock interstices and the relation of rock texture to porosity. (*a*) Well-sorted sedimentary deposit having high porosity; (*b*) poorly sorted sedimentary deposit having low porosity; (*c*) well-sorted sedimentary deposit consisting of pebbles that are themselves porous, so that the deposit as a whole has a very high porosity; (*d*) well-sorted sedimentary deposit whose porosity has been diminished by the deposition of mineral matter in the interstices; (*e*) rock rendered porous by solution; (*f*) rock rendered porous by fracturing. (After Meinzer, *U.S. Geological Survey Water-Supply Paper* 489, 1923, Fig. 1, p. 3.)

A simple relationship exists between Eqs. 3.1, 3.5, and 3.6:

$$S = c/n \tag{3.7}$$

SPECIFIC WEIGHT

It is defined as the weight per unit volume of porous medium, or unit weight. It is commonly designated by the Greek symbol γ and is the ratio of the total weight W to the total volume $\mathcal{V}_0$ of porous medium

$$\gamma = W/\mathcal{V}_0 \tag{3.8}$$

A subscript is added in order to distinguish the specific weight of water γ_w or that of the solid minerals γ_s from the overall specific weight γ.

SPECIFIC GRAVITY

It is defined as the ratio of the unit weight of a material to the unit weight of pure water at 4°C. Often the unit weight of pure water is replaced by the specific weight of water γ_w, so that the specific gravity of water is 1. In general

$$G = \gamma/\gamma_w \tag{3.9}$$

A subscript s may be added to refer to the specific gravity of the solid minerals G_s. It is easy to derive [4] the equation

$$\gamma = \frac{G_s + Se}{1 + e}\gamma_\omega \tag{3.10}$$

It has become customary, however, to drop the subscript s of G in Eq. 3.10. Hence,

$$\gamma = \frac{G + Se}{1 + e}\gamma_\omega \tag{3.10*}$$

DENSITY

It is defined as the mass per unit volume, and is commonly designated by the Greek symbol ρ. By definition, weight is equal to mass times gravitational acceleration, $W = Mg$, and after division by $\mathcal{V}_0$

$$\frac{W}{\mathcal{V}_0} = \frac{M}{\mathcal{V}_0}g$$

or

$$\gamma = \rho g \tag{3.11}$$

SPECIFIC SURFACE

It is defined as the ratio of the surface of all grain particles to the volume of all the grain particles. If the medium were one of spheres of uniform diameter d, the specific surface would be $6/d$. In general the specific surface is inversely proportional to a characteristic grain diameter of the medium, and therefore clays have a larger specific surface than sands.

3.6 *Pore Pressure and Intergranular Stress; Soil Moisture Tension; Piezometric Head; Water Table*

The pressure experienced by the water in the voids of a porous medium such as in the pores of the saturated soil of Fig. 3.7 is called pore pressure. It is measured as pressure head h_p in a point P by the height of water in a piezometer inserted in P as indicated in Fig. 3.7. This height is measured

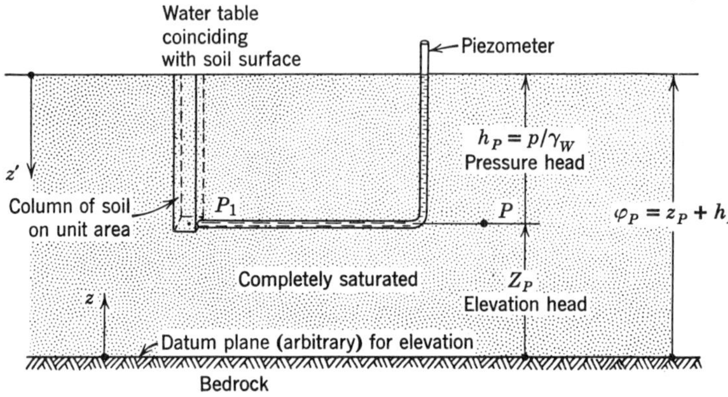

Fig. 3-7 Piezometric potential in the saturated zone.

with reference to point P, and is counted positive above P and negative below P. In Fig. 3.7 the pore pressure in P_1 is positive and equal to γh_p, while in Fig. 3.8 the pore pressure in P_2 is negative and equal to γh_t or $-\gamma h_p$. In the second case, where P_2 is situated in the unsaturated zone of the soil, the negative pore pressure is often called soil moisture suction or soil moisture tension, and the piezometer is called a tensiometer.

Pressures used in this textbook are gage pressures unless it is stated explicitly that absolute pressures are adopted. Gage pressure and absolute pressure are interrelated by the equation

$$\text{Absolute pressure} = \text{local atmospheric pressure} + \text{gage pressure}$$

Positive gage pressure is called pore pressure; negative gage pressure is called tension or vacuum. Absolute pressure is always positive (Fig. 3.9). In Fig. 3.9 this definition of the water table was implied: it is a phreatic surface at atmospheric pressure or at zero gage pressure.

The elevation z of a point P above an arbitrary datum plane is called the elevation head, positive above the datum plane and negative below the

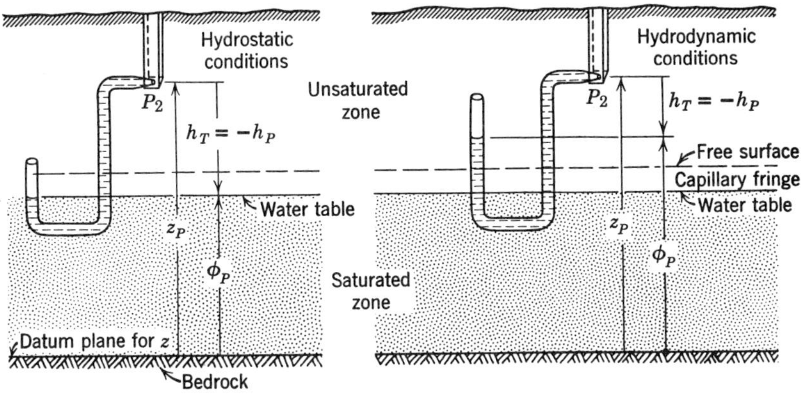

Fig. 3-8 Piezometric potential in the unsaturated zone $\varphi_P = z_P + h_T = z_P - h_P$. In hydrostatic conditions, $\varphi_P = \varphi$ at watertable. In hydrodynamic conditions, $\varphi_P > \varphi$ at watertable.

datum plane. The sum of the elevation head and the pressure head is called the piezometric head or piezometric potential φ. The concepts of head and potential are developed in sections 4.7 and 4.14.

The total pressure or stress in a point P of a porous medium is somewhat artificially defined as a macroscopic pressure [4], namely, as the weight of the overburden above P per unit area. It is possible to say that the total stress (pressure) or combined stress (pressure) is equal to the sum of the pore pressure and the intergranular stress. Intergranular stress is understood to be the stress in the granular skeleton. It would be difficult to define intergranular stress except by saying that it is the difference between total stress and pore pressure, as the intergranular forces are transmitted from

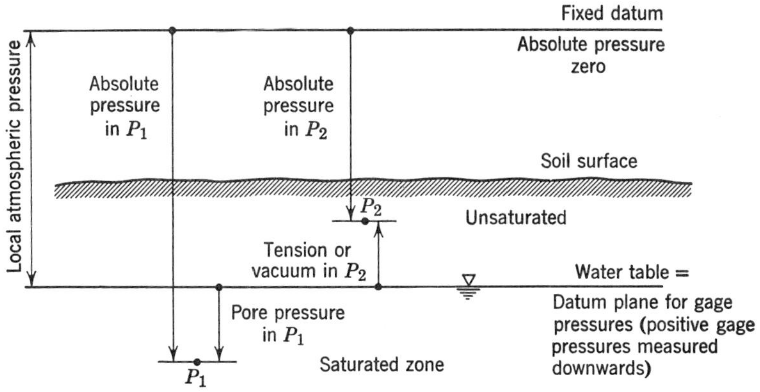

Fig. 3-9 Different kinds of pressure.

grain-to-grain contact, and it would be hard to define the surface over which the intergranular stress acts. However, the difference definition is completely satisfactory for our purposes. In the case of Fig. 3.7 where water table and soil surface coincide, the depth z' downward from the soil surface is measured and Eq. 3.10 is used to express

$$\text{Total stress} = \frac{\text{weight}}{\text{area}} = \frac{\gamma z' \text{ area}}{\text{area}} = \frac{G + Se}{1 + e} \gamma_w z' \tag{3.12}$$

Also,

$$\text{Pore pressure } p = \gamma_w z' \tag{3.13}$$

Therefore,

$$\text{Intergranular stress} = \left(\frac{G + Se}{1 + e} - 1\right) \gamma_w z' \tag{3.14}$$

and for $S = 1$, complete saturation,

$$\text{Intergranular stress} = \frac{G - 1}{1 + e} \gamma_w z' \tag{3.15}$$

Equation 3.15 states that the intergranular stress is the buoyant overburden per unit area.

Similar computations may be made when P is in the unsaturated zone as P_2 in Fig. 3.8. Here the intergranular stress in P_2 exceeds the total stress in that point by the amount $\gamma_w h_t$. In hydrostatic conditions no water flows from P_2 to the water table, and this is expressed by the equality of piezometric potential between P_2 and the water table. Therefore h_t assumes a large value as indicated on Fig. 3.8. When water flows from P_2 to the water table, the flow conditions are hydrodynamic, and therefore the piezometric potential in P_2 must be higher than that of the water table. Hence h_t is smaller than in hydrostatic conditions.

3.7 *Saturated and Unsaturated Zones of Flow—Capillary Fringe*

As a first approximation the water table is often taken as the boundary between the zone of saturation and the zone of aeration, of which capillary water is a part. Actually the capillary fringe may be defined as the zone of soil just above the water table which is still essentially saturated although under suction [5]. This zone, which is nearly saturated, may be of substantial thickness, and it is possible to define the free surface as the upper boundary of the capillary fringe and therefore as the surface which separates saturated from unsaturated flow.

The capillary fringe as defined here may be determined for a given soil from a desorption curve [6]. Such a curve is obtained from a column of soil (Fig. 3.10) by flooding the surface until water emerges freely at the bottom, removing the supply of water and allowing the column to drain (protected against evaporation) for a period of one week. The column is then sampled for the volumetric moisture content c at various depths. The graph of c versus h_t (Fig. 3.11) is called a desorption curve. At equilibrium, h_t is numerically equal to the height z of the sample above the outflow plane of Fig. 3.10. The desorption curve levels off to a horizontal asymptote, indicating the moisture content below which water cannot be drained by gravity. This moisture content is sometimes referred to as field capacity (soil physics) or specific retention (hydrology). The height h_c of the capillary fringe may be determined on Fig. 3.11 by the point on the curve for which c still has a value close to, say 99% of its maximum value at the beginning of the test. The desorption curve may be transposed to give a graphical relationship between soil moisture tension and drainable pore space, as in Fig. 3.12 [7]. This curve is idealized for low values of soil moisture tension because of errors involved in measurements of small water volumes. Figure 3.13 gives the soil moisture content and degree of saturation for a sand as a function of the

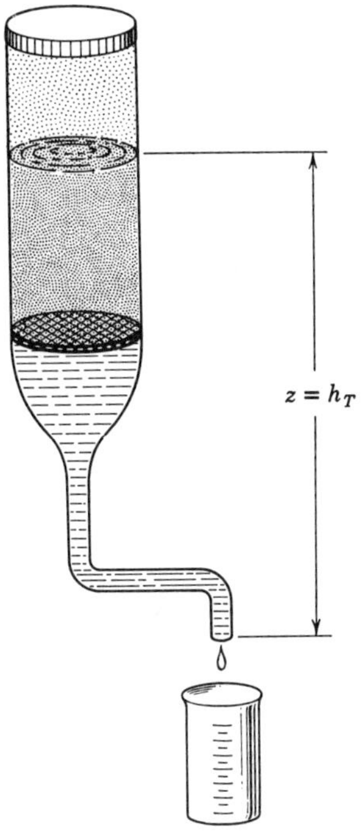

Fig. 3-10 Soil moisture tension. (The apparatus measures tension.)

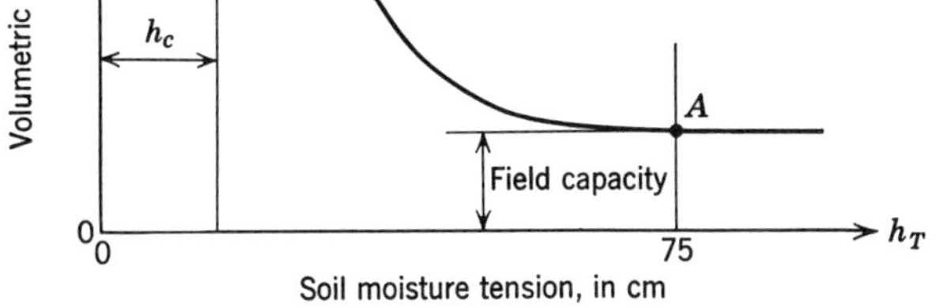

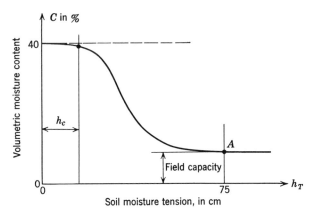

Fig. 3-11 Soil moisture tension versus volumetric moisture content. Desorption curve as measured in Fig. 3.10.

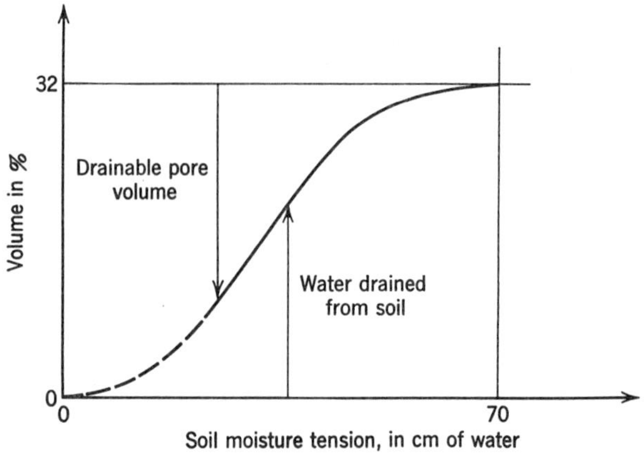

Fig. 3-12 Soil moisture tension versus percentage of drained volume.

height above the water table. The effective porosity or specific yield S_y [45, 46] is defined as the porosity minus the field capacity [8]. According to our definition the pressure at the free surface is equal to the atmospheric pressure minus $\gamma_w h_c$.

3.8 *Various Kinds of Subsurface Water*

There is no agreement on the nomenclature of the various kinds of subsurface water, and some terms are used in a contradictory sense. Thus Todd [16] refers to the zone of aeration as that occupied by suspended or

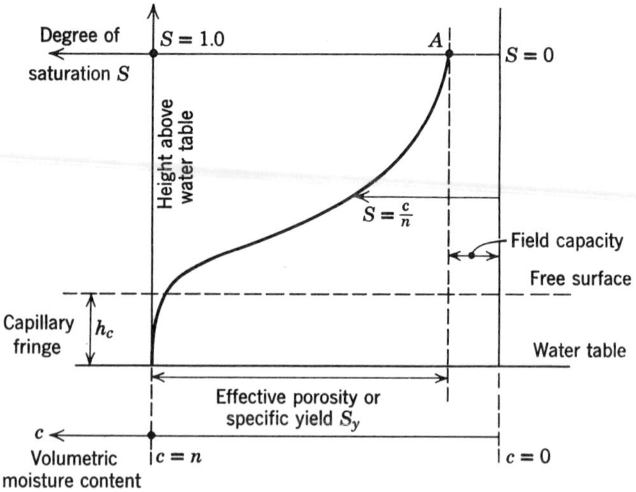

Fig. 3-13 Soil moisture content and degree of saturation above water table.

vadose water, while Schoeller [38] understands by vadose waters "all ground water that percolates essentially from the earth's surface to a certain depth and that again appears at the earth's surface after a relatively short time. Vadose waters are therefore moving waters, i.e., subjected to a certain hydrodynamic effect."

The term gravitational water in this country is applied to water that can be drained by gravity in a specified test and therefore is sometimes [16] classified in the zone of aeration. On the other hand gravitational water in the Russian [9] definition is all free water that escapes the action of the attraction forces towards the surface of the solid particles. It is simply called the ground water of the zone of saturation that moves under the influence of gravity and of hydrodynamical pressures.

Lebedev [9] considered four kinds of subsurface water:

1. *Water Vapor.* It fills completely the voids between the soil particles and shifts from regions of higher pressure to regions of much lower pressure.
2. *Hygroscopic Water.* This water condenses at the surface of the soil particles. When dry soil touches humid air the soil particles absorb moisture and the soil volume expands until the maximum hygroscopic effect is reached. Under these circumstances sands may increase in volume by about 1% of their total dry matter, while this figure is around 7% for silts and up to 17% for clays. Lebedev asserted that hygroscopic water can move in the ground only while transforming into the vaporous phase.
3. *Pellicular Water.* It is formed on soil particles under influence of the molecular forces of adhesion. This water holds together with a strong force and cannot be separated by centrifugal action with an acceleration exceeding that of gravity by seventy thousand times. Pellicular water is able to move as a liquid from dense films to much thinner ones. The gravitational force does not show any influence on the movement of pellicular moisture, and this moisture freezes only at minus 1.5°C. Pellicular water disappears under drying soil conditions.
4. *Gravitational Water.* As defined before, it is the ground water of the zone of saturation. It includes capillary water (see section 4.15). In the present textbook, the movement of ground water in the zone of saturation is studied exclusively.

◆ Interrelationship between Ground Water and Surface Water [42, 43, 44, 47]

3.9 *Introduction*

In the planning of engineering works to promote the conservation of water and to provide for its full use, it has become necessary to study the interaction of ground water and surface water—two categories of water resources which in the past were considered to be separate and distinct [10]. In the case of the development of a surface water supply, plans were usually based upon studies which utilized the records of runoff for a series of years and ignored the storage and release of ground water because until recently [10] no means were available for accounting for the time delays inherent in ground-water movements. Similarly [11] many ground-water problems involved only a determination of the occurrence and quality of water in geologic formations and estimates of the productivity of wells. Upson [11] considers three different kinds of problems involving the interrelationship between ground water and surface water:

1. Estimates of how changes in stream regimen affect ground-water conditions.
2. Relative feasibility of developing surface water and ground water for particular needs.
3. Measurement of surface discharge to indicate amounts of ground water available.

As an example of the first type of problem, Upson considers the Ipswich River drainage basin in Massachusetts and questions the effect on the public water-supply wells draining on the ground-water reservoir if the stream channel were lowered by dredging in amounts of several feet. It had been established previously that the ground-water reservoir is in hydraulic continuity with the river and, at least at times, sustains a large part of the river flow by efficient seepage.

For the second type of problem, Upson investigates the Blackstone River basin in Massachusetts and the feasibility of pumping the river water through wells, assuming again that there is hydraulic continuity between river and ground-water basin.

Finally, as an example of the third kind, the Pawcatuck River drainage basin in Rhode Island is cited. It is pointed out that mapping the geology of the basin, interpreting information from wells already drilled, and determining the nature, thickness, and extent of water-bearing deposits will locate sources of ground water and may even indicate the potential production of individual wells. However these techniques will not result in determining the total amount of ground water available for development and therefore quantitative methods, such as measurements of surface water flows, become imperative.

The interrelationship between ground water and surface water has also been studied by the author in his analysis of the Pennsauken aquifer [12].

In the sequel, two topics are treated which deal specifically with the conjunctive use of surface water and ground water—the determination of the sustained yield of a ground-water basin and the recharge or replenishment of aquifers.

3.10 *Sustained Yield* [39, 40]

Sustained yield or perennial yield from a ground-water basin is commonly determined in studies of a budgetary nature requiring no mathematical operations beyond arithmetic. The term sustained yield [13] is used as a synonym for safe yield, defined by Meinzer [14] as:

The rate at which water can be withdrawn from an aquifer for human use without depleting the supply to such an extent that withdrawal at this rate is harmful to the aquifer itself, or to the quality of the water, or is no longer economically feasible.

Conkling [15] modified this definition slightly by incorporating the limit of average annual recharge. Therefore.

Safe yield is the annual extraction from the ground-water unit which will not, or does not—(a) Exceed the average annual recharge; (b) So lower the water table that permissible cost of pumping is exceeded; or (c) So lower the water table as to permit intrusion of water of undesirable quality.

Many more changes have been suggested by various authors [13], among others, by Todd [16] who introduced the following broad definition:

Safe yield of a ground-water basin is the amount of water which can be withdrawn from it annually without producing an undesired result.

The undesired result may affect the water supply available to the basin, the economics of pumpage from the basin, the quality of the ground water, and the water rights in and near the basin.

In determining the sustained yield of a ground-water basin, the hydrologist should consider the fact that the hydrogeologic conditions are changing over a period of years and that the actual safe yield will also depend upon the hydraulic characteristics of production [17]. Safe yield is therefore not a constant, and its initially determined value will have to be revised in time. An accurate estimate of the sustained yield requires the making of the hydrologic budget for the ground-water basin [16, 18, 19, 20], expressing that a balance must exist between the quantity of water supplied to the basin and the amount stored within or leaving the basin. Over a period Δt of observation, change in volume of storage ΔS, inflow to and outflow from the basin are related as

$$\text{Rate of inflow} = \text{rate of outflow} + \frac{\Delta S}{\Delta t} \tag{3.16}$$

Among the contributions to inflow, the following items have to be considered [18, 20]:

1. Lateral inflow from adjacent ground-water basins.
2. Replenishment by deep percolation from rain water after deduction of evapo-transpiration losses.
3. Replenishment by surface runoff from rivers and overland flow from nearby plateaus.
4. Return flow from irrigation (around 30% of irrigation water).
5. Artificial recharge, if any.

Among the outflow components, the following are generally considered:

1. Pumpage, leading to a value of sustained yield.
2. Outflow to neighboring ground-water basins (natural discharge).
3. Streamflow passing out of discharge areas, springs.
4. Evaporation from watertable.
5. Water exported from the basin, if any.

The interaction between ground-water basins and evaporation from the water table are the most difficult outflow components to determine [18]. Baker [20] stresses the quality concept and the rate concept of safe yield linked to the transmissivity of the aquifer. In the quantity concept, it is assumed that the transmissivity of the aquifer is adequate to supply water at or approaching any rate of production desired. In the rate (i.e., the volume of water extracted during a day, week, month, or longer period rather than the instantaneous rate) concept, however, transmissivity becomes the limiting factor and determines the rate of production which will result in a safe lowering of the water level in the well or well field.

Safe yield may be increased by increasing recharge or decreasing the natural outflow. Increasing the unsaturated storage capacity in unconfined portions of a ground-water reservoir through increased production during years of low recharge, and storing increased recharge therein during wet years when streamflow would otherwise have been wasted for lack of underground storage capacity has been carried on with success, according to Baker [20]. The same principle, but applied on a seasonal rather than on an annual basis, also underlies the author's study of the Pennsauken aquifer near Princeton, N.J. [12].

Sutcliffe [19] has introduced the term long term yield in the investigation of a large aquifer at Tehran, Iran. In this case the main source of ground-water recharge is runoff from the mountains where precipitation is heaviest (orographic effect). Evapotranspiration losses due to high temperatures in the area overlying the ground-water basin are significant. Three methods paralleling Conkling's definition of safe yield are given to determine the long term safe yield.

1. The mean annual flow of the mountain rivers was estimated from a rainfall-elevation graph and from deduced losses from neighboring catchment areas. The percentage of this surface flow that recharges the aquifer was estimated from the topography.

2. Because of the marked dry season, the annual replenishment could be estimated from seasonal fluctuations of the watertable (rate of change in storage = rate of fall of water table times area of aquifer × specific yield).

3. An estimate of the flow through the aquifer is obtained by determining the value of the hydraulic conductivity (pumping test) measuring the slope of the water table and the cross section of the aquifer. (See sections 4.5 and 6.1.)

Sutcliffe then adopts for the long term safe yield about half of the estimated annual rate of recharge. A proper safety factor is applied to the concept of safe yield, as suggested by Baker, assuming that safe yield would be equal to recharge per year if natural discharge from the basin were reduced to zero.

For a recent discussion of the factors affecting the safe yield of ground-water basins, the reader is referred to reference [13].

3.11 *Artificial Recharge* [41]

Artificial recharge or replenishment of water-bearing strata that would otherwise be prematurely depleted has been practiced in Europe since the beginning of the nineteenth century, first in Scotland and France, later in England, Germany, and Sweden [21, 22]. In New Jersey, the Perth Amboy Water Works for more than fifty years have artificially recharged the Old Bridge sand of the upper Cretaceous Age [23] and have derived 4 to 5 mgd of water from this source.

The main purpose of artificial recharge is water conservation, often with improved quality as a second aim, e.g., when soft river water is used to reduce the hardness of the original or native ground water. The source of recharge may be storm runoff, river water, water used for cooling purposes, industrial waste water, and treated sewage water. In general recharge will be efficient only after some kind of pretreatment of the raw water, either of a chemical nature or through rapid sand filtration, or both. Recharge may be accomplished with the help of various surface-spreading methods which utilize basins, furrows, or flooded areas, or by injecting methods through vertical shafts, horizontal collector wells, pits, and trenches.

A major biochemical effect on the recharge water may be obtained in the case of filtration basins by building the basins on as high a level as possible

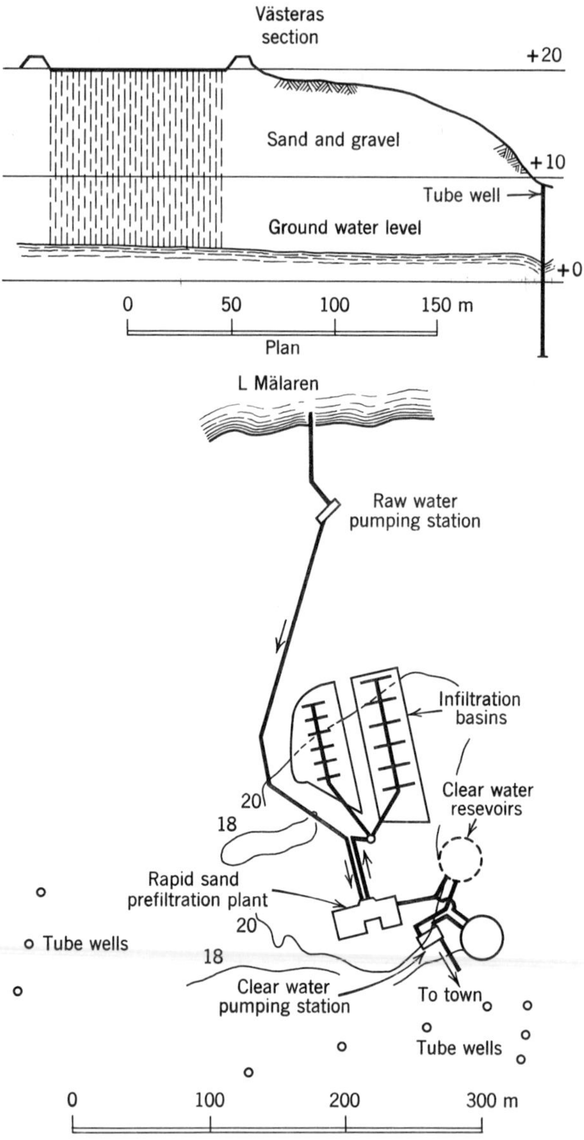

Fig. 3-14 Artificial recharge at Västeras, Sweden. (After Jansa, *International Association of Scientific Hydrology Publication* 33, p. 231, 1951.)

above ground-water surface, thus enabling the infiltrating water to percolate through a large volume of air-filled ground [21]. Also, soil has the outstanding capability of removing and destroying the abundant microorganisms always present in raw surface waters [22]. Figure 3.14 gives a sketch of the plant at Västeras, Sweden, which was completed in 1953 [21]. Raw water is pumped from Lake Mälaren to a prefiltration plant with sand filters and is conducted from there to infiltration basins, placed about

Table 3.1 Swedish Artificial Ground-Water Supplies

(After Jansa [22])

	Units	Min.	Ave.	Max.
Height of infiltration basin above ground-water level	m	2	8	30
Effective size of natural sand, d_{10}, rough estimate	mm	0.2	0.4	1.7
Rate of infiltration	m/day	1.5	6	16
Distance from infiltration basins to pump wells	m	100	400	1700
Raw water characteristics				
Colour, *Pt*	ppm	5	27	70
Oxygen consumed, $KMnO_4$	ppm	11	25	55
B. coli, in	100 cm³	10	650	2000
Gelatine bacteria, in	1 cm³	200	1,200	3300
Ground-water characteristics				
Colour, *Pt*	ppm	<5	6	15
Oxygen consumed, $KMnO_4$	ppm	4	9	21
B. coli in	100 cm³	0	0	8
Gelatine, bacteria	1 cm³	0	0	33

15 m above the water table. Purification of the surface water through seepage into the ground is assured because this water has to travel over some 700 ft before it reaches the service wells. Artificial recharge in the Scandinavian countries had become a necessity because of the depletion of the ground-water basins, which are only present where the glacial streams flowing under the ice of the glacial periods have formed a special kind of gravel deposit. A vast majority of the land consists of igneous rocks covered with almost impermeable moraine beds or clay and is worthless as far as ground water is concerned.

The main data from the Swedish artificial ground-water supplies are summarized in Table 3.1. This table shows the quality improvement of the water after recharge and a very high average rate of infiltration, of the order of 20 ft/day, as compared with a rate of 2 ft/day, reported by Groot

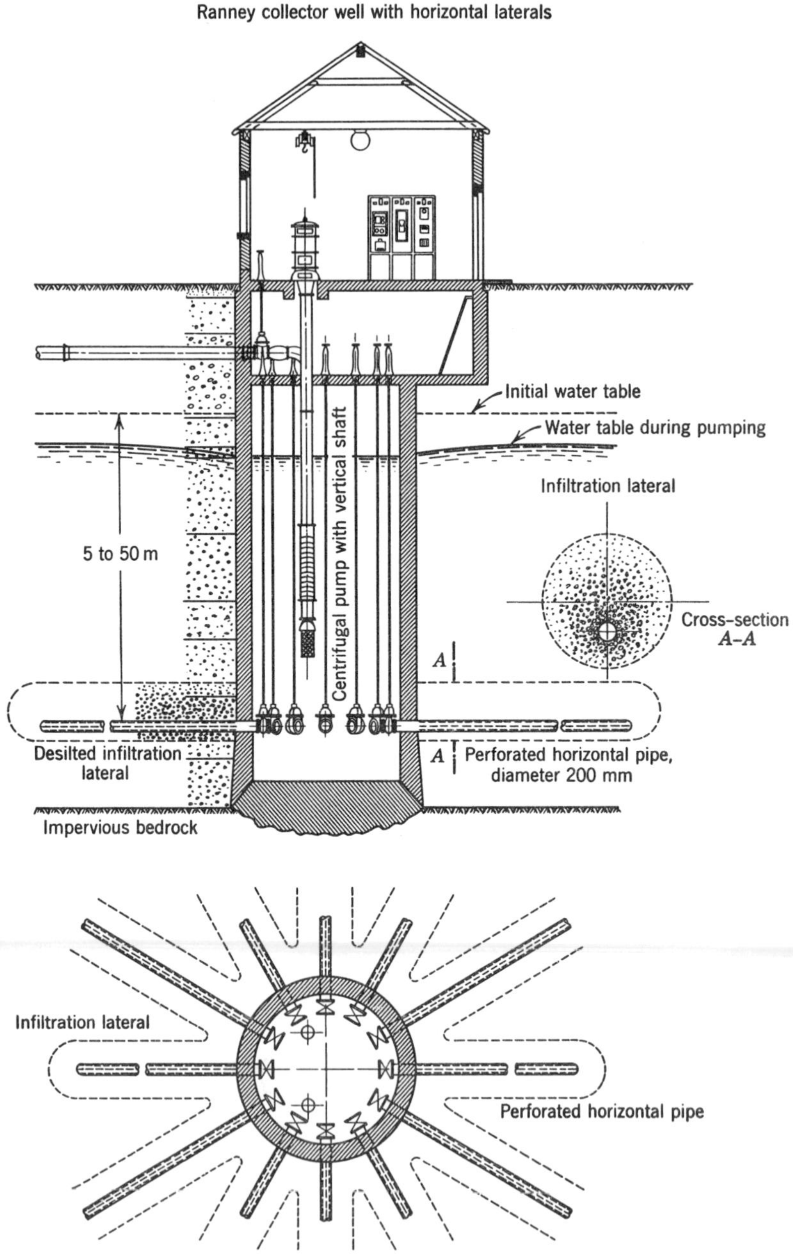

Fig. 3-15 Sketch of Ranney collector wells. (*IASH Publication* 37, p. 233, 1954.)

[24] as being the minimum acceptable, according to a study of nearly 40 published articles.

Artificial recharge depends, in the first place, on the ease of infiltration, or infiltrability [25], a phenomenon different from filtration, a term used in foreign literature as a synonym of seepage or ground-water flow. According to the U.S. Salinity Laboratory [24], the rate of infiltration is defined as the maximum rate at which a soil, in a given condition at a given time, can absorb water poured on the surface at shallow depth. Infiltration as such is the passage of water through the soil surface. However, ease of seepage is not unrelated to artificial recharge, as the rate of inflow, in the case of a recharge pit, for example [26], depends heavily on the facility to get water away from the pit into the adjacent strata. Suter [26] describes the operation of a pilot plant at Peoria, Illinois, consisting of a pit 40 × 62.5 ft with sloping sides dug to a depth of 10 ft below pool stage of the Illinois River and fed by gravity with screened and chlorinated river water. A recharge rate of 7 to 15 mgd per acre was obtained for a water table some 15 ft below the bottom of the pit. Most flow occurred through the sides of the pit, as the bed was more sensitive to clogging. Cleaning of the pit was accomplished through emptying and scraping silt from the bottom.

Horizontal collector wells [Fig. 3.15] of the Ranney type [27] offer great advantages in the cleaning or declogging of the well surfaces. Indeed, during the period of recharge, all valves mounted on the 10 to 20 horizontal screen pipes remain open, causing a minimum inlet velocity for the surface water entering the aquifer. During the cleaning period, however, when the direction of flow is reversed, only one valve is opened at a time, and the others remain closed. Since the whole discharge has to pass through the gallery pertaining to the one open valve, the scouring velocity is much larger than the recharge velocity. The recharge velocity must be smaller than the limit velocity for clogging, which has been estimated at 0.8 mm/sec [27]. Also, the scouring velocity must exceed the lower limit of 2.1 mm/sec of the drag velocity necessary to flush the particles deposited in the voids of the aquifer during the recharge process. Collector wells are installed 15 to 160 ft below the water level and their horizontal length may attain 250 ft. Scouring becomes necessary when the value of the hydraulic conductivity K (see section 4.5) of the material around the wells has dropped to 60% of its initial value. Polubarinova-Kochina [28] gives a quantitative analysis of the effect of clogging upon seepage and infiltration.

Information helpful in the classification of water-spreading sites is obtained by infiltrometer tests [24] and studies involving the determination of soil texture, soil structure, and chemical properties of the soil [29]. Schiff [29] examines different cases of infiltrability and ease of percolation in the light of Darcy's law (see section 4.4) and indicates the type of recharge method accordingly. When surface and subsurface soils permit rapid infiltration and percolation to the ground-water table, the basin method, furrow method, and flooding method may be used. When the surface soil restricts infiltration, treatment of the surface soil with vegetation, organic residues, and chemicals may increase infiltration rates from 0.5 ft/day to 1 to 1.5 ft/day. Treatments are most practical on large areas where the percolation rate of the subsurface soils is appreciably higher than the inherent rate of infiltration of the surface soil. The maximum allowable turbidity of the ponded recharge water is 25 cu ft of sediment per acre foot [29]. When shallow, less pervious soil layers restrict percolation and infiltration, the use of long narrow strips of land or long narrow trenches for water spreading is indicated. When deep, less pervious soil layers restrict percolation to the ground-water table, spreading over the entire area or in strips or trenches is possible. Perched water tables will develop and should be available for use or should ultimately reach the ground-water table [29]. Also according to Schiff [30], tests at Bakersfield, California, have indicated that filter materials are desired that will cause sedimentation through a certain depth of the filter and thus retain sufficient porosity to sustain reasonably high infiltration rates during a spreading season. The depth of sedimentation should permit cleaning by suction or by scraping for a reasonable cost.

Analytical [36, 37] work on the spreading of ground water has been helpful in determining optimum designs and in predicting relative benefits. Bittinger [36] developed an equation to compute the expected rise in water levels in observation wells located near Olds Reservoir, a spreading basin in northeastern Colorado for which field measurements were available. A total of 8,400 AF had been recharged during a 128-day period from December 3, 1959 to April 11, 1960 at a rate of infiltration of approximately 1 ft per day. Figure 3.16 gives a cross section of the system used by Bittinger, showing pertinent dimensions and assumed conditions below the recharge spreading basin. Radial symmetry is assumed; K is the hydraulic conductivity of the aquifer, S_y is the specific yield as defined in section 3.7, and D is the average saturated thickness. The decay of the ground-water mound h_0 at its center is given by

$$\frac{h_0}{H} = 1 - e^{-a^2/4\alpha t} \tag{3.17}$$

in which a is the radius of the spreading basin, H is the height of the idealized disc of ground water, t is the time, and $\alpha = KD/S_y$. A general formula for the shape of the mound as a function of time is also derived [36].

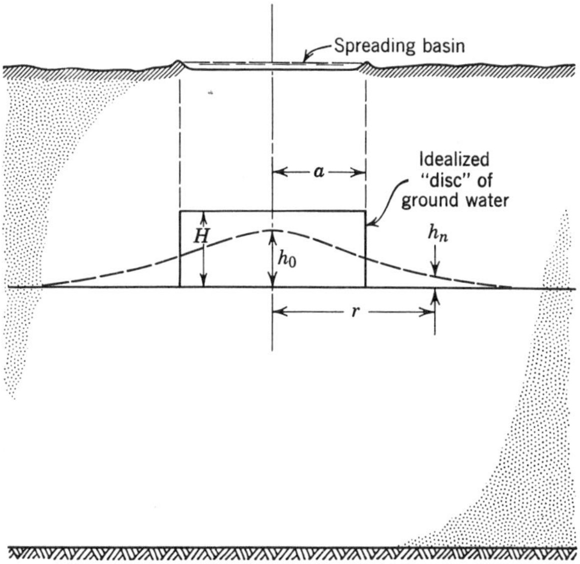

Fig. 3-16 Dissipation of a ground-water mound from a spreading basin. Cross section showing pertinent dimensions and assumed conditions below recharge spreading basin.(After Bittinger, reference 36, Chapter 3.) Decay in center

$$\frac{h_0}{H} = 1 - e^{-\frac{a^2}{4\alpha t}}$$

D = average saturated thickness

S_y = specific yield

$$\alpha = \frac{KD}{S_y}$$

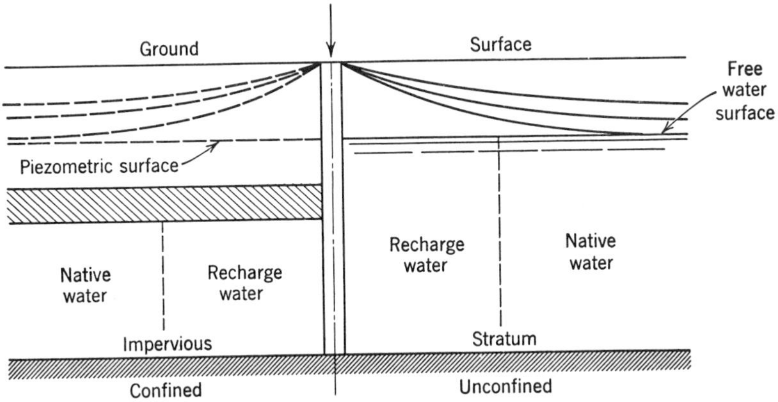

Fig. 3-17 Artificial recharge through a well. (After Baumann.)

Baumann [31] examines theoretical and practical aspects of well recharge. Figure 3.17 represents a recharge well. The left side applies to a confined aquifer, and the right side of the figure illustrates watertable conditions. Baumann points out that the study of unsteady flow to a recharge well is more difficult than is the study of similar flow to an ordinary well. The ground-water mound created through recharge of an open horizontal aquifer moves away from the center of recharge, that is, from the well, radially in the form of a wave. The celerity of this wave may greatly exceed the velocity of ground-water particles.

Recharge wells have been used by the Los Angeles County Flood Control District to stem sea-water intrusion into the West Basin Aquifer of Los Angeles County, California. A fresh water barrier [31] will give rise to a saline wave if located within the intruded sea-water wedge. Hence, if feasible, such a barrier should be located landward from the sea-water wedge, so as to prevent the initiation of a saline wave.

◆ Ground-Water Laws in the United States

The use of water in general has been regulated in this country by a variety of laws established since the second half of the nineteenth century at the state and federal level. To the engineer, however, these laws appear somewhat less defined than the single or even multivalued functions he is used to deal with, as they often leave room for alternate decisions to be made by local courts.

As the consumption of water increases in time with growing population and industrialization, it is only natural that occasionally there is more demand than available supply, and litigations are bound to arise more often in the future than they have in the past. Competition in the use (for example, irrigation versus wildlife) as well as in the quantity for identical consumptive use [32] is the basis of most conflicts. The problem is twofold —first, what function water will serve if the competing functions are incompatible and, second, how competition for use of the water for the same function shall be resolved. To this two fold problem, which besets every legislation concerning water, one has to add a jurisdictional problem stemming from the constitution of the United States. This entails the intricate and troublesome division of powers between the federal and state governments. The Desert Land Act (1877) and the Reclamation Act (1902) are among the most important pieces of United States legislation in water resources.

Courts in the United States generally divide ground waters into two classifications: well-determined underground streams and percolating waters [33]. The general tendency of the courts is to consider all subterranean water as percolating water unless the party alleging so can prove that it flows in a defined channel. Defined ground-water streams are subject to the same laws that are applicable to surface streams, i.e., the riparian and appropriative laws.

Three different concepts of rights are applied to percolating waters:

1. The common-law doctrine of riparian rights, based on the principle that the owner of the soil has absolute right to use all minerals found therein, including ground water.
2. The appropriative doctrine, analogous to the appropriation of surface waters, of which rights are acquired on a time-dependent priority basis regardless of landownership.
3. The reasonable-use rule, under which a land owner may make only a reasonable use of percolating waters underlying the land and must consider the equal right of all other owners of land overlying the same ground-water supply. As an adaptation of this rule, California follows the rule of correlative rights under which the rights of all owners of overlying lands are considered correlative and coequal.

DOCTRINE OF RIPARIAN RIGHT [34]

This doctrine for ground water is equivalent to that of land ownership, as the possession of land overlying water-bearing strata becomes the governing factor. Riparian means adjacent to the river and describes a right which is given to the owner of land because of its position and its contact with the river. The concept of riparian right originated in England in the first half of the nineteenth century. Because there was no water shortage in England at that time, the major benefit derived by the riparian owner was the continuation of the streamflow in its natural state past his land, undiminished in quantity and unimpaired in quality. A riparian right, although usufructuary[3] by nature, could not be lost through nonuse.

DOCTRINE OF APPROPRIATION

This doctrine contains two elements, one involving time, according to the principle "first in time is first in right," the second implying the beneficial use of the appropriated water. No rights can be acquired for nonbeneficial use of water and rights are lost if use of the water is discontinued.

The doctrine of appropriation originated in the gold mining days of California and was recognized by the California Supreme Court in 1855 [35]. Today all of the seventeen Western states recognize this doctrine to some extent.

[3] From the Latin, meaning "using the fruits of the property."

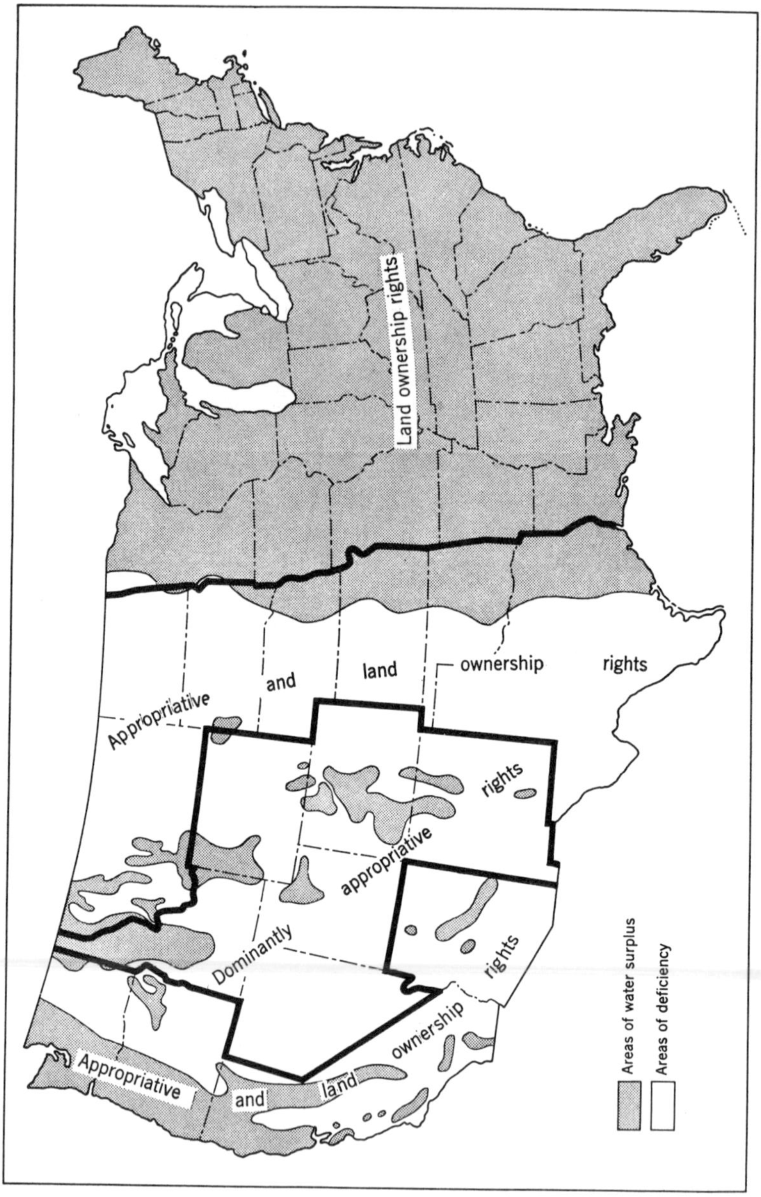

Fig. 3-18 Water rights doctrines by states of the United States, and areas of normal water surplus and deficiency. (After Thomas.)

The Desert Land Act (1877) recognized and gave sanction, insofar as the United States is concerned, to the state and local doctrine of appropriation, leaving to each state the right to determine for itself to what extent the rule of appropriation or the rule of riparian rights should apply. The right of appropriation was subsequently upheld by the U.S. Reclamation Act (1902), Section 2339, as follows: "Vested rights to the use of water from a stream or public land for mining, shall be maintained and protected. While an appropriator of water may not waste the same, he cannot be required to change his system of husbandry or devote his land to other purposes because it will require less water or leave more for subsequent appropriators."

In times of shortages those who were last to appropriate water will be first without water supply.

Prescriptive Rights

Rights acquired by landownership or through prior appropriation may be lost by adverse use or prescription. To satisfy the legal elements of prescription, it must be open and notorious and with the knowledge of the holders of the existing rights; it must be actual, and exclusive under claim of right and during a specified period.

In Pasadena vs. Alhambra (1938–1949), the court considered three kinds of rights for ground water: overlying, appropriative, and prescriptive. The court held that appropriative rights may apply only to surplus ground water, surplus water is not subject to prescription as against an overlying owner, and indeed, no right of the overlying owner is invaded in the case of surplus, that surplus water may be exported from the basin for non-overlying uses, and that the appropriative rights in such surplus are, in time of shortage, subject to the rights of overlying owners which are paramount to appropriative rights except as invaded by prescription. The court limited all withdrawals from the Raymond Basin, which had been considerably overdrafted, to the amount of the estimated safe yield.

Conclusion

In the Eastern states having a surplus of water there is a prevalence of the land ownership doctrine, although in densely populated areas such as northern New Jersey, where the supply of ground water is not ample, there is a tendency for the courts to limit the land owner to reasonable use, or to put the burden upon a new user by having him prove that pumpage from his well will not cause excessive drawdown in neighboring wells.

In the arid West, the doctrine of appropriation is predominant. The Pacific Coast states and the Great Plains states, however, accept both the riparian and appropriative doctrines.

Figure 3.18 [35] shows that the doctrines of water rights as applied in different states follow a climatic pattern.

REFERENCES

1. Graton, L. C. and H. J. Fraser, "Systematic Packing of Spheres with Particular Relation to Porosity and Permeability, and Experimental Study of the Porosity and Permeability of Clastic Sediments," *Journal of Geology*, Vol. 43, p. 785 (1935).
2. Pirson, S. J., *Oil Reservoir Engineering*, 2nd Ed. pp. 30–44, McGraw-Hill, New York, 1958.
3. Scheidegger, A. E., *The Physics of Flow through Porous Media*, pp. 8–12, Macmillan, New York, 1960.
4. Taylor, D. W., *Fundamentals of Soil Mechanics*, Chapter 7, Wiley, New York, 1948.
5. Brutsaert, W. H., G. S. Taylor, and J. N. Luthin, "Predicted and Experimental Water Table Drawdown during Title Drainage," *Hilgardia*, Vol. 31, pp. 389–418 (November 1961).
6. Day, P. R. and J. N. Luthin, "A Numerical Solution of the Differential Equation of Flow for a Vertical Drainage Problem," *Proceedings American Society of Soil Science*, Vol. 20, pp. 443–447 (October 1956).
7. Luthin, J. N. and R. V. Worstell, "The Falling Water Table in Tile Drainage, A Laboratory Study," *Proceedings American Society of Soil Science*, Vol. 21, pp. 580–584 (1957).
8. Smith, W. O., "Mechanism of Gravity Drainage and its Relation to Specific Yield of Uniform Sands," *U.S. Geological Survey Professional Paper* 402-A, 1961.
9. Polubarinova-Kochina, P. Ya., *Theory of Ground-Water Movement*, Chapter I, Princeton University Press, Princeton, 1962.
10. Glover, R. E., *Ground Water-Surface Water Relationships*, Western Resources Conference, Boulder, Colorado, August 24, 1960.
11. Upson, J. E., "Ground-Water Problems in New York and New England," *Journal of the Hydraulics Division ASCE*, pp. 1–12 (June 1959).
12. De Wiest, R. J. M., "Replenishment of Aquifers Intersected by Streams," *Journal of the Hydraulics Division ASCE*, November 1963.
13. Mann, J. F., "Factors Affecting the Safe Yield of Ground-Water Basins," *Journal of the Irrigation and Drainage Division ASCE*, pp. 63–69 (September 1961).
14. Meinzer, O. E., *Outline of Ground Water Hydrology with Definitions, Water Supply Paper* 494, U.S. Geological Survey, 1923.
15. Conkling, H., "Utilization of Ground-Water Storage in Stream System Development," *ASCE Transactions*, Vol. 111, p. 275, 1946.
16. Todd, D. K., *Ground Water Hydrology*, Wiley, New York, 1959 (336 pp.).
17. Bogomolov, G. V. and N. A. Plotnikov, "Classification des ressources d'eaux souterraines et évaluation de leurs réserves," *IASH*, Vol. 2, pp. 263–271, Symposia Darcy, Dijon, September 20–26, 1956.
18. Mortier, F., "Eléments pour l'établissement du Bilan de la nappe phréatique de la plaine des Triffa," *IASH Publication* 44, pp. 115–134, General Assembly of Toronto, 1957.
19. Sutcliffe, J. V. and W. R. Rangeley, "An Estimation of the Long Term Yield of a Large Aquifer at Tehran," *IASH Publication* 52, pp. 264–271, General Assembly of Helsinki, 1960.
20. Baker, D. M., "Safe Yield of Ground-Water Reservoirs, *IASH*, Vol. 2, pp. 160–164, Assemblée Générale de Bruxeiles, 1950.

21. Jansa, O. V., "Artificial Ground-Water Supplies of Sweden," *IASH Publication* 33, pp. 227–237 (1951).
22. Jansa, O. V., "Artificial Ground-Water Supplies of Sweden," *IASH Report* 2, *Publication* 37, pp. 269–275 (1954).
23. Barksdale, H. C. and G. D. De Buchananne, "Artificial Recharge of Productive Ground-Water Aquifers in New Jersey," *Economic Geology*, Vol. 41, No. 7, pp. 726–737 (1946).
24. Groot, C. R., "Feasibility of Artificial Recharge at Newark, Delaware, *Journal American Water Works Association* Vol. 52, No. 6, pp. 749–755 (1960).
25. Pioger, R., "Etude quantitative de l'infiltration," *IASH Publication* 37, pp. 248–258 (1954).
26. Suter, M., High-Rate Artificial Ground-Water Recharge at Peoria, Illinois," *IASH Publication* 37, pp. 219–224. (1954).
27. Wegenstein, M., "La recharge de Nappes souterraines au moyen de puits centraux et galeries d'alimentation horizontales," *IASH Publication* 37, pp. 232–237 (1954).
28. Polubarinova-Kochina, P. Ya., *Theory of Ground-Water Movement*, pp. 142, 495, Princeton University Press, Princeton, 1962.
29. Schiff, L., "The Darcy Law in the Selection of Water-Spreading Systems for Ground-Water Recharge," *IASH Publication* 41, pp. 99–110 (1956).
30. Schiff, L., "The Use of Filters to Maintain High Infiltration Rates in Aquifers for Ground-Water Recharge," *IASH Publication* 44, pp. 217–221 (1957).
31. Baumann, P., "Theoretical and Practical Aspects of Well Recharge," *ASCE Transactions*, Vol. 128, Part I, pp. 739–764 (1963).
32. Ely, Northcutt, *Legal Problems in Development of Water Resources*, 57th Meeting of Princeton University Conference, pp. 89–94 (May 1963).
33. Report of "President's Water Resources Policy Commission," *Water Resources Law*, Vol. III, 1950.
34. Baker, D. M. and H. Conkling, *Water Supply and Utilization*, Chapters 8 and 12, Wiley, New York, 1930.
35. Thomas, H. E., *Water Rights in Areas of Ground-Water Mining, U.S. Geological Survey Circular* 347, Washington, D.C., 1955 (16 pp.).
36. Bittinger, M. W. and F. J. Trelease, *The Development and Dissipation of a Ground-Water Mound beneath a Spreading Basin*, 1960 Winter Meeting of American Society Agricultural Engineers, Memphis, Tenn., December 4–7, 1960.
37. Baumann, P., "Ground-Water Movement Controlled Through Spreading," *ASCE Transactions*, Vol. 117, pp. 1042–1074 (1952).
38. Schoeller, H., *Les eaux souterraines*, Masson, Paris, 1962 (642 pp.).
39. Walton, W. C., "Potential Yield of Aquifers and Ground Water Pumpage," *Journal American Water Works Association*, Vol. 56, pp. 172–186 (February 1964).
40. Walton, W. C., "Future Water-level Declines in Deep Sandstone Wells in Chicago Region," *Ground-Water Journal of NWWA*, Vol. 2, pp. 13–20 (1964).
41. Todd, D. K., "Economics of Ground Water Recharge by Nuclear and Conventional Means," *Report UCRL-7850, Lawrence Radiation Laboratory*, pp. 1–135, Livermore, California, February 1964.
42. Stern, W., M. Jacobs, and S. Schmorak, *Hydrogeological Investigations in the Southern Coastal Plain of Israel*, Hydrological Service of Israel, Jerusalem, Israel, 1956 (12 pp.).
43. Goldschmidt, M. J. and M. Jacobs, *Precipitation over and Replenishment of the Yarqon and Nahal Halteninim Underground Catchments*, Hydrological Service of Israel, Jerusalem, Israel, 1958 (8 pp.).

44. Goldschmidt, J. J., *On the Water Balances of Several Mountain Underground Water Catchments in Israel and their Flow Patterns*, Hydrological Service of Israel, Jerusalem, Israel, 1959 (10 pp.).
45. Smith, W. O., "Mechanism of Gravity Drainage and its Relation to Specific Yield of Uniform Sands," *USGS Professional Paper* 402-A, 1961 (12 pp.).
46. Johnson, A. J., R. C. Prill, and D. A. Morris, "Specific Yield-Column Drainage and Centrifuge Moisture Content," *USGS Water Supply Paper* 1662-A, pp. 1–60 (1963).
47. Toth, J., "A Theory of Groundwater Motion in Small Drainage Basins in Central Alberta, Canada," *Journal of Geophysical Research*, Vol. 67, pp. 4375–4387 (October 1962).

CHAPTER FOUR

Elementary Theory of Ground-Water Movement

4.1 *Introduction*

Ground water in any stratum or basin is subject to a continuous process of natural and artificial discharge into rivers, springs, and wells, and of replenishment through deep percolation of precipitated or spread water, recharge from wells, rivers, and irrigation ditches. This process causes ground water to have different energies in various locations. As a result of these differences in energy or potential, water moves from levels of higher energy to levels of lower energy. It tries to achieve a state of minimum energy for a given system, obeying a well-known principle of physics. The state of minimum energy corresponds to hydrostatic conditions, which can only prevail theoretically and for a short duration. While flowing from a level of higher to one of lower energy, water experiences a loss in energy due to friction against intergranular walls of the porous medium along its seepage path.

Experiments were first conducted to obtain some quantitative idea of the velocity at which water travels through the ground, and hence of the discharge of ground water through given geological cross sections. Later, laws were derived analytically to interpret the fundamental characteristics of ground-water flow. In the present book only linear laws of seepage are studied in detail, i.e., laws which describe a linear relationship between seepage velocity and loss due to friction per unit length of distance traveled, or hydraulic gradient. These laws correspond to such natural phenomena as laminar flow of water in sandy aquifers or seepage through earth embankments. They do not predict accurately enough the turbulent flow in the vicinity of a well, nor the flow in fractured rocks or cavernous limestone, although in the case of cavernous limestone they are often used as a first approximation. From the foregoing considerations of linearity between loss due to friction, commonly called head loss, and seepage velocity, the similarity existing between ground-water flow and laminar pipe flow is made apparent. Experiments with pipe flow were first conducted by Hagen (1839) and almost simultaneously by Poiseuille (1841), after whom the law for laminar flow in pipes was named. However, it remained for Henri Darcy [1] in 1856 to discover, also by experiment, the fundamental law of ground-water movement. In fact, Poiseuille's law and Darcy's law are so similar that the analytical derivation of the first law, which was achieved in 1858–1860 by F. Neumann and E. Hagenbach [2], contributes to the understanding of the latter.

As in classical isothermal fluid mechanics, three velocity components, pressure, and density at any point of the fluid are the five unknown quantities in problems of ground-water flow. Water will be assumed to be incompressible, unless explicitly stated to the contrary, as for example in the calculation of the storage coefficient of a confined aquifer. This means that the equation of state reduces to the expression of constant density, and that only four unknowns (three velocity components and the pressure) exist in single-phase flow with a three-dimensional pattern. Theoretically it would be possible to solve for these unknowns if three equations of motion of the Navier-Stokes type, and one continuity equation expressing conservation of mass were available. Hydrodynamically, such a problem would be tractable if the granular skeleton were a simple geometrical assembly of prismatic, unconnected tubes. The seepage path, far from being a prismatic channel however, is tortuous, branching into a multitude of tributaries and recombining several of them as the flow proceeds.

Darcy's law in its original form avoids the insurmountable difficulties of the hydrodynamic microscopic picture by introducing a doubly averaging macroscopic concept. First, it considers a fictitious flow velocity, the Darcy velocity or specific discharge,[1] through a given cross section of porous medium rather than the true velocity between the grains. Second, it treats average hydraulic values rather than local hydrodynamic values of this velocity.

The basic reason for the introduction of this simplifying concept lies in the nature of Darcy's experiment: it utilized a sand-filled cylindrical pipe which permitted a measurement of only the average hydraulic values.

4.2 *Poiseuille Flow*

Consider the steady laminar flow of an incompressible fluid through a cylindrical pipe of circular cross-sectional area A with piezometer taps a

[1] The Russian literature uses the term "seepage velocity" (see reference 7, p. 11), but this is not compatible with United States nomenclature for soil mechanics (see reference 4, p. 100).

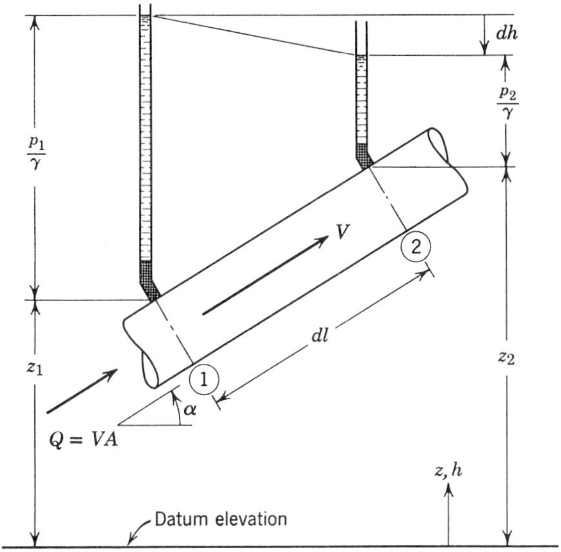

Fig. 4-1 Steady laminar flow in a pipe. (Poiseuille flow.)

distance dl apart, as shown in Fig. 4.1. The average velocity V is the same in sections 1 and 2. The streamlines in this case are parallel to the axis of the cylinder, but the velocity along these streamlines is not the same. Intuitively one feels that the fluid layers close to the wall will be slowed down because of the friction against the wall, and that at the wall the velocity of the fluid is zero. In fact τ, the effort or resistance per unit area that adjacent fluid layers, dy apart, have to overcome in order to attain a difference in velocity dv, is expressed by Newton's law of friction

$$\tau = \mu \, dv/dy \tag{4.1}$$

Here τ is called the shear stress and μ is the dynamic viscosity of the fluid. The distance y is measured away from the wall, as indicated in Fig. 4.2.

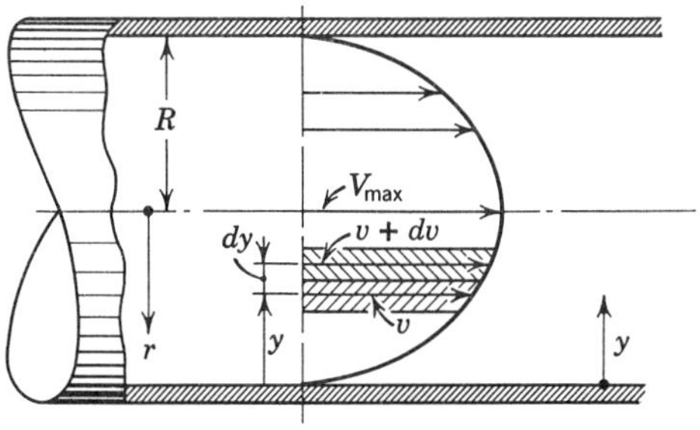

Fig. 4-2 Velocity distribution for Poiseuille flow.

The velocity picture is identical in any cross section. Moreover, the flow has radial symmetry so that a longitudinal section through the axis of the cylinder suffices to illustrate the velocity distribution. If the y-axis is replaced by an r-axis emanating from the center of the cylinder, then $dy = -dr$, and

$$\tau = -\mu \, dv/dr \tag{4.1*}$$

In Fig. 4.1, the generalized Bernoulli equation is applied along a streamline between 1 and 2 per unit weight of fluid:[1]

$$z_1 + p_1/\gamma + v_1^2/2g = z_2 + p_2/\gamma + v_2^2/2g + h_{L,1-2} \tag{4.2}$$

where z is the elevation measured as positive above an arbitrary datum plane, p is the pressure in the fluid, v is the local flow velocity, $h_{L,1-2}$ is the head loss between sections 1 and 2, and γ is the unit weight of the fluid. The sum $z + p/\gamma + v^2/2g$ represents the total energy or head per unit weight of fluid at any point of the flow. Bernoulli's law expresses the conservation of energy. It states that a mass of fluid traveling from section 1 to section 2 has lost energy in accomplishing this travel. The toll is accounted for as a loss $h_{L,1-2}$, due to friction. The different terms of the sum are commonly designated as elevation head, pressure head, and velocity head. In the present example $v_1 = v_2$, so that the velocity head terms drop out of Eq. 4.2. This would not be true if the pipe were not prismatic. On the other hand, as will be seen later, velocity head terms are in general negligible in ground-water flow. Because of the absence of velocity head terms in Bernoulli's equation for laminar flow in prismatic pipes, a bundle of thin prismatic pipes has been used as a model for ground water flow. With the notation

$$h = z + p/\gamma + \text{arbitrary constant} \tag{4.2*}$$

for the piezometric head, Eq. 4.2 may be written

$$h_2 - h_1 = dh = -h_{L,1-2} \tag{4.3}$$

Because the head loss is positive, dh is negative, indicating a loss of head. In this example, the flow proceeds from higher to lower pressure. Clearly, to have flow in the direction indicated on Fig. 4.1, the fluid must be delivered to the pipe by a pump at a higher pressure than when it leaves the pipe. But it is evident that, by tilting the pipe in the opposite direction, i.e., with section 2 lower than section 1, flow could take place predominantly under the influence of gravity and in this case from lower to higher pressure. In elementary hydraulics the absolute value of dh/dl is commonly

[1] The subscript w for the specific weight of water γ is dropped in the rest of the book, as no confusion is possible.

called the hydraulic gradient. In American literature, the symbol S is generally reserved for this quantity, but in view of the storage coefficient and the drawdown which often occur simultaneously in ground water problems and for which the use of the symbols S and s is deeply rooted, the symbol S^* will be used for the hydraulic gradient

$$S^* = -dh/dl \tag{4.4}$$

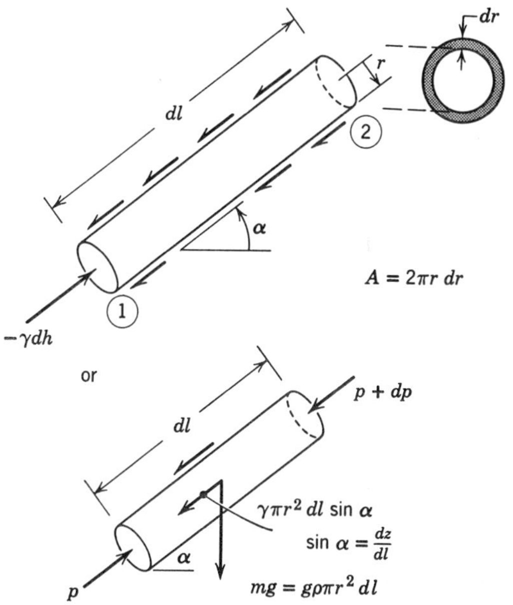

Fig. 4-3 Free body diagram for element of fluid in Poiseuille flow.

The use of the minus sign should be defended if dh/dl is to be inserted in differential equations because dh is always negative and because the proper meaning of a differential (i.e., end value of quantity — begin value of quantity) should always be respected. The omission of the minus sign may lead to errors even in the formulation of very elementary differential equations. For example, consider the free body diagram of Fig. 4.3, resulting from Fig. 4.1, by taking out a cylinder of radius r and length dl. In steady flow there must be an equilibrium of forces acting on the body as follows: (1) The unbalance in the head, dh, at cross sections 1 and 2 gives rise to a corresponding difference in pressure $-\gamma\, dh$ and hence to a resulting force $-\gamma\, dh\, \pi r^2$, from left to right in Fig. 4.3. (2) The shearing stresses add up to a resulting force $\tau 2\pi r\, dl$, with τ given by Eq. 4.1*.

The force equilibrium requires

$$-\gamma\, dh\, \pi r^2 = -\mu(dv/dr)2\pi r\, dl \tag{4.5}$$

The same result follows if gravity and pressure forces are treated separately. Indeed a projection of the force equilibrium on the l-axis gives:

$$p\pi r^2 - (p + dp)\pi r^2 - \tau 2\pi r\, dl - \gamma\pi r^2\, dl \sin\alpha = 0$$

and with $\sin\alpha = dz/dl$,

$$-\pi r^2(dp + \gamma\, dz) = \tau 2\pi r\, dl$$

which is essentially Eq. 4.5.

The symbol S^* of Eq. 4.4 is introduced in Eq. 4.5 and the variables are separated to render

$$dv = -(S^*\gamma/2\mu)r\, dr$$

This ordinary first-order differential equation can immediately be integrated. The result is

$$v = (-S^*\gamma/4\mu)r^2 + C$$

C is determined by the boundary condition $v = 0$ at $r = R$. It follows that

$$v = (S^*\gamma/4\mu)(R^2 - r^2) \tag{4.6}$$

This is a parabola with its vertex on the axis of the cylinder. There $r = 0$ and $v = V_{max} = (S^*\gamma/4\mu)R^2$.

The discharge or flowrate Q may be obtained by first calculating the flowrate through an annular section of width dr, say $v2\pi r\, dr$, and then integrating over the entire area.

$$Q = \int_0^R v2\pi r\, dr = (S^*\gamma\pi/2\mu)\int_0^R (R^2 - r^2)r\, dr$$
$$= (S^*\gamma\pi R^4)/(8\mu) \tag{4.7}$$

Equation 4.7 may be transformed by means of the general formulas

$$Q = AV$$

and

$$A = \pi R^2$$

It follows that

$$Q = (S^*\gamma R^2/8\mu)A \tag{4.8}$$

and that the average velocity V is expressed as

$$V = (S^*\gamma R^2)/(8\mu) = (1/2)V_{max} \tag{4.9}$$

If the symbol

$$K' = \gamma R^2/8\mu \tag{4.10}$$

is introduced, Eq. 4.9 reduces to

$$V = S^*K' \tag{4.11}$$

It is evident that the inclination α is immaterial in the proofs and that a correct use of Eq. 4.4 has led to a positive Q, which was physically required.

From the foregoing derivations and from section 4.4 the similarity between Poiseuille's and Darcy's laws will become evident at once.

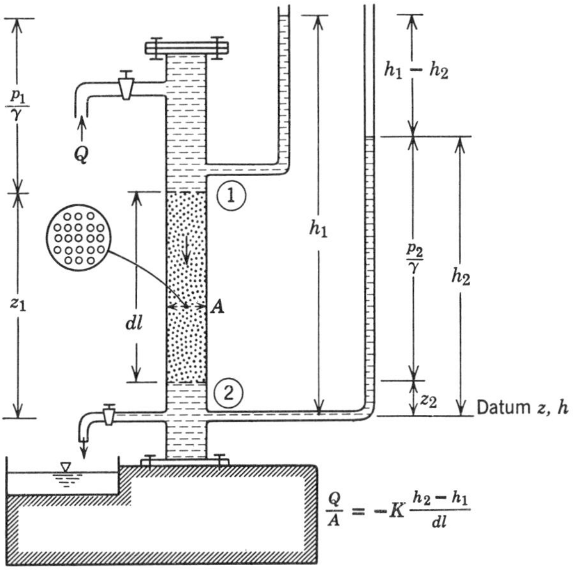

Fig. 4-4 Apparatus to demonstrate Darcy's law. (After Hubbert.)

4.3 *Porosity*

The study of laminar flow in a porous medium is more complicated than that of laminar flow in a prismatic tube because of the complexity of the cross section of the actual flow and of the variation of the cross section along the path of flow. If a granular material fills the space of a cylinder between two screens, 1 and 2 in Fig. 4.4, and if a difference of head is applied at these points, flow will take place through the pores of the medium. At any elevation z between 1 and 2, the cross section A_p of actual flow is different, i.e., A_p is a function of z, say $A_p(z)$, the sum of the cross sections of all the pores cut by a plane at elevation z. The superficial porosity

$$n(z) = \frac{A_p(z)}{A} \tag{4.12}$$

is defined as the ratio of $A_p(z)$ and A, the cross section of the cylinder, and varies with z in a very complicated way. Fortunately, the average value of $n(z)$ between sections 1 and 2 is a constant [7], namely, the (volumetric) porosity n,

$$n = \frac{\mathcal{V}_v}{\mathcal{V}_0} = \frac{\mathcal{V}_v}{\mathcal{V}_v + \mathcal{V}_s} = \frac{1}{1 + \dfrac{1}{e}} \tag{4.13}$$

where $\mathcal{V}_v$ is the volume of all the pores or voids, $\mathcal{V}_s$ is the volume of all the grains or solids, $\mathcal{V}_0$ is the total volume containing the porous medium, and $e = \mathcal{V}_v/\mathcal{V}_s$ is the void ratio. Because only average values of the velocities and hence of the porosity are of interest in this elementary course, the average value of $n(z)$, i.e., the porosity n, will be used in the sequel. The specific discharge V is defined as the quotient of the flow rate Q and the total cross section A,

$$V = Q/A$$

On the other hand, the true velocity in the pores V_p times the area A_p is also equal to the flowrate Q. Therefore

$$Q = VA = V_p A_p$$

and

$$V = nV_p \tag{4.14}$$

In general, the specific discharge V is used in the mathematical formulation of the laws of ground-water flow, and the true velocity follows from Eq. 4.14 after V has been found.

4.4 *Darcy's Law*

Darcy [1] ran his experiment on a vertical pipe filled with sand, under conditions simulated by Fig. 4.4 [3]. From his investigations of the flow through horizontally stratified beds of sand, Darcy concluded that the flowrate was proportional to the head loss, inversely proportional to the length of the flow path, and proportional to a coefficient K, depending on the nature of the sand. Darcy's law may be expressed as

$$\begin{aligned} Q &= KA(h_1 - h_2)/dl \\ &= -KA\, dh/dl = KAS^* \end{aligned} \tag{4.15}$$

where

$$h = z + \frac{p}{\gamma} + \text{arbitrary constant} \tag{4.2*}$$

and where S^* is the hydraulic gradient as defined by Eq. 4.4. From Eq. 4.15 and the considerations of section 4.3 it follows that

$$V = \frac{Q}{A} = S^*K = -K\frac{d}{dl}\left(z + \frac{p}{\gamma}\right) \tag{4.16}$$

must be considered to be a fictitious velocity, called specific discharge. A comparison of Eqs. 4.11 and 4.16 shows that the specific discharge is mathematically analogous to the average velocity V of the Poiseuille flow, provided K, the hydraulic conductivity of the sand, can be expressed similarly to K'. This similarity will be derived by dimensional analysis.

4.5 *Hydraulic Conductivity K*

From Eq. 4.15 it follows that K has the dimensions of a velocity, $[L]/[T]$. Furthermore, an analogy with K' would imply that K is a function not only of the nature of the sand, but also of the fluid properties γ and μ. The presence of R in Eq. 4.10 suggests naturally that K is a function of some characteristic length, say the average pore size d of the sand. From inspection of Eq. 4.10 it may be concluded only that K is a function of γ, μ, d, and the correct dimensions of the functional relationship remain to be determined. In this relationship, there will remain a dimensionless constant or shape factor C which takes into account effects of stratification, packing, arrangement of grains, size distribution, and porosity [4]. To find the value of K by dimensional analysis, let

$$K = C\gamma^{x_1}\mu^{x_2}d^{x_3} \tag{4.17}$$

and express all quantities in one of the two fundamental systems: mass length, time, i.e., $[M]$, $[L]$, $[T]$, or force, length, time, i.e., $[F]$, $[L]$, $[T]$. Both fundamental systems are related through Newton's first law (force = mass × acceleration), and it is easy to convert from one to the other. Here the use of the first fundamental system in Eq. 4.17 renders

$$[M]^0[L][T]^{-1} = C[M/L^2T^2]^{x_1}[M/LT]^{x_2}[L]^{x_3}$$

This functional relationship is dimensionally correct if the exponents of $[M]$, $[L]$, $[T]$ on both sides are the same. Hence three equations in x_1, x_2, and x_3 are obtained:

$$\begin{aligned} [M] &\rightarrow \quad 0 = x_1 + x_2 \\ [L] &\rightarrow \quad 1 = -2x_1 - x_2 + x_3 \\ [T] &\rightarrow -1 = -2x_1 - x_2 \end{aligned}$$

This system of equations has a unique solution, namely, $x_1 = 1$, $x_2 = -1$, $x_3 = 2$. Therefore

$$K = Cd^2\gamma/\mu \tag{4.18}$$

A variety of names has been given to K, such as effective permeability [5], coefficient of permeability [4], [6], seepage coefficient [7], and hydraulic conductivity. The term hydraulic conductivity, in view of the analogy with thermal and electrical conductivity, is gaining widespread use. It is used throughout this text. The name intrinsic permeability for

$$k = Cd^2 \tag{4.19}$$

is now generally accepted as being characteristic of the medium alone, and K may be rewritten as

$$K = k\frac{\gamma}{\mu} \tag{4.18*}$$

As was pointed out before, K has the dimensions of a velocity. It may therefore be expressed in a variety of consistent units, generally different from one discipline to another (soil mechanics, petroleum engineering, hydrogeology) and from continent to continent (Europe, the United States). In soil mechanics, centimeter-gram-second units are universally used, whereas water supply studies in this country are carried out with K in gallons per day and per ft² (gpd/ft²).

The darcy unit, which is used in petroleum engineering, has been preempted as a measure of the intrinsic permeability k because of an incomplete definition of Darcy's law. Indeed, for horizontal flow, Darcy's law may be written as

$$V = Q/A = -(k/\mu)(dp/dl)$$

in view of Eqs. 4.16 and 4.18*, or in absolute value,

$$k = \frac{\mu(Q/A)}{dp/dl} \tag{4.20}$$

From Eq. 4.20, the darcy has been defined as

$$1 \text{ darcy} = \frac{\dfrac{1 \text{ centipoise} \times 1 \text{ cm}^3/\text{sec}}{1 \text{ cm}^2}}{1 \text{ atmosphere/cm}}$$

The darcy unit has the dimensions of $[L]^2$. It may be expressed as such by replacing in the above formula:

$$1 \text{ centipoise} = 0.01 \text{ poise} = 0.01 \text{ dyne sec/cm}^2$$

and

$$1 \text{ atmosphere} = 1.0132 \times 10^6 \text{ dynes/cm}^2$$

Table 4.1 Average Values of K and k

Soil Class	K cm/sec	K darcys	K gpd/ft^2
Gravel	1—10^2	10^3—10^5	10^4—10^6
Clean sands (good aquifers)	10^{-3}—1	1—10^3	10—10^4
Clayey sands, fine sands (poor aquifers)	10^{-6}—10^{-3}	10^{-3}—1	10^{-2}—10

This leads to

$$1 \text{ darcy} = 0.987 \times 10^{-8} \text{ cm}^2$$

or

$$1 \text{ darcy} = 1.062 \times 10^{-11} \text{ ft}^2$$

Because Eq. 4.18 may be written as

$$K = kg/\nu \tag{4.21}$$

in which $\nu = \mu/\rho$ is the kinematic viscosity of the fluid, and ρ is the density of the fluid. It is easy to convert from darcy units to gpd/ft^2. The knowledge of the temperature of the fluid makes it possible to read ν from tables of physical properties of fluids [11]. The U.S. Geological Survey uses the meinzer unit to measure the hydraulic conductivity, honoring the late Dr. O. E. Meinzer. The meinzer unit is defined as the flow of water in gallons per day through a cross-sectional area of 1 square foot under a hydraulic gradient of one at a temperature of 60°F.

A second coefficient, the field coefficient of hydraulic conductivity, is also used by the Water Resources Division of the Geological Survey.

Table 4.2 Representative Values for k and K

Geologic Classification		Darcys (k)	Meinzers (K)
Argillaceous limestone	2% porosity	1.0×10^{-4}	1.80×10^{-3}
Limestone	16% porosity	1.4×10^{-1}	2.50
Sandstone, silty	12% porosity	2.6×10^{-3}	4.74×10^{-2}
Sandstone, coarse	12% porosity	1.1	19.90
Sandstone	29% porosity	2.4	43.60
Very fine sand	well sorted	9.9	18.00×10
Medium sand	very well sorted	2.6×10^2	4.60×10^3
Coarse sand	very well sorted	3.1×10^3	5.80×10^4
Gravel	very well sorted	4.3×10^4	7.88×10^5
Montmorillonite clay		10^{-5}	10^{-4}
Kaolinite clay		10^{-3}	10^{-2}

For the clays only the order of magnitude is indicated.

This unit is defined as the flow of water in gallons per day through a cross section of aquifer 1 ft thick and 1 mi wide under a hydraulic gradient of 1 ft per mile at field temperature [27].

In Eq. 4.18 it is assumed that both porous material and water are chemically and mechanically stable. This may never be true. Ion exchange on clay and colloid surfaces will cause changes of mineral volume which in turn will change the pore size and shape. Extreme changes in pressure will cause dilatation or compaction of aquifers. Moderate to high ground-water velocities will move colloids and small clay particles. Also, all water movement will facilitate solution or deposition of dissolved constituents. Relatively small changes in pressure or temperature may cause gases to come out of solution and clog pore space, thus reducing the hydraulic conductivity.

Table 4.3 Equivalence Between K and k Values

1 darcy	$= 9.87 \times 10^{-9}$ cm^2 $= 1.062 \times 10^{-11}$ ft^2
10^{-10} cm^2	$= 1.012 \times 10^{-2}$ darcys
0.1 cm/day	$= 1.15 \times 10^{-6}$ cm/sec $\approx 1.18 \times 10^{-11}$ cm^2 for water at 20°C
1.0 cm/sec	$\approx 1.02 \times 10^{-5}$ cm^2 for water at 20°C
1 darcy	$\approx$ 18.2 meinzer units for water at 60°F
1 meinzer	$= 0.134$ ft/day $= 4.72 \times 10^{-5}$ cm/sec $\approx 5.5 \times 10^{-2}$ darcys for water at 60°F

Tables 4.1, 4.2, and 4.3 contain useful data about K and k. In Table 4.3, $\approx$ means equivalent to.

4.6 *Measurement of K* [37, 38, 39, 40, 41, and 42]

K may be determined by a variety of laboratory methods as well as by field measurements [46, 47, 48]. The determination of K by field measurements will be treated further. Field measurement is the more significant method, as it renders average values of K for the aquifer. As to laboratory methods, the hydraulic conductivity may be determined by means of so-called permeameters, where small samples taken at different points of the aquifer are subject to flow under constant head or variable head. These samples are generally disturbed, and, moreover, even if they were undisturbed, they might not be representative of the average K of the aquifer. In both constant head and falling head permeameters water may be admitted to the top as well as to the bottom of the sample (downward and upward flow types). Figures 4.5*a* and *b* show respectively a constant head and a falling head permeameter. The specimens must be completely

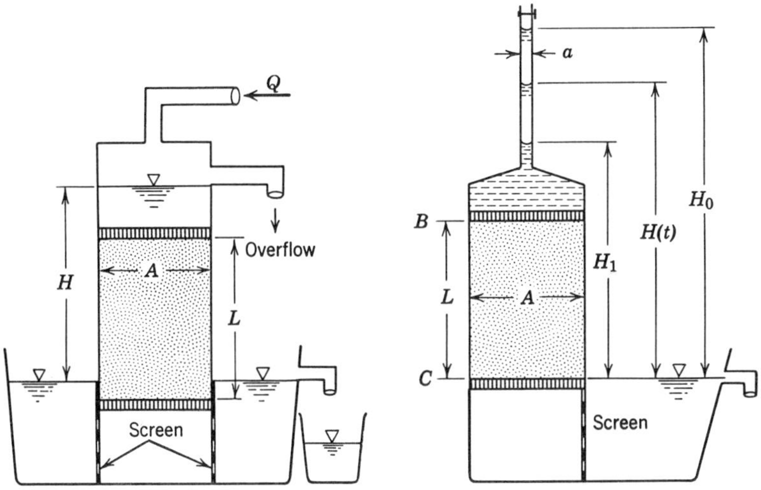

Fig. 4-5 (*a*) Constant head permeameter. (*b*) Falling head permeameter.

saturated with gas-free distilled water before testing in order to minimize the effects of entrapped air and foreign matter in the voids. Under normal circumstances water has a definite amount of air in solution (about 20 liter per m^3 water at room temperature and atmospheric pressure). An increase in temperature or a decrease in pressure results in a release of air from solution. This shows a disadvantage of the variable-head permeameter where changes in the hydraulic conductivity are due to variations of entrapped air resulting from differences in pore pressures as the head on the sample declines. (See reference 4, p. 121.) On the other hand, constant head devices commonly have a much larger amount of entrapped air due to the larger volume of water which must be used. In the constant head permeameter, water flows under constant head through a sample of area A and length L. The flowrate Q is measured by weighing the amount of water caught in a beaker during a given time t. The head loss dh over the sample in absolute value is equal to the difference H between headwater and tailwater. From Darcy's law it follows that

$$K = \frac{QL}{AH} \tag{4.22}$$

This method is satisfactory only for relatively pervious soils because it is difficult to measure small discharges in the manner just described. To avoid this difficulty the falling head permeameter was designed. Here discharge measurements are replaced by observations of the rate of fall of the water level in the standpipe of Fig. 4.5*b*. During a time length t_1 the water level falls from its initial level H_0 to H_1. If the datum plane for elevation is chosen through C, the head in C is always zero while the head in B is given by H:

$$\underset{\text{elevation head}}{L} + \underset{\text{pressure head}}{(H - L)}$$

so that the specific discharge may be expressed as

$$V = \frac{-K(O - H)}{L} = \frac{KH}{L}$$

The flow rate through A is $Q = AKH/L$. By continuity Q must have the same value in the standpipe. Here the velocity at which the water level falls is

$$v = \frac{-dH}{dt}$$

since H measures the water level upward from the datum plane and hence dH is negative. It follows that

$$AK\frac{H}{L} = -a\frac{dH}{dt}.$$

Separation of variables and integration render

$$-a\int_{H_0}^{H_1}\frac{dH}{H} = K\frac{A}{L}\int_0^{t_1} dt$$

or

$$K = 2.3\frac{aL}{At_1}\log_{10}\frac{H_0}{H_1} \tag{4.23}$$

Many researchers have investigated the various factors which influence the intrinsic permeability and hence the hydraulic conductivity, both in experimental and analytical work. Among others, Kozeny [12] has studied the influence of porosity, Hazen [13], grain size, and Zunker [14], void ratio. Other important contributions were made by Slichter [15], [16], Fair and Hatch [17], Rose [18], [19], and Bakhmeteff and Feodoroff [20].

The influence of the fluid properties on the value of the hydraulic conductivity is evident from Eq. 4.18. Hence hydraulic conductivity may be measured by using different fluids. It is possible, at least in principle, to measure the hydraulic conductivity using a gas, say air, and to compute the intrinsic permeability k. Once k is known, it suffices to multiply k by g/ν to find the hydraulic conductivity for any fluid. However, Klinkenberg [30] has shown that the permeability of a porous medium to a liquid and to a gas is not the same. Because of the slip phenomenon, whereby the velocity of a gas layer in the immediate vicinity of the surface of the grains has a finite (instead of zero) velocity, the permeability to a gas is higher than to a liquid. Klinkenberg derived the equation

$$k_a = k(1 + b/\bar{p}) \tag{4.19*}$$

where k_a is the apparent permeability and k the intrinsic permeability when gas is used in the measurement, b a constant, and $\bar{p}$ the mean pressure over the sample. From this equation it follows that the intrinsic permeability may be obtained by an extrapolation of the gas-permeability data for low pressure to those for an infinitely high mean pressure.

Also, samples containing clay and silt change hydraulic conductivity in response to the type of fluid used. In general permeabilities determined with air have the highest values, those determined with brines are next, and those determined with distilled water are the lowest. The difference between air and water values is explained by the fact that hydration of clays and other minerals causes swelling of the grains and hence clogging of pores. Air permeabilities and water permeabilities may differ by a factor of more than 100 in clay rich sediments. Brine permeabilities and distilled water permeabilities may also differ by a factor of 100. The latter difference is due also to the effects of partial hydration. The water molecules on the clay surface tend to be in osmotic equilibrium with the surrounding pore fluids. When the pore fluid is a brine, only a few water molecules cluster about the clay particles, and the effective volume of the clay particle is small. With a decrease in ion concentration in the pore fluid, a new equilibrium will be established in which a much larger cluster of water molecules will surround each clay particle. Thus the effective volume of the clay is increased with a resultant decrease in pore space and permeability.

4.7 *Velocity Potential* Φ

In Fig. 4.4 water flows from lower to higher pressure. The datum plane for elevation has been chosen arbitrarily. It shows that the energies in sections 1 and 2 are determined only within the same arbitrary constant, so that their difference is independent of this constant. The pressures however are uniquely determined throughout the flow system from the knowledge of the heads in sections 1 and 2. The heads indicated on Fig. 4.4 consist of elevation head z and pressure head p/γ. The kinetic energy $V^2/2g$, which is the same in both sections and therefore would not affect the difference in total energies for this particular experiment, is nevertheless in general negligible compared to $h = z + p/\gamma$. The piezometric head h is a scalar quantity, i.e., at any point it can be expressed and defined solely by a number in contrast to a vectorial quantity which requires in addition a direction. A special kind of vector that occurs in the present theory is that derived from a scalar quantity Φ, namely gradient Φ or grad Φ. The vectorial form of Eq. 4.15 generalized to more than one dimension is

$$\mathbf{V} = -K \operatorname{grad} h \tag{4.24}$$

which is valid for two- and three-dimensional flow. If K is a constant, i.e., if the medium is homogeneous and isotropic, and if the fluid is of constant density and viscosity, Eq. 4.24 becomes

$$\mathbf{V} = -\operatorname{grad} \Phi \tag{4.25}$$

in which

$$\Phi = Kh = K(z + p/\gamma) + \text{constant} \tag{4.26}$$

is called the velocity potential, and has the dimensions $[L]^2/[T]$. It will be shown later that Φ may be combined in a particular way with a stream function Ψ having the same dimension. In two-dimensional problems, aquifers and porous media of unit width (measured perpendicularly into the plane of the paper) are commonly treated. For these problems Ψ may be expressed in discharge units per unit width. The arbitrary additive constant in Eq. 4.26 is a result of the arbitrariness in the choice of the datum plane for elevation. This constant is often omitted, but sometimes it is convenient to assign a particular value to it. The velocity potential concept is very useful for the solution of problems in which K is constant, as is the case in most simplified civil engineering applications. Its continued use may be defended for this reason alone. In general, however, the theory of ground-water flow is more complete when the specific discharge is derived via the gradient of a force potential, as will be shown later. Again it is emphasized that Eq. 4.25 is valid only for constant K and therefore is not a correct expression of Darcy's law in general.

4.8 *Derivation of Darcy's Law*

Darcy's law states that macroscopically, the Darcy velocity or specific discharge $\mathbf{V}$ is equal to the negative gradient of Kh when K is constant. The true velocity in the pores of the medium is $\mathbf{V}/n$, in which n is the porosity of the medium. Although Darcy's law was first discovered by experiment, several researchers, among whom Hubbert [8] was outstanding, have derived it from the general Navier-Stokes equations for viscous flow. Such derivations invariably stem from statistical considerations and simplifications of the complicated microscopic flow picture. Although they do not contribute to the formulation of a new law, they confirm the early belief that Darcy's law is of the nature of a statistical result giving the empirical equivalent of the Navier-Stokes equations.

The following simple proof, more heuristic than rigorous, was devised by De Wiest [9]

Consider the Navier-Stokes equations in two-dimensional form:

$$(1/n)(\partial u/\partial t) + (u/n^2)(\partial u/\partial x) + (v/n^2)(\partial u/\partial z) = -(1/\rho)(\partial p/\partial x) + (\mu/\rho n)\nabla^2 u \quad (4.27)$$

and

$$(1/n)(\partial v/\partial t) + (u/n^2)(\partial v/\partial x) + (v/n^2)(\partial v/\partial z) = -(1/\rho)(\partial p/\partial z) + (\mu/\rho n)\nabla^2 v - g \quad (4.28)$$

In Eqs. 4.27 and 4.28 u/n and v/n are the components of the true velocity $\mathbf{V}/n$. The statistical averages may essentially be summarized by the following assumptions, dimensionally correct;

$$\nabla^2 u = -\frac{1}{c}(u/L^2)$$

and

$$\nabla^2 v = -\frac{1}{c}(v/L^2) \quad (4.29)$$

where L may be thought of as being a characteristic length, for example the pore size d of the medium, and where c is a dimensionless parameter. If it is further assumed that

$$u = -\partial\Phi/\partial x$$

and

$$v = -\partial\Phi/\partial z \quad (4.30)$$

the Navier-Stokes equations become

$$-(1/n)\partial/\partial x(\partial\Phi/\partial t) + (1/n^2)\partial/\partial x[1/2(\partial\Phi/\partial x)^2 + 1/2(\partial\Phi/\partial z)^2] = -(1/\rho)(\partial p/\partial x) + (\mu/c\rho d^2 n)(\partial\Phi/\partial x) \quad (4.31)$$

and

$$-(1/n)\partial/\partial z(\partial\Phi/\partial t) + (1/n^2)\partial/\partial z[1/2(\partial\Phi/\partial x)^2 + 1/2(\partial\Phi/\partial z)^2] = -(1/\rho)(\partial p/\partial z) + (\mu/c\rho d^2 n)(\partial\Phi/\partial z) - g$$

If μ and ρ are constant, both equations may be integrated directly and their integrals should be the same, i.e.,

$$-(1/n)(\partial\Phi/\partial t) + n^2(1/2)[(\partial\Phi/\partial x)^2 + (\partial\Phi/\partial z)^2] + p/\rho - \mu\Phi/c\rho n d^2 + gz = F(t) \quad (4.32)$$

For steady flow, provided the inertia terms here represented by the kinetic energy of the liquid may be neglected, Eq. 4.32 reduces to

$$\Phi = \left(k\frac{\gamma}{\mu}\right)(p/\gamma + z) + \text{constant} \quad (4.33)$$

in which

$$k = cnd^2 \quad (4.34)$$

Equation 4.34 may be identified with Eq. 4.19 because it is always possible to factor out the porosity from the shape factor C, independently of the functional relationship between C and n. As a matter of comparison, Eq. 4.34 is in agreement with Taylor's Eq. 6.20 for the intrinsic permeability. With the symbols of this textbook, Taylor's equation (see reference 4, p. 111) may be written as

$$k = C'e^2nd^2$$

in which e is the void ratio of the medium. Obviously both formulas are compatible. Hence Eq. 4.33 may be identified with Eq. 4.26, and Darcy's law has been derived.

4.9 *Range of Validity of Darcy's law*

In pipeflow the transition from laminar to turbulent flow is characterized by well-known values of Reynolds' number which expresses the ratio of inertial to viscous forces. Thus there is a lower critical number, around 2,100 below which the flow in pipes is always laminar. By analogy, in flow through porous media a Reynolds number has been established as

$$N_R = VD/\nu \tag{4.35}$$

in which V is the specific discharge, ν the kinematic viscosity of the fluid, and D a characteristic length. For D some have used the average grain diameter and others the effective diameter of the particles, d_{10} (equal to the diameter of the aperture of the sieve through which 10% by weight of the soil sample is screened). In Darcy's law the inertia forces have been neglected, as was demonstrated in section 4.8. If the flow at low velocities is laminar, it is expected to become turbulent at higher velocities. Turbulent flow, if the analogy with pipe flow holds true, would require a nonlinear, near quadratic or quadratic, relationship between velocity and head loss. Hence Darcy's law would not be valid any longer. However, experiments conducted by Lindquist [21, 28] have shown that digressions from Darcy's law occur even in the laminar flow regimen when inertial forces become effective. Lindquist defined a value of $N_R{}^*$ at which digression from Darcy's law starts because the inertia forces become important. He found $N_R{}^*$ to be of the order of 4 in the case of a medium of uniform grain size, with diameters ranging from 1 to 5 mm and with a porosity n of 38%. He also found that for this medium the upper limit of N_R, above which there is always turbulence (called $N_{R,\text{crit}}$), was greater than 180.

These experiments were confirmed by Schneebeli [29], who used the same definition for N_R as did Lindquist and also the same value for D, namely, the average grain diameter. For a medium of spheres of uniform diameter $d = 27$ mm, $k = 6.13 \times 10^{-3}$ cm², $n = 39\%$ and water at 20°C, Schneebeli found $N_R{}^*$ to be around 5, whereas $N_{R,\text{crit}}$ was around 60. For a medium of crushed stone of equilavent diameter $d_e = 37$ mm, $k_e = 11.5 \times 10^{-3}$ cm², $n = 47\%$ and water at 20°C, he found $N_R{}^*$ to be around 2, and $N_{R,\text{crit}}$ around 60 as in the case of the spheres. The equivalent diameter d_e used in N_R for D is defined by

$$\frac{d_e^{\,2}}{d^2} = \frac{k_e}{k} \tag{4.19*}$$

where k and d are the intrinsic permeability and grain size of a medium of uniform spheres and k_e the intrinsic permeability of the medium of crushed stone. Actually, Eq. 4.19* may be rewritten as

$$\frac{d_e^{\,2}}{k_e} = \frac{d^2}{k} = \frac{1}{C}$$

where C is the constant of Eq. 4.19. Here, to compare the two tests, Schneebeli used for K and d the values of his first test.

From these and other similar experiments one concludes that there are transition zones first from laminar flow where resistance (viscous) forces are predominant to laminar flow where inertial forces govern, and then finally to turbulent flow. In the case of a typical discharging well, departures from laminar flow are to be expected in the immediate vicinity of the well, where the velocities of the water are maximal. A study of turbulent flow near well screens has been made by Engelund [22]. Considerable departures from Darcy's law occur if clay and colloidal material are abundant in the medium [33].

4.10 *Main Equation for Conservation of Mass*

Along with the principle of conservation of energy, the principle of conservation of mass has to be observed in flow through porous media. The following derivations are essentially borrowed from Jacob [23]. Consider an elemental parallelepipedum of porous medium completely saturated with fluid of density ρ. Let the vectorial specific discharge $\mathbf{V}(x, y, z)$, in the center P of the volume element, have the components $u(x, y, z)$, $v(x, y, z)$, $w(x, y, z)$. In order to find the flow rate through the sides of the volume element, the specific discharge at these sides must be referred to that at the center P or to its components. This is done by an expansion of the functions $u(x, y, z)$ $v(x, y, z)$, $w(x, y, z)$ in a Taylor series about the values of u, v, w, in P. Since these functions contain more than one variable, it is necessary to introduce the concept of a partial derivative. This derivative differs from the ordinary derivative in that only one of the variables is considered as such in the process of differentiation, and the others are kept constant. For example, to find the value of u through the plane at $x + \Delta x/2$. Taylor's expansion gives:

$$u\left(x + \frac{\Delta x}{2}, y, z\right) = u(x, y, z) + \frac{\Delta x}{2}\frac{\partial u}{\partial x} + \frac{I}{2}\left(\frac{\Delta x}{2}\right)^2 \frac{\partial^2 u}{\partial x^2} + \cdots$$

Note that the derivatives of all orders are evaluated at the point P. It is common practice to break off the series after the linear term in Δx and to neglect the higher order terms in Δx. The mass flowrate through an elemental area ΔA is expressed as $(\rho V_n \Delta A)$, where V_n is the normal component of the velocity to the area ΔA. Since ρ may be a function $\rho(x, y, z)$, one has to consider the products ρu, ρv, ρw to compute the mass flow rate through the sides of the volume element. This is accomplished,

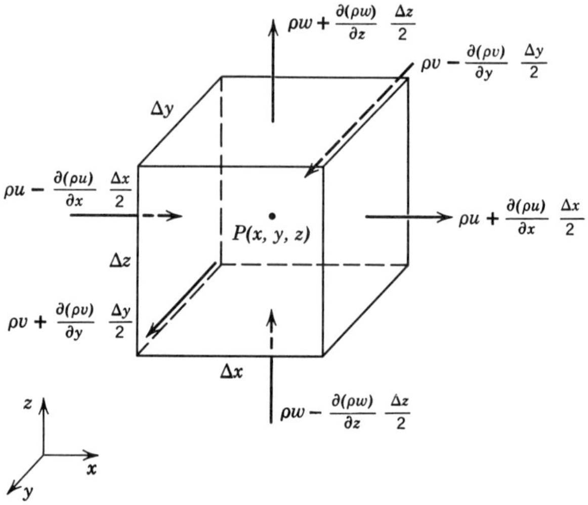

Fig. 4-6 Elemental parallelepipedum of porous medium completely saturated.

for the flow in the x-direction for example, by replacing u by ρu in the above Taylor series expansion. The terms that are kept in the different Taylor series are written in Fig. 4.6. The principle of conservation of mass for the volume element requires:

$$\text{Mass inflow rate} = \text{mass outflow rate} + \text{change of mass storage in time} \tag{4.36}$$

The contributions for the mass inflow rate are:
In the

x-direction: $$\left[\rho u - \frac{\partial(\rho u)}{\partial x}\frac{\Delta x}{2}\right] \Delta y \, \Delta z$$

y-direction: $$\left[\rho v - \frac{\partial(\rho v)}{\partial y}\frac{\Delta v}{2}\right] \Delta x \, \Delta z$$

z-direction: $$\left[\rho w - \frac{\partial(\rho w)}{\partial z}\frac{\Delta z}{2}\right] \Delta x \, \Delta y$$

These contributions must be added to obtain the total mass inflow rate. Similarly, the contributions for the mass outflow rate are:
In the

x-direction: $$\left[\rho u + \frac{\partial(\rho u)}{\partial x}\frac{\Delta x}{2}\right] \Delta y \, \Delta z$$

y-direction: $$\left[\rho v + \frac{\partial(\rho v)}{\partial y}\frac{\Delta y}{2}\right] \Delta x \, \Delta z$$

z-direction: $$\left[\rho w + \frac{\partial(\rho w)}{\partial z}\frac{\Delta z}{2}\right] \Delta x \, \Delta y$$

These contributions must be added to obtain the total mass outflow rate. From this it follows that

Mass inflow rate − mass outflow rate

$$= -\left[\frac{\partial(\rho u)}{\partial x} + \frac{\partial(\rho v)}{\partial y} + \frac{\partial(\rho w)}{\partial z}\right] \Delta x \, \Delta y \, \Delta z \tag{4.37}$$

This must be equal to the change of mass storage in time. The mass of fluid ΔM stored in the volume element $\Delta \mathcal{V}_0$ is

$$\Delta M = n\rho \, \Delta x \, \Delta y \, \Delta z$$

where n is the porosity defined in section 4.3. In the considerations of the change of ΔM in time, n and Δz may vary due to vertical compression or expansion of the medium, and ρ may change in time as well as in place. But variations of the lateral dimensions of aquifers subject to variable flow are negligible because of the constraints of these aquifers by their surroundings. It follows that

$$\frac{\partial(\Delta M)}{\partial t} = \left[n\rho \frac{\partial(\Delta z)}{\partial t} + \rho \, \Delta z \frac{\partial n}{\partial t} + n \, \Delta z \frac{\partial \rho}{\partial t}\right] \Delta x \, \Delta y \tag{4.38}$$

It remains now to express the three terms in the right member of Eq. 4.38 in terms of the compressibility α of the aquifer, the compressibility β of the fluid, and the pore pressure p.

1° $$n\rho \frac{\partial(\Delta z)}{\partial t}$$

Here the concept of vertical compressibility α of the granular skeleton of the medium, treated as a continuum, is introduced. $\alpha = 1/E_s$, where E_s is the bulk modulus of elasticity of this skeleton. The stress σ_z on the intergranular skeleton in the vertical direction is called intergranular pressure or stress [24]. By definition

$$E_s = -\frac{d\sigma_z}{\dfrac{d(\Delta z)}{\Delta z}} = \frac{1}{\alpha}$$

so that

$$d(\Delta z) = -\alpha \, \Delta z \, d\sigma_z$$

and

$$\frac{\partial(\Delta z)}{\partial t} = -\alpha(\Delta z)\frac{\partial \sigma_z}{\partial t} \tag{4.39}$$

2°

$$\rho \, \Delta z \frac{\partial n}{\partial t}$$

The volume of solid grains $\Delta \mathcal{V}_s = (1 - n)\,\Delta x\,\Delta y\,\Delta z$ may be considered as constant because the compressibility of the individual grains is considerably smaller than that of their skeleton as considered above and is also smaller than the compressibility of water. The total derivative of this quantity is zero, or

$$d(\Delta \mathcal{V}_s) = d[(1 - n)\,\Delta x\,\Delta y\,\Delta z] = 0$$

Again Δx and Δy, the lateral dimensions of the volume element, do not change in comparison to the change in the vertical dimension Δz. Therefore $\Delta x\,\Delta y$ in the above total derivative is treated as a constant, and only two terms remain:

$$\Delta z\, d(1 - n) + (1 - n)\, d(\Delta z) = 0$$

or

$$dn = \frac{1 - n}{\Delta z} d(\Delta z)$$

and

$$\frac{\partial n}{\partial t} = \frac{1 - n}{\Delta z}\frac{\partial(\Delta z)}{\partial t} \tag{4.40}$$

This expression may be modified somewhat if $\partial(\Delta z)/\partial t$ from Eq. 4.39 is inserted in Eq. 4.40:

$$\frac{\partial n}{\partial t} = -(1 - n)\alpha \frac{\partial \sigma_z}{\partial t} \tag{4.41}$$

This shows that the terms $n\rho \dfrac{\partial(\Delta z)}{\partial t}$ and $\rho(\Delta z)\dfrac{\partial n}{\partial t}$ are really not independent but that they express the effects of the same cause, namely, the vertical compression of the medium:

3°

$$n \, \Delta z \frac{\partial \rho}{\partial t}$$

To introduce the compressibility β of the fluid or the reciprocal of its bulk modulus of elasticity,

$$\beta = -\frac{\dfrac{d(\Delta V_v)}{\Delta V_v}}{dp}$$

the equation of conservation of mass is written as

$$\rho \, \Delta V_v = \rho_0 \, \Delta V_{v_0} = \text{constant}$$

in which ρ_0, ΔV_{v_0} are constant reference values of density and elemental volume of fluid. Total differentiation of this equation gives

$$\rho d(\Delta V_v) + (\Delta V_v) \, d\rho = 0$$

or

$$-\rho(\Delta V_v)\beta \, dp + (\Delta V_v) \, d\rho = 0$$

and

$$\rho\beta \frac{\partial p}{\partial t} = \frac{\partial \rho}{\partial t}. \tag{4.42}$$

Here p is the pressure experienced by the water in the pores and is called pore pressure or neutral pressure.

At any depth intergranular pressure and pore pressure add to render the total or combined pressure. The combined pressure is numerically equal to the weight of all the matter that rests above per unit area if arching effects of the overlying strata are neglected. It follows that

$$\sigma_z + p = \text{constant} \tag{4.43}$$

and

$$d\sigma_z = -dp$$

Equation 4.38 can be rewritten as

$$\frac{\partial(\Delta M)}{\partial t} = [n\rho\alpha + \rho(1 - n)\alpha + n\rho\beta] \, \Delta x \, \Delta y \, \Delta z \frac{\partial p}{\partial t}$$

and the final continuity equation becomes

$$-\left[\frac{\partial(\rho u)}{\partial x} + \frac{\partial(\rho v)}{\partial y} + \frac{\partial(\rho w)}{\partial z}\right] = \rho(\alpha + n\beta) \frac{\partial p}{\partial t} \tag{4.44}$$

4.11 *Ground-Water Flow in Homogeneous Isotropic Medium*

The expansion of the left side of Eq. 4.44 leads to

$$-\rho\left(\frac{\partial u}{\partial x} + \frac{\partial v}{\partial y} + \frac{\partial w}{\partial z}\right) - \left(u \frac{\partial \rho}{\partial x} + v \frac{\partial \rho}{\partial y} + w \frac{\partial \rho}{\partial z}\right) \tag{4.45}$$

The second term of expression in Eq. 4.45 is in general quite small in comparison with the first [23] and may therefore be neglected, especially for low-angle flow with $\partial h/\partial z$ very small. Indeed from

$$h = z + \frac{p}{\gamma} + \text{constant} \tag{4.2*}$$

The following partial derivatives follow

$$\frac{\partial p}{\partial x} = \rho g \frac{\partial h}{\partial x} + \frac{p}{\rho}\frac{\partial \rho}{\partial x}$$
$$\frac{\partial p}{\partial y} = \rho g \frac{\partial h}{\partial y} + \frac{p}{\rho}\frac{\partial \rho}{\partial y}$$
$$\frac{\partial p}{\partial z} = \rho g \left(\frac{\partial h}{\partial z} - 1\right) + \frac{p}{\rho}\frac{\partial \rho}{\partial z}$$

They may be inserted in the partial derivatives derived from

$$d\rho = \rho\beta\, dp \tag{4.42}$$

to give[1]

$$\frac{\partial \rho}{\partial x} = \frac{1}{1-\beta p}\beta\rho^2 g \frac{\partial h}{\partial x} \approx \beta\rho^2 g \frac{\partial h}{\partial x}$$
$$\frac{\partial \rho}{\partial y} = \frac{1}{1-\beta p}\beta\rho^2 g \frac{\partial h}{\partial y} \approx \beta\rho^2 g \frac{\partial h}{\partial y}$$
$$\frac{\partial \rho}{\partial z} = \frac{1}{1-\beta p}\beta\rho^2 g \left(\frac{\partial h}{\partial z} - 1\right) \approx \beta\rho^2 g \left(\frac{\partial h}{\partial z} - 1\right)$$

Furthermore, Darcy's law (Eq. 4.24) may be expressed as

$$u = -K\frac{\partial h}{\partial x}$$
$$v = -K\frac{\partial h}{\partial y} \tag{4.24}$$
$$w = -K\frac{\partial h}{\partial z}$$

Also, from Eq. 4.2*,[1]

$$\frac{\partial p}{\partial t} = \frac{1}{1-\beta p}\rho g \frac{\partial h}{\partial t} \approx \rho g \frac{\partial h}{\partial t}$$

[1] For practical values of p in ground-water flow and for $\beta = 3.3 \times 10^{-6}$ square inch per pound, the ratio $\frac{1}{1-\beta p}$ may be replaced by one, as a first approximation, indicated by the symbol $\approx$.

Equation 4.44 may now be rewritten, for K constant, as

$$K\rho\left(\frac{\partial^2 h}{\partial x^2} + \frac{\partial^2 h}{\partial y^2} + \frac{\partial^2 h}{\partial z^2}\right) + K\beta\rho^2 g\left[\left(\frac{\partial h}{\partial x}\right)^2 + \left(\frac{\partial h}{\partial y}\right)^2 + \left(\frac{\partial h}{\partial z}\right)^2 - \frac{\partial h}{\partial z}\right]$$
$$= \rho^2 g(\alpha + n\beta)\frac{\partial h}{\partial t} \quad (4.46)$$

The second term of the left side of Eq. 4.46 may be neglected for low-angle flow with $\partial h/\partial z$ very small, whereas the first term may be written as $K\rho\nabla^2 h$ where

$$\nabla^2 = \frac{\partial^2}{\partial x^2} + \frac{\partial^2}{\partial y^2} + \frac{\partial^2}{\partial z^2}$$

is the Laplacean operator. The final form of Eq. 4.46 is

$$\nabla^2 h = \frac{S_s}{K}\frac{\partial h}{\partial t} \quad (4.47)$$

where

$$S_s = \rho g(\alpha + n\beta) \quad (4.48)$$

is the specific storage. It has the dimensions $1/L$, as may be seen from Eq. 4.47, and therefore it may be conceived of as the amount of water in storage that is released from a unit volume of aquifer per unit decline of head. Its two parts may be interpreted as:

$\rho g\alpha$ = water in storage released due to the compression of the intergranular skeleton per unit volume and per unit decline of head

$\rho g n\beta$ = water in storage released due to the expansion of the water per unit volume and per unit decline of head

It may indeed be observed that $\beta > 0$ for a decrease in pore pressure and a resulting expansion of the water. In the special case of a confined aquifer of thickness b, the storage coefficient S

$$S = S_s b \quad (4.49)$$

and the transmissivity T of the aquifer

$$T = Kb \quad (4.50)$$

may be introduced. Also, $S_s/K = S/T$, and therefore Eq. 4.47 becomes

$$\nabla^2 h = \frac{S}{T}\frac{\partial h}{\partial t} \quad (4.51)$$

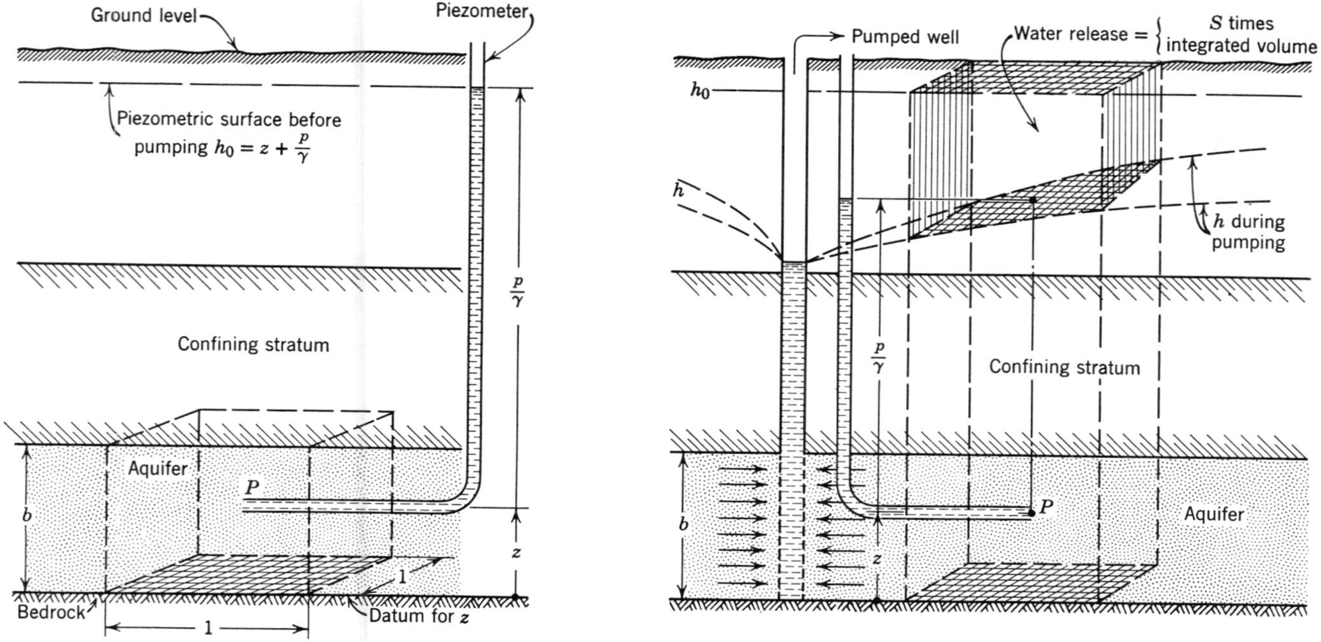

(a) (b)

Fig. 4-7 Physical interpretation of storage coefficient S. (*a*) Before pumping. (*b*) During pumping.

S and T are called the formation constants of the confined aquifer. S is dimensionless and may be interpreted as the amount of water in storage released from a column of aquifer with unit cross section under a unit decline of head. In the computation of S, the compressibility of water cannot be neglected because the term $\rho g n \beta$ may be of the same order of magnitude as the term $\rho g \alpha$ [23]. The significance of S may be visualized by means of Fig. 4.7. The decrease of head h in a given point of a confined aquifer is generally a result of water withdrawal through pumpage and the accompanying decrease in pore pressure.

Equations 4.47 and 4.51 are the general equations for unsteady flow respectively in an unconfined and in a confined aquifer.

In fact, Eq. 4.47 for the unconfined aquifer may be still more simplified. Indeed the compressibility of a sand is relatively important only when the sand is completely saturated with water and confined between impervious strata. When the flow is unconfined the compressibility of the sand and of the water are relatively unimportant compared to unsteady perturbations or vertical displacements of the free surface which affect the flow pattern. For unconfined flow the term $(S_s/K)(\partial h/\partial t)$, may be neglected in Eq. 4.47, and formally the so-called Laplace equation

$$\nabla^2 h = 0 \tag{4.52}$$

is obtained both for steady and unsteady unconfined flow. Treatment of problems of the latter type in which the solution is a function of time as well as of the space coordinates [25, 43, 44, and 45] is beyond the scope of this elementary textbook.

Equation 4.52 is also the equation for confined steady flow as follows immediately from Eq. 4.51 when h is independent of t. The above derivations were made on a hydrodynamic basis. It is further shown how the hydraulic theory modifies Eq. 4.52 for steady unconfined flow through the Dupuit assumptions. The formation constants S and T are generally determined from pumping tests, as will be shown further.

4.12 *Extension of Main Equation of Conservation of Mass*

Equation 4.47 may be altered to take into account the time dependency of external forces that are acting on the elemental mass of Fig. 4.6, such as the effect of variation of barometric pressure, of river stage in an estuary due to tidal fluctuations, and of moving load of a passing train. In the derivation of Eq. 4.47, atmospheric pressure was assumed to be constant and therefore it was convenient to use the relative pore pressure p. However, when the variations in atmospheric pressure p_a are important and when the effect of the time dependency of the atmospheric pressure cannot be neglected or is studied as such, the absolute pore pressure p'

$$p' = p + p_a \tag{4.52)*$$

must be used. Time dependency of p_a may be expressed conveniently by multiplying the average atmospheric pressure p_{a0} by $[1 + f(t)]$, an arbitrary function of time, so that

$$p_a = p_{a0}[1 + f(t)] \qquad (4.53)$$

Equation 4.43 must now be replaced by

$$\sigma_z + p' = \text{static overburden} + p_{a0}[1 + f(t)] \qquad (4.54)$$

or, if $p' - p_{a0} = p$ as a first approximation, in view of Eq. 4.52,*

$$\sigma_z + p = \text{static overburden} + p_{a0}f(t) \qquad (4.55)$$

Instead of Eq. 4.2, let

$$h = z + \frac{p'}{\gamma} - \frac{p_{a0}}{\gamma}[1 + f(t)] + \text{constant} \qquad (4.56)$$

or

$$h = z + \frac{p}{\gamma} - \frac{p_{a0}}{\gamma}f(t) + \text{constant} \qquad (4.57)$$

From Eq. 4.55, after differentiation with respect to t, it follows that

$$\frac{\partial\sigma_z}{\partial t} = -\frac{\partial p}{\partial t} + p_{a0}\frac{\partial f(t)}{\partial t} \qquad (4.58)$$

If this value of $\partial\sigma_z/\partial t$ (from Eq. 4.58) is inserted in the right side of the equation of mass conservation, the result is

$$\left[\frac{\partial(\rho u)}{\partial x} + \frac{\partial(\rho v)}{\partial y} + \frac{\partial(\rho w)}{\partial z}\right] = \rho(\alpha + n\beta)\frac{\partial p}{\partial t} - \alpha\rho p_{a0}\frac{\partial f}{\partial t} \qquad (4.59)$$

Partial differentiation of Eq. 4.57 with respect to t gives, for $\rho\beta \ll 1$,

$$\rho g\frac{\partial h}{\partial t} = \frac{\partial p}{\partial t} - p_{a0}\frac{\partial f}{\partial t} \qquad (4.60)$$

Equation 4.59 may now be rewritten as

$$-\left[\frac{\partial(\rho u)}{\partial x} + \frac{\partial(\rho v)}{\partial y} + \frac{\partial(\rho w)}{\partial z}\right] = \rho\alpha\left(\frac{\partial p}{\partial t} - p_{a0}\frac{\partial f}{\partial t}\right) + \rho n\beta\frac{\partial p}{\partial t} - \rho n\beta p_{a0}\frac{\partial f}{\partial t} + \rho n\beta p_{a0}\frac{\partial f}{\partial t}$$

or

$$-\left[\frac{\partial(\rho u)}{\partial x} + \frac{\partial(\rho v)}{\partial y} + \frac{\partial(\rho w)}{\partial z}\right] = \rho^2 g(\alpha + n\beta)\frac{\partial h}{\partial t} + \rho n\beta p_{a0}\frac{\partial f}{\partial t} \qquad (4.61)$$

The left side of Eq. 4.61 may now be expanded in the same way as that of Eq. 4.44, and if the same assumptions are made as for the derivation of Eq. 4.47,

$$\nabla^2 h = \frac{S_s}{K}\frac{\partial h}{\partial t} + \frac{n\beta}{K}p_{a0}\frac{\partial f}{\partial t} \qquad (4.62)$$

replaces Eq. 4.47. Equation 4.62 may be solved by superposition. Assume that the solution is of the form

$$h(x, y, z, t) = h_1(x, y, z, t) + h_2(x, y, z, t) \tag{4.63}$$

Then $h_1(x, y, z, t)$ must satisfy

$$\nabla^2 h_1 = \frac{S_s}{K} \frac{\partial h_1}{\partial t} \tag{4.64}$$

for the boundary conditions and initial conditions imposed for the solution of Equation 4.62.

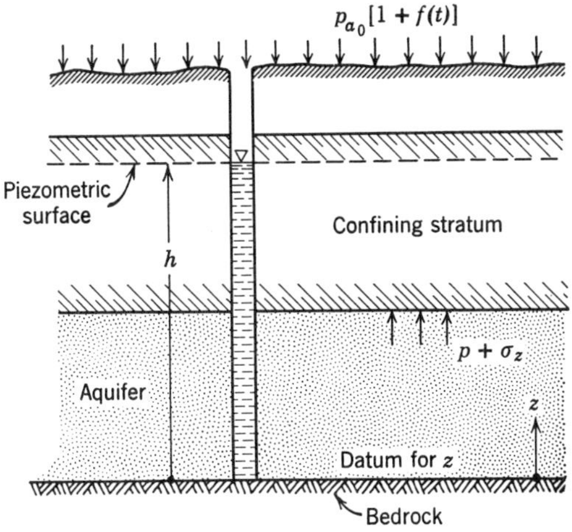

Fig. 4-8 Variation in atmospheric pressure. Barometric efficiency.

At the same time h_2 (x, y, z, t) must satisfy

$$\nabla^2 h_2 = \frac{S_s}{K} \frac{\partial h_2}{\partial t} + \frac{n\beta}{K} p_{a0} \frac{\partial f}{\partial t} \tag{4.65}$$

with all boundary conditions and initial conditions equal to zero.

In Eq. 4.65, $\frac{n\beta}{K} p_{a0} \frac{\partial f}{\partial t}$ plays the role of a forcing function. The solution of Eq. 4.65 may be found by the method of variation of parameters. For an example of this method, the reader is referred to reference 31. The same approach may be used to study the influence of variations in time of other external loads, such as those induced by passing trains or fluctuating tides.

BAROMETRIC EFFICIENCY

Readings of water level in wells are affected by variations of external loads as has been observed and also as follows from the preceding theory. Sometimes, for example in a pumping test of long duration and during which the atmospheric pressure fluctuates from a value p_a^* to a value $p_a^* + dp_a^*$, it is sufficient to apply a correction dh to the record. This correction may be computed in the following way [23] (see Fig. 4.8).

Replace Eq. 4.57 by an equivalent expression

$$h = z + \frac{p}{\gamma} - \frac{p_a{}^*(t)}{\gamma} + \frac{p_{a0}}{\gamma}, \tag{4.66}$$

where p_{a0}/γ has been chosen for the value of the constant. Also

$$\sigma_z + p = \text{constant overburden} + p_a{}^*(t) \tag{4.67}$$

Differentiation of Eqs. 4.66 and 4.67 gives

$$dp = dp_a{}^* + \gamma\, dh \tag{4.68}$$

and

$$dp = dp_a{}^* - d\sigma_z \tag{4.69}$$

From Eqs. 4.68 and 4.69 it follows that

$$\frac{\gamma\, dh}{dp_a{}^*} = \frac{dp - dp_a{}^*}{dp_a{}^*} = \frac{-d\sigma_z}{dp + d\sigma_z} = -\frac{d\sigma_z/dp}{1 + (d\sigma_z/dp)} \tag{4.70}$$

Consider now the elemental volume ΔV_0 of Fig. 4.6:

$$\Delta V_0 = \Delta V_v + \Delta V_s \tag{4.71}$$

since the assumption of complete saturation of the aquifer is made. Hence differentiation of Eq. 4.71 renders

$$d(\Delta V_0) = d(\Delta V_v) \tag{4.72}$$

and if both sides are divided by $n(\Delta V_0) = \Delta V_v$, the result is

$$\frac{d(\Delta V_0)}{n(\Delta V_0)} = \frac{d(\Delta V_v)}{\Delta V_v} \tag{4.73}$$

The definitions of the vertical compressibility α of the granular skeleton of the medium and of the compressibility β of the fluid are now recalled.

$$\alpha = -\frac{d(\Delta z)/\Delta z}{d\sigma_z} = -\frac{d(\Delta V_0)/\Delta V_0}{d\sigma_z} \tag{4.74}$$

where the second equality sign is only valid because Δx and Δy are considered to be constant in the product $\Delta V_0 = \Delta x\, \Delta y\, \Delta z$.

$$\beta = -\frac{d(\Delta V_v)/\Delta V_v}{dp} \tag{4.75}$$

Hence

$$\frac{d(\Delta V_0)}{\Delta V_0} = -\alpha\, d\sigma_z$$

and

$$\frac{d(\Delta V_v)}{\Delta V_v} = -\beta\, dp$$

These ratios may be inserted in Eq. 4.73 to render

$$-\alpha \frac{d\sigma_z}{n} = -\beta\, dp$$

and
$$\frac{d\sigma_z}{dp} = \frac{n\beta}{\alpha} \tag{4.76}$$

which may be introduced in Eq. 4.70 for the final result

$$\frac{dh}{dp_a{}^*} = -\frac{1}{\gamma}\frac{n\beta}{\alpha + n\beta} \tag{4.77}$$

Equation 4.77 shows that for $dp_a{}^* > 0$, the correction $dh < 0$. This is confirmed by observations in nature. A physical interpretation of this phenomenon is given by Zhukovsky (see reference 7, pp. 15–16). However, Eq. 4.77 does not explain the faster flow of a well under decreasing atmospheric pressure as stated by Zhukovsky, because dh in Eq. 4.77 is uniform.

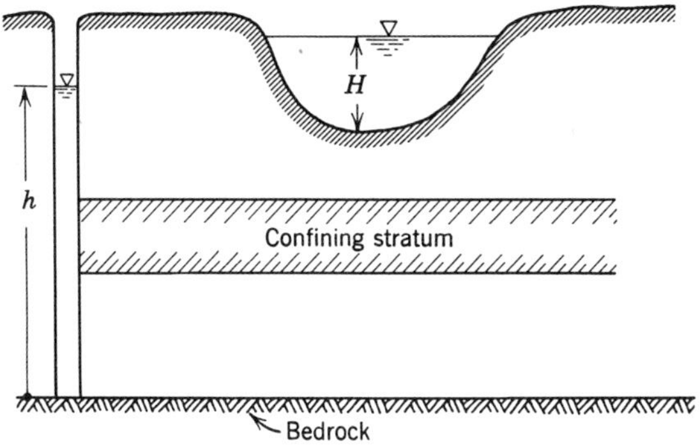

Fig. 4-9 Effect of tidal changes on well levels.

TIDAL FLUCTUATIONS

If it is assumed that the pressure γH at the bottom of the estuary is transmitted according to Pascal's law to the confined aquifer of Fig. 4.9, then

$$p + \sigma_z = \text{static overburden} + \gamma H \tag{4.78}$$

so that for a change dH in river stage,

$$dp + d\sigma_z = \gamma \, dH \tag{4.79}$$

Also, from Eq. 4.2*, $dp = \gamma \, dh$. Hence

$$\frac{\gamma \, dh}{\gamma \, dH} = \frac{dp}{dp + d\sigma_z}$$

and by means of Eq. 4.76, the final result is

$$\frac{dh}{dH} = \frac{\alpha}{\alpha + n\beta} \tag{4.80}$$

A comparison of Eq. 4.80 with Eq. 4.77 shows that

$$\left|\frac{\gamma \, dh}{dp_a{}^*}\right| + \frac{dh}{dH} = 1$$

so that it is possible to compute one correction dh from the knowledge of the other.

4.13 *Streamlines and Equipotential Surfaces—Hubbert's Force Potential*

To derive Eq. 4.24, the directional derivative dh/dl in Eq. 4.16 was replaced by (grad h), and a vector meaning was given to the specific discharge by printing **V** in boldface type. This step was easy because the problem was unidirectional. In fact, the transition from scalar equation

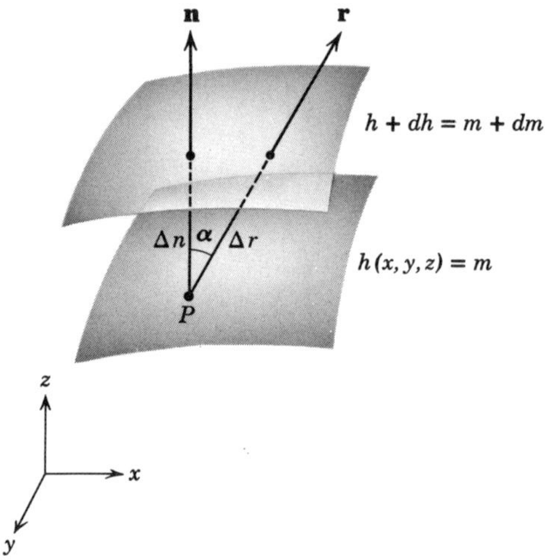

Fig. 4-10 Physical interpretation of grad $h(x, y, z)$.

to vectorial equation was made possible by multiplication of both sides by a unit vector $\mathbf{u}_l$[1] in the l direction. Thus

$$V\mathbf{u}_l = -K\frac{dh}{dl}\mathbf{u}_l$$

became

$$\mathbf{V} = -K\,(\text{grad}\ h) \tag{4.24}$$

In unidirectional flow, gradient and directional derivative are overlapping concepts. Their physical and mathematical extensions to two- and three-dimensional problems follow. In three dimensions, $h(x, y, z) = m$ may be represented by a surface as sketched in Fig. 4.10. If the parameter m assumes an increment Δm, then h assumes an increment Δh, and

[1] A unit vector is a vector of unit length.

$h + \Delta h = m + \Delta m$ is represented by another surface. Let Δn be the intercept between the two surfaces along the normal n in point P of $h(x, y, z) = m$ perpendicular to h. The gradient of h is then defined as $\text{grad } h = \lim\limits_{\Delta n \to o} \dfrac{\Delta h}{\Delta n} \mathbf{u}_n$ where $\mathbf{u}_n$ is the unit normal. It is the directional derivative of h in the direction $\mathbf{n}$ normal to the surface $h(x, y, z) = m$ in the point P. The special feature of the gradient is that it represents the

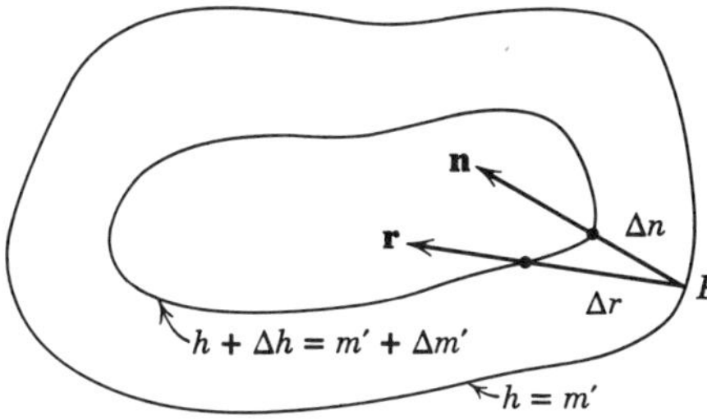

Fig. 4-11 Physical interpretation of grad $h(x, y)$.

maximal value of the directional derivative. To see this, one takes the directional derivative D in any other direction, say $\mathbf{r}$,

$$\underset{\mathbf{r}}{D}h = \lim_{\Delta r \to o} \frac{\Delta h}{\Delta r} \mathbf{u}_r \tag{4.81}$$

But since $|\Delta r| > |\Delta n|$, the latter limit is smaller than the gradient. From Fig. 4.7, it follows that the magnitude of the component of the gradient in any direction is equal to the magnitude of the directional derivative in that direction. Indeed $\Delta n = \Delta r \cos \alpha$, and if this is inserted in Eq. 4.81 it follows that

$$(\text{grad } h) \cos \alpha = \underset{\mathbf{r}}{D}h$$

The surfaces $h(x, y, z) = m$, for various values of the parameter m, are called level surfaces.

If these level surfaces are intersected with a plane $z =$ constant, then contour curves $h(x, y) = m'$ are generated. A well-known case is that of topographic contours, where h simply is the elevation of different points of the earth's surface (Fig. 4.11). The gradient again could be defined exactly as in the three-dimensional case. Here it would be oriented along the path of steepest descent perpendicular to $h = m'$ in P.

Any vector may be decomposed into its components in a coordinate system. If $\mathbf{u}_x$, $\mathbf{u}_y$, $\mathbf{u}_z$ are unit vectors along the x-, y-, and z-axis, decomposition of grad h renders

$$\text{grad } h = (\text{grad } h)_x \mathbf{u}_x + (\text{grad } h)_y \mathbf{u}_y + (\text{grad } h)_z \mathbf{u}_z$$

But according to what was stated before, the components of the gradient in any direction are the directional derivatives in that direction. Therefore $(\text{grad } h)_x = \partial h/\partial x$, etc., and

$$\text{grad } h = \frac{\partial h}{\partial x}\mathbf{u}_x + \frac{\partial h}{\partial y}\mathbf{u}_y + \frac{\partial h}{\partial z}\mathbf{u}_z \tag{4.82}$$

With these elements of vector calculus, one may proceed to find a relationship between equipotential surfaces and streamlines in three-dimensional flow or between equipotential lines and streamlines in two-dimensional flow.

Let it be remembered that the piezometric head for water

$$h = z + \frac{p}{\gamma} + \text{constant} \tag{4.2*}$$

may be thought of as energy or potential per unit weight. Furthermore, in Eq. 4.26,

$$\Phi = Kh = K\left(z + \frac{p}{\gamma}\right) + \text{constant}$$

has been called the velocity potential because its gradient has the dimensions of a velocity. Hubbert [3] uses the concept of force potential, written for water as

$$\Phi^* = gh = g\left(z + \frac{p}{\gamma}\right) + \text{constant} = \frac{g}{K}\Phi + \text{constant} \tag{4.83}$$

because grad Φ^* represents a force per unit mass of fluid. In fact, Φ^* is the potential per unit mass of fluid, derived from the general formula

$$\Phi^* = g\int_{z_0}^{z} dz + \int_{p_0}^{p} \frac{dp}{\rho}$$

where the kinetic energy $v^2/2$ has been neglected and where ρ must be a function only of p in order to have a uniquely defined Φ^*. This is the case for water, where ρ = constant, so that the second integral can be performed. For $z_0 = p_0 = 0$ as reference values, Eq. 4.83 follows immediately.

If we take the gradient of Eq. 4.83, then

$$\text{grad}\,\Phi^* = g(\text{grad}\,z) + \frac{1}{\rho}(\text{grad}\,p)$$

or

$$-\text{grad}\,\Phi^* = \mathbf{g} - \frac{1}{\rho}(\text{grad}\,p) \qquad (4.85)$$

because grad $z = \mathbf{u}_z$ according to Eq. 4.82 and because $g\mathbf{u}_z = -\mathbf{g}$, $\mathbf{u}_z$ being oriented upward against the direction of gravity.

Equation 4.85 is a force equation per unit mass of fluid. The left side of this equation states that the unit mass of fluid is subject to a force oriented from surfaces of higher Φ^* to surfaces of lower Φ^* and perpendicular to these surfaces. The right side of the equation states that this force has a gravity component and a component caused by the gradient of the fluid pressure. In Eq. 4.85 only the direction of $\mathbf{g}$ is fixed, but the vectors $\left(\frac{1}{\rho}\,\text{grad}\,p\right)$ and ($-$ grad Φ^*) may have arbitrary directions (Fig. 4.12).

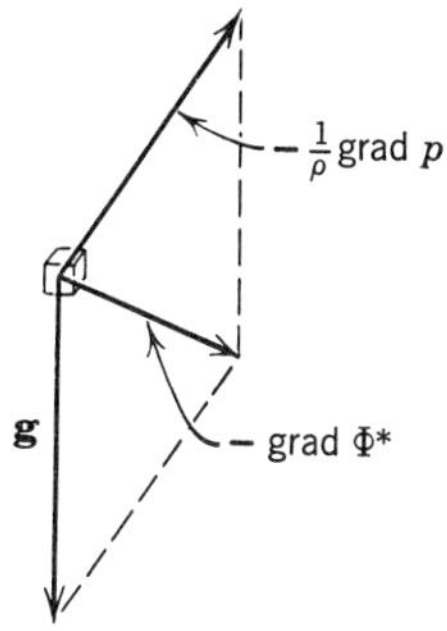

Fig. 4-12 Graphical representation of forces on unit mass of fluid (Eq. 4.85).

The magnitude of $-$ grad Φ^* in most cases of ground-water flow is only a small fraction of $\mathbf{g}$ so that $-\frac{1}{\rho}(\text{grad}\,p)$ becomes nearly vertical which means that the surfaces of constant pressure are nearly horizontal. As may be seen from Fig. 4.12 only in the special case that $-\frac{1}{\rho}(\text{grad}\,p) = -\mathbf{g}$ do we find that

$$\text{grad}\,\Phi^* = 0$$

and this means that there is hydrostatic equilibrium. The vectors $-$ grad p generate a vector field everywhere normal to the isobaric surfaces $p =$ constant. The vectors $-$grad Φ^* generate a family of field lines or a vector field everywhere normal to the equipotential surfaces $\Phi^* =$ constant. How may the latter vector field be related to the streamlines? Let it be recalled that a streamline is a line tangent in each of its points to the velocity vector of a particle inserted at these points. The relationship asked for is Darcy's law, which may be written according to Hubbert [3] as

$$\mathbf{V} = -\frac{K}{g}\,\text{grad}\,\Phi^* \qquad (4.86)$$

If K is constant, as for a homogeneous and isotropic medium and for water, then $\mathbf{V}$ will have the same direction as $-\text{grad}\ \Phi^*$, i.e., streamlines are normal to surfaces or lines of $\Phi^* = \text{constant}$. But for anisotropic media, where K has preferential properties in given directions, flowlines will in general be oblique to the direction of $-\text{grad}\ \Phi^*$ and hence not normal to equipotential surfaces or — lines. With the value

$$\Phi = \frac{K}{g}\Phi^* + \text{constant}$$

taken from Eq. 4.83, Eq. 4.25 may be expanded as

$$\mathbf{V} = -\frac{K}{g}\,\text{grad}\ \Phi^* - \frac{\Phi^*}{g}\,\text{grad}\ K \qquad (4.25)$$

which proves that Eq. 4.25 is only compatible with Eq. 4.86, and therefore only valid when $\text{grad}\ K = 0$ or $K = \text{constant}$. Unless stated to the contrary, this is the case throughout this book. For general conditions, Eq. 4.25 is not correct and Eq. 4.86 should be applied. In this elementary book, certain advantages are derived by using the velocity potential in analogy to classical hydrodynamics, as stated before in section 4.7. A discussion of the differential form of Darcy's law which is helpful in the understanding of the different potential concepts is given by Jones [32].

As will be shown further, orthogonality of streamlines and equipotential lines in two-dimensional problems facilitate the construction of flow nets, a graphical method of solving Laplace's equation. From hydrodynamics it is known that the satisfaction of this equation by a potential function is associated with irrotational flow. The fact that a ground-water potential has been derived seems to be in contradiction to the predominance of viscous resistance forces in ground-water movement and the invariably rotational character of viscous flow. The paradox is explained because only the specific discharge has been derived from a potential. In the individual porous channels the flow is truly viscous and perfectly analogous to Poiseuille flow. It is assumed, however, that when a great number of channels are assembled the rotations in the individual interstices balance out and the average resultant flow is irrotational [23].

4.14 *Other Forces Causing Water Movement*

We have previously examined the flow of ground water under influence of gravity and pressure forces. These are the principal forces acting in saturated media, and they are of main interest in this book. Water in the unsaturated zone is of greater concern to soil physicists and agronomists than to hydrogeologists. It is subjected to other forces, a few of which are briefly discussed here.

Differences of temperature will cause movement of water in unsaturated material. This action is commonly referred to as thermo-osmosis. Water does not move unless the water content is considerably below the quantity needed to fully saturate the pores. In some medium-grained sands the optimum moisture content for water migration is 1 to 6% of the total weight of the sand [34]. When air in soil is saturated with water vapor and then subjected to cooling, condensation and local depletion of water vapor result. Water vapor will then migrate by molecular diffusion towards the area of condensation, which causes, in turn, vaporization of soil water in the warmer part of the soil. Lebedev [26] considered the condensation of water vapors moving in the ground to be one of the reasons for the formation of ground water, especially in deep strata. However, it has been demonstrated since then that vapor coming from deep origins is not a significant source of ground water except in the vicinity of recent volcanoes.

A second cause of water movement is the increase of surface tension which accompanies a decrease of temperature. The increase of surface tension causes greater capillary attraction in the direction of heat flow. Actual flow of water under a temperature gradient is probably a combination of capillary and vapor transport. The fact that salts are not transported indicates that some movement is in the vapor phase. On the other hand, many researchers believe that vapor transport uses much more energy than capillary transport; therefore as much water as possible moves as capillary water, and vapor phase transport is only through short distances.

If temperatures below freezing occur ice forms from interstitial water. A number of different explanations have been given for this phenomenon. An important factor is probably that the water film on the growing ice crystals has a lower temperature and hence higher surface tension, and capillary water in contact with the ice will be drawn towards the growing ice. Water movement in connection with ice growth can take place in a medium much nearer to saturation than in the capillary vapor phase transport. Salts can be moved to a zone of freezing through a temperature gradient, shus indicating that a connected system of capillary water exists in the region of accumulating subsurface ice.

Thermo-osmosis through combined capillary-vapor phase movement is generally of minor importance as far as most ground-water production problems are concerned. The accumulation of moisture under asphalt pavements and troublesome changes in thermal conductivity of soils adjacent to electrical conductors are two causes for engineering problems related to thermo-osmosis. If the water table is near the surface, some water may be transported to the ground surface during cold periods to be later lost to the atmosphere by evaporation. Thermo-osmosis in connection with ice accumulations is of the utmost importance in many engineering problems of cool-temperate, sub-Arctic, and Arctic regions. Also, in the northern latitudes accumulation of ice may deplete water from aquifers as well as change the hydraulic head of the aquifers.

If an electrical potential from a direct-current source is imposed on a water-saturated sample of fine-grained material, such as kaolinitic clay, a flow of water from the positive to the negative electrode will be induced. This phenomenon is known as electro-osmosis. Electro-osmotic conductivity, expressed in terms of volume of water moved per second across a unit area under an electrical gradient of one volt per centimeter, is many times the hydraulic conductivity of fine-grained silts and clays. To date (1964), electro-osmosis has been of greatest interest as a method of dewatering fine-grained sediments. Minor amounts of water may move by electro-osmosis in natural environments which have considerable changes in natural electrical potentials; for example, in the vicinity of oxidizing sulfide minerals.

If two types of water have different dissolved solids concentrations and if they are separated by a semipermeable membrane, water will move from the side of low concentration to the side of high concentration. The movement will continue until the concentrations become equal or until some other counteracting force, such as hydrostatic pressure, equals the chemical force. The tendency of water to move in the direction of increasing chemical concentration is called osmosis. It is explained by the fact that water molecules continuously pass through the semipermeable membrane in both directions but the higher concentration of ions on one side slows the return of water so that there is a net movement towards the side with the greater concentration. The effectiveness of osmosis in moving normal ground water is largely a matter of speculation. Certainly the action of plants in absorbing subsurface water is partly an osmotic process, but, in addition, important hydraulic head differences may be caused by osmotic pressures. In fresh-water aquifers which are separated from sea water by clay deposits, osmotic pressures of more than 10 atmospheres are possible. These pressures undoubtedly increase the amount of seepage of water from the aquifer into the ocean. If the hydraulic head in the aquifer is reduced below sea level by pumping, osmotic pressure through the clay membrane may prevent sea-water intrusion into the aquifer.

4.15 *Capillary Forces*

The tendency of water molecules to adhere to surfaces of solids combined with surface tension of water gives rise to capillary forces. Surface tension is due to molecular attraction forces in the surface of a liquid when the liquid is exposed to a fluid with which it does not mix (water and air are immiscible). As a result of this exposure the free surface tends to assume a minimum area. It is a skin effect that is expressed as a force per unit length of contour made by the free liquid surface with its solid boundaries. When a circular tube of small diameter $2r$, wetted by the liquid to be tested, is partly immersed in a vessel containing the liquid, the liquid will rise to a height h_c above the free liquid surface in the vessel.

The formulas for capillary rise given in most textbooks make certain simplifying assumptions concerning the physics and geometry of the water and tube. First, a finite time is assumed during which the capillary rise takes place. Actually, the viscosity of the water in the tube exerts a restraining force which causes the upward movement of the capillary water to diminish exponentially with time so that an infinite length of time is needed to reach a theoretically stable condition. Second, an angle of contact is assumed between the water surface and the solid. The water surface, nevertheless, approaches the solid surface as a continuously curving surface, and the so-called angle of contact is formed by a line tangent to that surface at the point of contact. Inasmuch as most natural materials have irregular surfaces in minute detail, the concept of a microscopic angle of contact is somewhat artificial as applied to problems in hydrology.

If surface tension is designated by σ, the maximum capillary rise by h_c, the tube radius by r, the angle of contact by θ, the water density by ρ, and the acceleration due to gravity by g, then the vertical component of the upward force on the inside of a tube is $\sigma 2\pi r \cos\theta$ and the downward pull is the weight of the water, or $\pi r^2 h_c \rho g$ (Fig. 4.13*a*). If these forces are equated, the following expression results for h_c:

$$h_c = \frac{2\sigma \cos\theta}{r\gamma} \tag{4.87}$$

If for water the approximate values $\sigma = 75$ dynes/cm, $\cos\theta \approx 1.0$, and $\gamma = 1$ gram weight/cm^3 are used, and in the presence of solids such as glass and quartz at about 20°C, Eq. 4.87 may be simplified for most purposes to

$$h_c = \frac{0.153}{r} \tag{4.88}$$

in which h_c and r are measured in centimeters. If a glass tube has a radius of 0.01 mm the theoretical rise of water will be slightly more than $1\frac{1}{2}$ meters. The openings in many fine-grained unconsolidated sediments average about this radius, and the maximum capillary rises observed in these sediments range from about 1 to 2 meters. If the material is a sand the capillary rise will be between 0.1 and 1.0 meter. In gravel the capillary rise will be less than 0.1 meter. The actual capillary rise will be a function of water temperature, mineral composition, orientation, shape, and packing of the component grains. In fine-grained clay the capillary rise is on the average 2 to 4 meters. The permeability of such clay is, nevertheless,

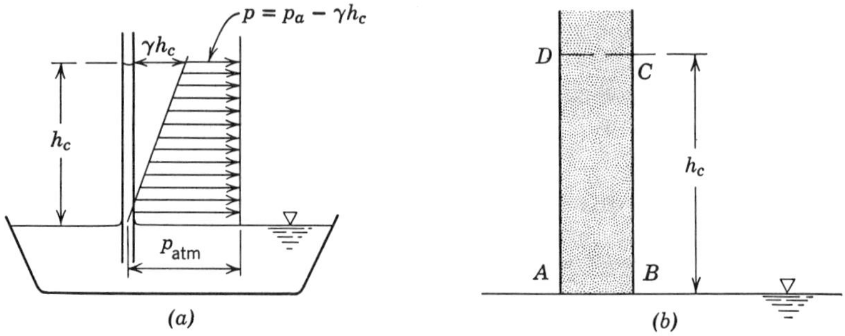

Fig. 4-13 (*a*) Rise in a capillary tube. (*b*) Capillary rise in a pipe filled with sand.

so small that the length of time needed for the rise is measured in terms of many years. The phenomenon observed in Fig. 4.13*a* also occurs when the open end of a pipe filled with sand is immersed in water (Fig. 4.13*b*). The pressure at the bottom *AB* is atmospheric so that in the pipe pore pressures lower than atmospheric prevail. All water within the capillary zone between *AB* and *CD* is under tension (negative pressure, if atmospheric pressure is assumed to be zero). Actually the granular skeleton is under pressure in the same way as the wall of the capillary tube is under vertical compression. In the capillary zone the intergranular pressure is larger than the combined pressure, owing to the weight of the overlying strata, by an amount equal to the capillary tension in the water. In *CD*, the pore pressure is minimum:

$$p = p_a - \gamma h_c \tag{4.89}$$

This condition must be satisfied at the free surface of ground-water flow, according to some Russian authors [35]. Averjanov [36] used a modified form of Eq. 4.89, namely,

$$p = p_a - \alpha\gamma h_c$$

where α is a number smaller than one, depending on the saturation of the soil. Averjanov proposed to take $\alpha = 0.3$ for most cases because the capillary rise in laboratory tests is exaggerated in comparison with the actual capillary rise in the field. (See reference 37, Chapter 8.)

REFERENCES

1. Darcy, H., *Les fontaines publiques de la ville de Dijon*, V. Dalmont, Paris, 1856 (674 pp.).
2. Rouse H. and S. Ince, *History of Hydraulics*, p. 161, Iowa Institute of Hydraulic Research, Iowa City, 1957.
3. Hubbert, M. K., "Entrapment of Petroleum under Hydrodynamic Conditions," *Bulletin of American Association Petroleum Geologists*, Vol. 37, pp. 1954–2026 (1953).
4. Taylor, D. W., *Fundamentals of Soil Mechanics*, p. 108, Wiley, New York, 1948.
5. Muskat, M., *The Flow of Homogeneous Fluids through Porous Media*, p. 72, McGraw-Hill, New York, 1937.
6. Todd, D. K., *Groundwater Hydrology*, p. 50, Wiley, New York, 1959.
7. Polubarinova-Kochina, P. Ya., *Theory of Ground-Water Movement*, p. 14, Princeton University Press, Princeton, 1962. (English translation by J. M. R. De Wiest.)
8. Hubbert, M. K., "Darcy's Law and the Field Equations of the Flow of Underground Fluids," *Transactions American Institute Mining and Metallurgical Engineers*, Vol. 207, pp. 222–239 (1956).
9. De Wiest, J. M. R., "Unsteady State Phenomena in Flow through Porous Media," *Technical Report* 3, p. 5, Civil Engineering Department, Stanford University, 1959.
10. Taylor, D. W., *Fundamentals of Soil Mechanics*, p. 111.
11. Hatschek, *The Viscosity of Liquids*, D. Van Nostrand, Princeton, 1928.
12. Kozeny, J., *Wasserkraft und Wasserwirtschaft*, Vol. 22, p. 86, 1927.
13. Hazen, A., "Discussion of Dams on Sand Foundations by A. C. Koenig," *Transactions American Society Civil Engineers*, Vol. 73, p. 199 (1911).
14. Zunker, F., "Das Verhalten des Bodens zum Wasser," Vol. 6 of *Handbuch der Bodenlehre* by E. Blanck, p. 152, Springer, Berlin, 1930.
15. Slichter, C. S., "Theoretical Investigation of the Motion of Ground Waters," "*USGS 19th Annual Report*, Part 2, pp. 295–384, Washington, D.C., 1899.
16. Slichter, C. S., "The Motion of Underground Waters," *USGS Water Supply Paper* 67, Washington, D.C., 1902 (106 pp.).
17. Fair, G. M. and L. P. Hatch, "Fundamental Factors Governing the Streamline Flow of Water through Sand," *Journal American Water Works Association*, Vol. 25, pp. 1551–1565 (1933).
18. Rose, H. E., "An Investigation into the Laws of Flow of Fluids through Beds of Granular Materials," *Proceedings of Institute of Mechanical Engineers*, Vol. 153, pp. 141–148 (1945).
19. Rose, H. E., On the Resistance Coefficient—Reynolds Number Relationship for Fluid Flow through a Bed of Granular Material," *Proceedings of Institute of Mechanical Engineers*, Vol. 153, pp. 154–168 (1945).
20. Bakhmeteff, B. A. and N. V. Feodoroff, "Flow through Granular Media," *Journal Applied Mechanics*, Vol. 4A, pp. 97–104; discussion Vol. 5A, 86–90, 1937.
21. Lindquist, E., "On the Flow of Water through Porous Soil," Premier Congrès des grands barrages, pp. 81–101, Stockholm, 1933.

22. Engelund, F., *Transactions of Danish Academy of Technical Sciences No.* 3, 1953 (105 pp.).
23. Jacob, C. E., "Flow of Ground Water," in *Engineering Hydraulics*, pp. 321–386, H. Rouse (Ed.), Wiley, New York, 1950.
24. Taylor, D. W., *Fundamentals of Soil Mechanics*, p. 126.
25. De Wiest, J. M. R., "Unsteady Flow through an Underdrained Earth Dam," *Journal of Fluid Mechanics*, Vol. 8, pp. 1–9 (1960).
26. Lebedev, A. F., in Polubarinova-Kochina's *Theory of Ground-Water Movement*, p. 9, Princeton University Press., Princeton, 1962.
27. Wenzel, L. K., "Methods for Determining Permeability of Water-Bearing Materials," *USGS Water Supply Paper* 887, pp. 7–11, 1942.
28. Hubbert, M. K., "The Theory of Ground-Water Motion," *The Journal of Geology*, Vol. XLVIII, No. 8, pp. 819–822 (November, December 1940).
29. Schneebeli, G., "Expériences sur la limite de validité de la loi de Darcy et l'apparition de la turbulence dans un écoulement de filtration," *La Houille Blanche*, Vol. 10, No. 2, pp. 141–149 (1955).
30. Klinkenberg, L. J., "The Permeability of Porous Media to Liquids and Gases," *Drilling and Production Practice*, pp. 200–214, American Petroleum Institute, 1941.
31. De Wiest, J. M. R., "On the Theory of Leaky Aquifers," *Journal of Geophysical Research*, Vol. 66, pp. 4257–4262 (December 1961).
32. Jones, K. R., "On the Differential Form of Darcy's Law," *Journal of Geophysical Research*, Vol. 67, No. 2, pp. 731–732 (1962).
33. Swartzendruber, D., "Non-Darcy Flow Behavior in Liquid-Saturated Porous Media," *Journal of Geophysical Research*, Vol. 67, pp. 5205–5213 (December 1962).
34. Hadley, W. A. and R. Eisenstadt, "Thermally Acutated Moisture Migration in Granular Media," *Transactions American Geophysical Union*, Vol. 36, No. 4, pp. 615–623 (August 1955).
35. Polubarinova-Kochina, P. Ya., *The Theory of Ground-Water Movement*, p. 18.
36. Polubarinova-Kochina, P. Ya., *The Theory of Ground-Water Movement*, p. 159.
37. Bouwer, H., "Variable Head Technique for Seepage Meters," *Journal of the Irrigation and Drainage Division ASCE*, pp. 31–44 (March 1961).
38. Bouwer, H., L. E. Myers, and R. C. Rice "Effect of Velocity on Seepage and its Measurements," *Journal of Irrigation and Drainage Division ASCE*, pp. 1–14 (September 1962).
39. Bouwer, H., "In-Place Measurement of Soil Hydraulic Conductivity in the Absence of a Water Table," *Proceedings, International Society of Soil Mechanics and Foundation Engineering*, Vol. 1, pp. 130–133, 2nd Asian Regional Conference, Tokyo, Japan, May 1–4, 1963.
40. Bouwer, H. and R. C. Rice, "Simplified Procedure for Calculation of Hydraulic Conductivity with the Double-Tube Method," *Proceedings, Soil Science Society of America*, Vol. 28, pp. 133–134 (1964).
41. Bouwer, H., "Measuring Horizontal and Vertical Hydraulic Conductivity of Soil with the Double-Tube Method," *Proceedings, Soil Science Society of America*, Vol. 28, pp. 19–23 (1964).
42. Irmay, S., J. Bear and D. Zaslavsky, *Determination of Hydraulic Conductivity of Soils by Means of Infiltration Rings*, Technion, Haifa, Israel, 1962 (68 pp.).
43. Krayenhoff van de Leur, D. A., "A Study of Nonsteady Groundwater Flow With Special Reference to a Reservoir Coefficient," *Ingenieur*, Vol. 70, pp. 87–94, 1958.
44. Krayenhoff van de Leur, D. A., "A Study of Nonsteady Ground-Water Flow, II, Computation Methods for Flow to Drains," *Ingenieur*, Vol. 74, pp. 285–292, Utrecht, 1962.

45. Krayenhoff van de Leur, D. A., "Some Effects of the Unsaturated Zone on Non-steady Free-Surface Groundwater Flow as Studied in a Scaled Granular Model," *Journal of Geophysical Research*, Vol. 67, pp. 4347–4362 (October 1962).
46. Bentall, R., "Methods of Collecting and Interpreting Ground-Water Data," *USGS Water Supply Paper* 1544-H, pp. 1–97 (1963).
47. Bentall, R., "Methods of Determining Permeability, Transmissibility and Drawdown," *USGS Water Supply Paper* 1536-I, pp. 243–341 (1963).
48. Donnan, W. W. and V. S. Aronovici, "Field Measurement of Hydraulic Conductivity," *Journal of the Irrigation and Drainage Division ASCE*, pp. 1–13 (June 1961).

CHAPTER FIVE

Steady State Flow

5.1 *Introduction*

Ground-water movement is steady or time-independent when its characteristics are not affected by the lapse of time. In order to preserve the steadiness of the flow inside a region, specific conditions along the boundaries of this region must remain unchanged in time. If these conditions, called boundary conditions, are not a function of time, the flow will ultimately become steady even though initially it might have developed in the unsteady state (for example, owing to a disturbance in the interior of the flow region). A typical example of the transition from unsteady to steady flow is given by the withdrawal of water from a well in an island aquifer. Water, indicating the piezometric head in that point of the aquifer, fills the borehole when it is completed. When pumping of the well starts, the water level in the well drops until a final value of the drawdown is reached for a constant flowrate. For water bodies at rest around the island, the flow is unsteady from the beginning of pumping until the ultimate drawdown at the well for a given flowrate is established; from then on the flow is steady. The disturbance here is the sudden withdrawal of water at a point in the interior of the bounded flow region or aquifer. The property of steadiness may also depend on the length of time during which ground-water flow is observed or analyzed. The assumption of at-rest conditions for the water bodies surrounding the island aquifer (above) is valid only for the period during which tidal fluctuations may be neglected. A steady state analysis is often made as a first approximation everywhere in a ground-water basin where the rate of change of head in time is small or negligible.

In some problems it suffices to obtain average values of such dependent variables as pressure and velocity of flow. Therefore the boundary conditions are averaged over a given time interval and a steady state solution is found for these conditions. The elimination of time as an independent variable simplifies the equation of flow presented in Chapter 4. Derivatives with respect to time vanish, and Eq. 4.51 becomes

$$\nabla^2 h = 0 \tag{5.1}$$

The study of Eq. 5.1 in a given domain and for specified boundary conditions is the subject of courses in potential theory and is beyond the scope of this textbook But it is worthwhile knowing that Eq. 5.1 may be solved analytically when the geometry of the boundaries is relatively simple and when the analytical conditions imposed along these boundaries are also simple. To make a solution of Eq. 5.1 possible, it is mandatory that the value of h or its normal derivative $\partial h/\partial n$ [n indicates the normal to the boundary of the flow region], or a linear combination of both, be known along the entire boundary containing the flow region. It is sufficient to know h along some parts of this boundary and to have the value of $\partial h/\partial n$, or a linear combination of both, along the remaining parts.

The geometry of the region is simple when the region is rectangular or circular; the boundary conditions are simple when $h =$ constant and when $\partial h/\partial n =$ constant. In general these restrictions on the geometry of the region and the boundary conditions are not compatible with the majority of ground-water flow problems and analytical solutions become impossible or unwieldy. Approximate graphical and semigraphical methods, relaxation techniques, and a variety of model studies are substituted for analytical methods when these methods become too complex. A widely used graphical method consists of the fitting of a flow net to the boundary conditions as will be exposed in section 5.3. This method has definite advantages in speed of execution and range of applicability, especially when geometrically complicated boundaries occur. The construction of a flow net to solve Laplace's equation graphically is relatively easy in problems with fixed boundaries and for confined flows, but it is slightly more complicated for free surface flows or unconfined flows where the location of the water table is not known beforehand. A flow net is a two-dimensional graph composed of two families of curves of a special nature: flowlines or streamlines which indicate how water travels and equipotential lines which join points of the same potential. Its use is therefore limited to the investigation of two-dimensional cross sections of porous medium which are representative of the main flow and to the analysis of three-dimensional problems with either axial or radial symmetry. The first use of a flow net (to study seepage underneath a concrete dam) was made by Forchheimer [16].

5.2 *Flow Nets—Boundary Conditions*

The most commonly occurring boundaries and the conditions which have to be satisfied along these boundaries are described in various textbooks [1, 2, 15]. They are briefly mentioned hereafter.

BOUNDARIES OF CONSTANT HEAD OR CONSTANT POTENTIAL

In two-dimensional flow, boundaries of constant potential are represented by a line separating the region of porous medium from that

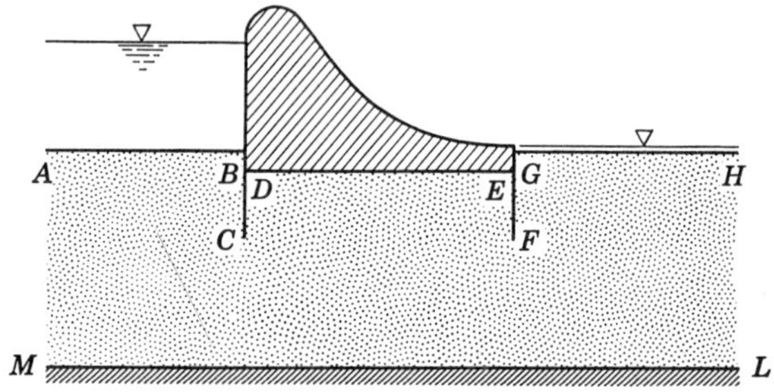

Fig. 5-1 Flow with fixed boundaries.

occupied by the water, as indicated by AB and GH on Fig. 5.1. In the first case, the water reservoir behind the concrete dam is so large and the seepage so slow that we may assume the hydrostatic law to hold for the water body. In the second case, water seeps to the surface and remains quasi stagnant above GH, so that again the assumption of hydrostatic law is valid. These boundaries correspond to the analytical condition $h =$ constant, and for $K =$ constant, also to $\Phi =$ constant as follows from Eq. 4.26.

IMPERVIOUS BOUNDARIES

Impervious boundaries are typified by the underground contour $BCDEFG$ of the hydraulic structure and the bedrock ML of Fig. 5.1. These boundaries are streamlines and correspond to the analytical condition $\partial h/\partial n = 0$, or to $\Psi =$ constant, where Ψ is the stream function (see Eq. 5.7).

Constant potential and impervious boundaries are the only types that are present in Fig. 5.1, and they enclose the domain completely if it is assumed that parallel lines intersect at infinity. These types of boundaries would also occur exclusively in the flow to a well in an island aquifer (Fig. 5.2), provided the flow remains completely confined even in the

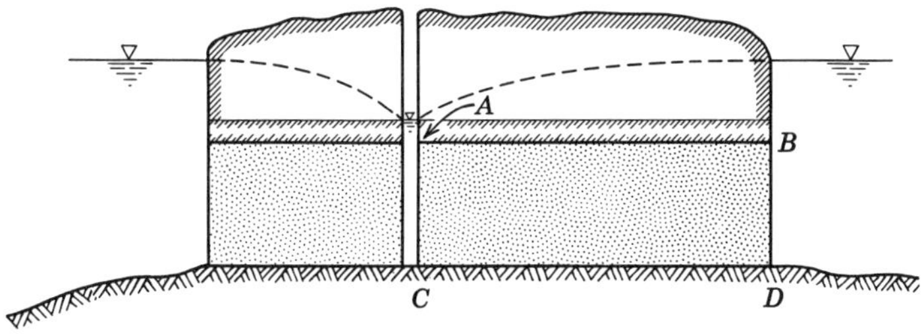

Fig. 5-2 Flow to well in confined aquifer.

vicinity of the well. In case the well is established in the center of the island, the flow has radial symmetry and it suffices to analyze the region *ABCD*, which is delineated by two streamlines and two equipotential lines.

FREE SURFACES

When the contribution by the capillary fringe to the flowrate may be neglected, as in the flow of water through large earth embankments or towards a well in an unconfined aquifer, the existence of the capillary fringe is often disregarded. In many civil engineering problems, as in this chapter, the positions of free surface and water table are assumed to coincide. In general, the equation of the free surface is not available and is one of the unknowns of the problem. The difficulty inherent in the unknown location of the free surface is somewhat compensated for by the fact that along this kind of boundary two conditions must be satisfied: constant pressure and the requirement that the free surface is a streamline in steady flow without recharge to or evaporation from the free surface. An example of a free surface is the line *AD* of Fig. 5.3.

SURFACES OF SEEPAGE

In Fig. 5.3 the free surface intersects the downward slope in point *D* above the toe of the dam, and water at this point leaves the porous medium to enter the free space and to trickle down along the surface *DC*.

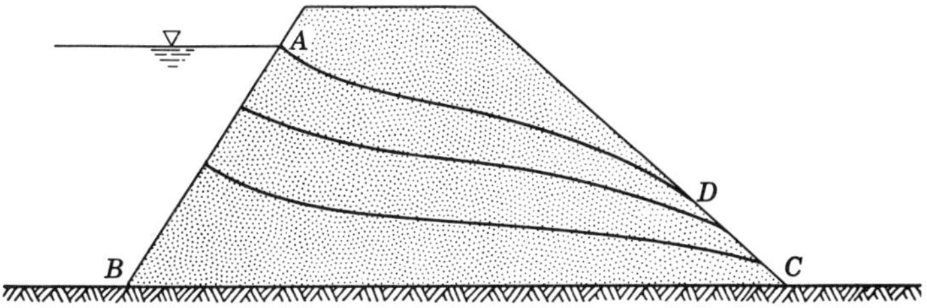

Fig. 5-3 Free surface flow.

This surface is called a seepage surface. The line DC is not a streamline as it is intersected by a great many streamlines which terminate upon it between D and C, and because streamlines by definition do not intersect each other. In Fig. 5.3 two intermediate streamlines are sketched between free surface and bedrock. The basic property of a seepage surface is that pressure along it is always atmospheric. A seepage surface also exists along the well casing between the water level in the well and the intersection of the water table with the well.

5.3 *Properties of Flow Nets*

In section 4.13 it was shown that streamlines are orthogonal to equipotential lines in a homogeneous and isotropic medium for which Darcy's law is valid. In the case of constant K, streamlines are everywhere tangent to the velocity vector or to $-\text{grad}\ \Phi$ because of Eqs. 4.25 and 4.86. If the velocity potential Φ is used in two-dimensional flow the velocity components u, w may be expressed as

$$u = -\frac{\partial \Phi}{\partial x}$$
$$w = -\frac{\partial \Phi}{\partial z} \tag{5.2}$$

Consider first a family of equipotential lines $\Phi = C_1, C_2, C_3, \cdots$. The tangent along any line $\Phi = C$ has a slope $\left(\frac{dz}{dx}\right)_{\Phi = C}$, which may be found by taking the total differential of $\Phi = C$.

$$d\Phi = \frac{\partial \Phi}{\partial x}\, dx + \frac{\partial \Phi}{\partial z}\, dz = 0$$

or

$$\left(\frac{dz}{dx}\right)_{\Phi} = -\frac{\partial \Phi / \partial x}{\partial \Phi / \partial z} = -\frac{u}{w} \tag{5.3}$$

If two adjacent streamlines are considered, say a and b of Fig. 5.4a, and if it is assumed that there is a flowrate dq in the stream tube which they form, then continuity requires that this flowrate pass through sections 1–2 and 2–3, which are chosen so that only the first has contributions due to w and the second only to u. As a first approximation, if the contributions $(\partial u/\partial x)\, dx\, dz$ and $(\partial w/\partial z)\, dz\, dx$ are neglected as second-order terms, continuity renders

$$dq = w\, dx = u\, dz$$
$$w\, dx - u\, dz = 0 \tag{5.4}$$

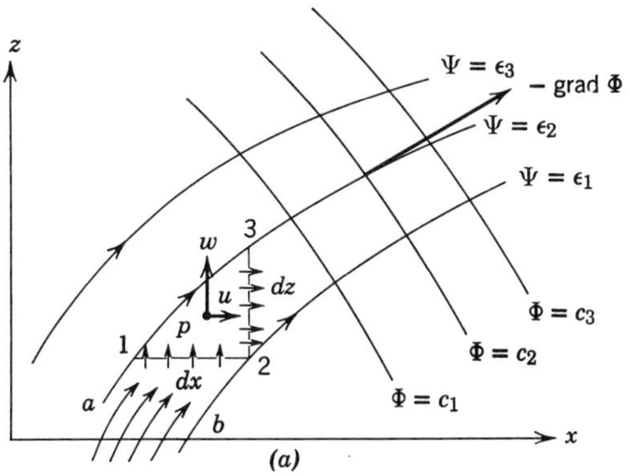

Fig. 5-4*a* Streamlines and equipotential lines.

in which dx and dz are the components of the chord of an elemental arc of length ds along a streamline. On the other hand, continuity equation (4.44) for two-dimensional incompressible steady flow reduces to

$$\frac{\partial u}{\partial x} + \frac{\partial w}{\partial z} = 0 \tag{5.5}$$

Equation 5.5 means that Eq. 5.4 is derived from a total differential as is shown in elementary calculus [3]. Let

$$w = \frac{\partial \Psi}{\partial x}, \qquad u = -\frac{\partial \Psi}{\partial z} \tag{5.6}$$

These values clearly satisfy Eq. 5.5 and when they are inserted in Eq. 5.4 the result is

$$\frac{\partial \Psi}{\partial x}\,dx + \frac{\partial \Psi}{\partial z}\,dz = d\Psi = 0$$

Hence

$$\Psi(x, z) = \text{constant} \tag{5.7}$$

and the existence of a stream function has been derived. The streamlines may now be labeled in the same way as the equipotential lines. Equation 5.4 may now be written as

$$\left(\frac{dz}{dx}\right)_{\Psi} = \frac{w}{u} \tag{5.8}$$

This equation compared with Eq. 5.4 again shows the orthogonality of streamlines and equipotential lines.

A comparison of Eqs. 5.1 and 5.6 and elimination of u and w leads to the relationships

$$\frac{\partial \Phi}{\partial x} = \frac{\partial \Psi}{\partial z}$$
$$\frac{\partial \Phi}{\partial z} = -\frac{\partial \Psi}{\partial x} \tag{5.9}$$

These are the Cauchy-Riemann equations showing that the complex potential $W(z_1) = \Phi(x, z) + i\Psi(x, z)$ is an analytical function of $z_1 = x + iz$. Φ and Ψ are conjugate harmonic functions. It could be shown that Ψ satisfies Laplace's equation by using the condition of irrotational flow. The condition [4] in this case is

$$\frac{\partial u}{\partial z} - \frac{\partial w}{\partial x} = 0 \tag{5.10}$$

Insertion of the values of u, w from Eq. 5.6 renders

$$\nabla^2\Psi = 0 \tag{5.11}$$

The general limitations on the use of flow nets in ground-water flow are:

1. Steady flow.
2. The medium must be piecewise homogeneous.
3. Incompressible fluid, of constant density and viscosity.
4. Darcy's law must be valid.

The above considerations about streamlines and equipotential lines are particularly useful when observations of static water levels are made in three closely spaced noncolinear wells. Here the piezometric heads are equal to the elevation of the water level in the wells, and it is possible to sketch the ground-water contours from an observation of these water levels (Fig. 5.4*b*). This permits us to estimate the direction of the ground-water flow, orthogonal to the family of contour lines and represented as straight lines in Fig. 5.4*b* as a first approximation. The direction of the ground-water flow may also be visualized by a series of lines, and a flow net would arise naturally.

5.4 *Construction of Flow Net*

A flow net, as demonstrated in the above paragraphs, is composed of two families of mutually orthogonal lines established in such a way that, as a rule, streamlines terminate upon potential lines delineating in part the flow domain, and vice versa. The exception is when a seepage surface is present. There is no unique way to construct a flow net because the number of streamlines and equipotential lines to choose from is extremely large, as both Ψ and Φ are continuously varying functions. A few representative lines of each family are picked. The ratio of the number of

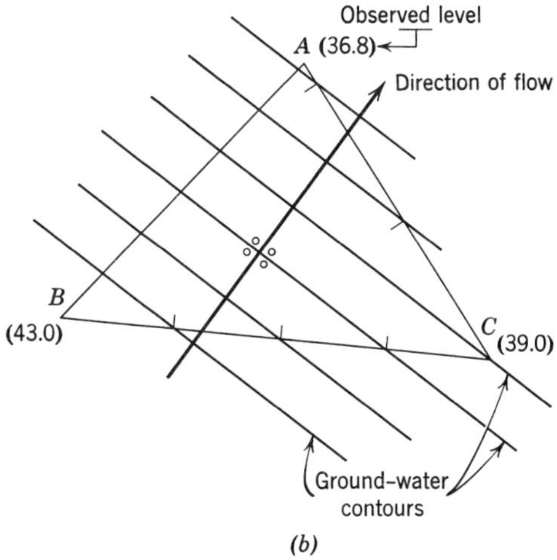

Fig. 5-4*b* Observation of water table.

streamtubes to equipotential drops, however, is a constant for each problem, as will follow from the formula for the seepage flowrate. It follows that the flow net becomes uniquely determined once the number of streamlines or equipotential lines is imposed. A. Casagrande [5] advises the limiting of flow tubes to four or five. Actually fractional flow tubes or potential drops may be used, and experienced students may prefer to use them. In a beginner's attempt, however, fractional flow tubes should be avoided. Only by coincidence will both the number of flow tubes and potential drops be an integer.

A cross section of a spillway or overflow weir with cut-off walls on both ends, is shown in Fig. 5.5*a*. This hydraulic structure is supposedly very long perpendicular to the paper so that the flow may be treated as two-dimensional. This assumption is acceptable at a reasonable distance, say 1.5 times the transversal dimension of the structure, away from the abutments of the weir. To draw this flow net the author decided to use four stream tubes. It is easily discovered that the fluid-filled porous medium has symmetry about the axis *X–X*, so that one half of the flow net must be the mirror image of the other. The first line to draw is the upper flowline *U–U*. Its first position is a trial one, but it should be kept in mind at this stage that there must be room for three more stream tubes. They could be laid out tentatively in the vicinity of *X–X* and an attempt could be made to construct curvilinear[2] squares (why squares and not

[2] A curvilinear element is square if its sides intersect orthogonally and if the line segments joining centers of opposite sides are of equal length.

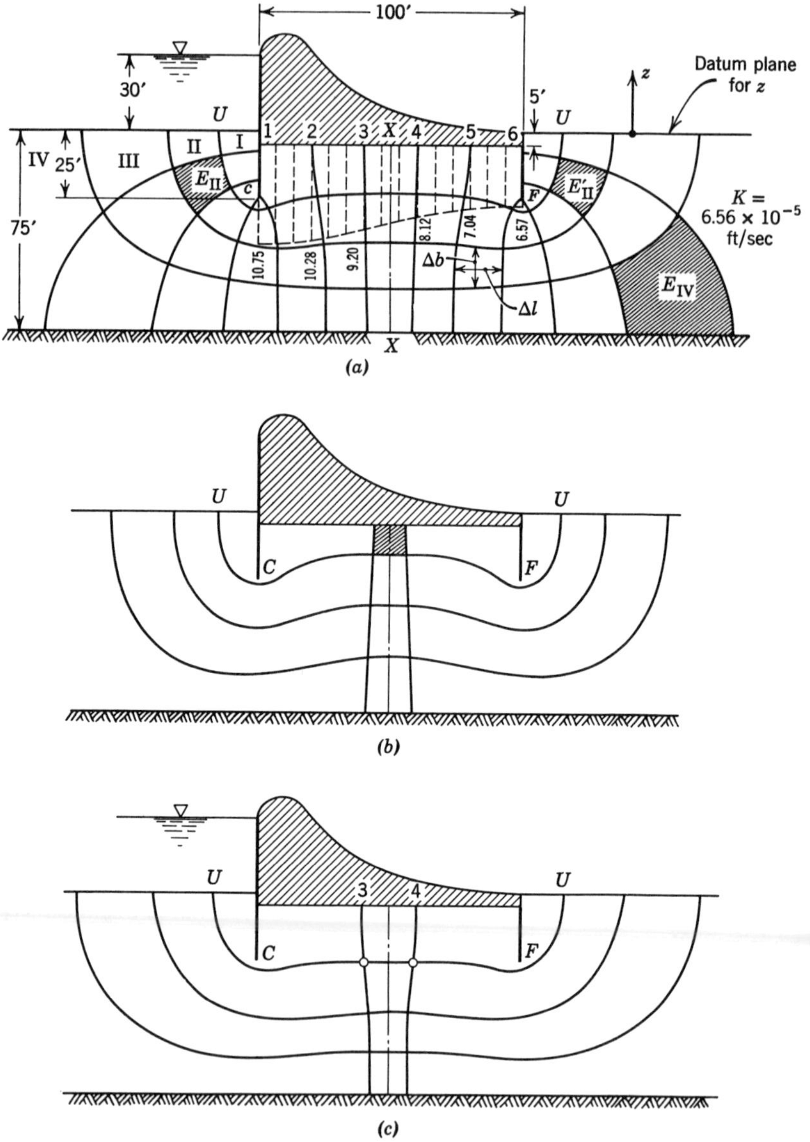

Fig. 5-5(*a*) Seepage under dam with cut-off walls. (*b*) Seepage under dam, trial line *U–U* too high. (*c*) Seepage under dam, trial line *U–U* too low.

rectangles are used will be explained) bounded by the equipotential lines 3 and 4 and the proposed streamlines. If this trial is successful, a significant step has been made towards the correct completion of the entire net. Figure 5.5*b*, for example, shows that if the upper flow line is located too high the construction of curvilinear squares centered around *X–X* becomes impossible, at least if the number of stream tubes is limited to four. On the other hand, part *c* of this figure shows that if the upper flow line is sketched too low, the construction of curvilinear squares around *X–X* requires the convergence of the equipotential lines through 3 and 4, making it difficult to observe orthogonality with the upper flowline. The

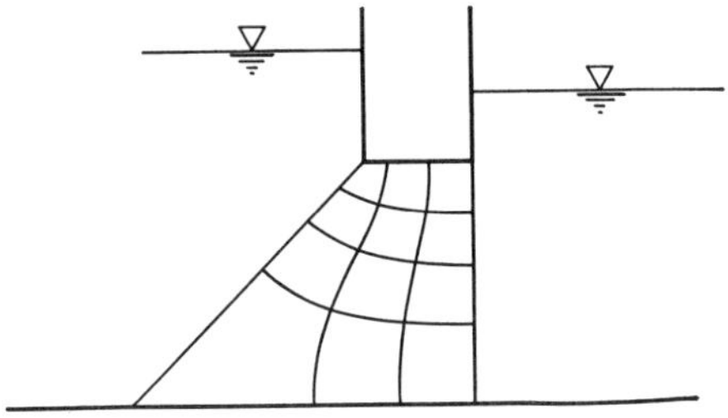

Fig. 5-6 Asymmetric flow net.

next problem is the proximity of the flowlines near the tips *C* and *F* of the cut-off walls. Actually these points are singular points in the flow net. In hydrodynamics it is proved that for potential flow about a corner forming an obtuse angle the velocity near the corner point tends to infinity. Infinite velocities would of course not occur in ground-water flow because microscopically, as was pointed out before, the flow is not irrotational. However, large velocities may be expected in the vicinity of the tips where Darcy's law would break down. Hence the flowline may pass very close to the tips. The squares around the tips are likewise singular.

Some hints on how to start the construction of a symmetric flow net have been given above. Its completion may involve relocation, by trial and error, of parts of the first set of streamlines. It is somewhat more difficult to draw a flow net without symmetry, as sketched in Fig. 5.6. For further explanation about the construction of flow nets, the reader is referred to Polubarinova-Kochina [6] and Taylor [7].

5.5 *Seepage Rate*

The computation of the seepage rate or flowrate underneath the hydraulic structure of Fig. 5.5*a* shows the result of a choice of squares or rectangles with constant width-length ratios on the properties of the flow net [7]. If the stream tubes of Fig. 5.5*a* are labeled I, II, III, IV, it suffices to consider three curvilinear elements E_{II}, E'_{II}, and E_{IV} and to write down the discharge through these elements:

$$E_{II} \qquad \therefore\ \Delta q_{II} = K\,\Delta h_{II}\left(\frac{\Delta b}{\Delta l}\right)_{II}$$

$$E'_{II} \qquad \therefore\ \Delta q'_{II} = K\,\Delta h'_{II}\left(\frac{\Delta b}{\Delta l}\right)'_{II}$$

$$E_{IV} \qquad \therefore\ \Delta q_{IV} = K\,\Delta h_{IV}\left(\frac{\Delta b}{\Delta l}\right)_{IV}$$

Provided $\Delta b/\Delta l$ = constant, it follows from these three relationships that Δq = constant and Δh = constant for the three elements. Since the choice of these elements was arbitrary and since it is possible to cover the entire net by appropriate successive choices, it follows that Δq = constant and Δh = constant for all curvilinear elements. For the construction of the flow net, the simplest constant for $\Delta b/\Delta l$ is one. In this case, if

$$n_s = \text{number of stream tubes}$$

$$n_d = \text{number of equipotential drops} = \frac{H_t}{\Delta h}$$

then the flowrate Q per unit width of hydraulic structure may be expressed as

$$Q = n_s\,\Delta q = n_s K\,\Delta h$$

or

$$Q = \frac{n_s}{n_d}\,KH_t \tag{5.12}$$

The flow net makes it possible to determine the pore pressure and the velocity in the porous medium from the knowledge of h. Actually discrete average values of h are obtained for every curvilinear square instead of a continuously varying h which would be the result of the analytical solution. Discrete values of h for the same quality of the flow net are more accurate the further the subdivision of elements is carried out. To find h at any point, advantage is taken of the fact that the total head is lost along any streamline and over the total length of the streamline between headwater and tailwater. The head at the origin of the streamline is determined after the datum plane for elevation is chosen. The loss between this point and the point where the values of pressure and velocity are sought is computed in terms of the number of equipotential drops between the two points and the loss per drop, say H_t/n_d. After h is computed, the values of p follow immediately from the knowledge of z and the definition of $h = z + p/\gamma$.

NUMERICAL EXAMPLE (Fig. 5.5*a*)

$n_s = 4 \qquad n_d = 12.2 \qquad K = 2 \times 10^{-3}\ \text{cm/sec} = 6.56 \times 10^{-5}\ \text{ft/sec}$

$H_t = 30\ \text{ft} \qquad \Delta h = 30\ \text{ft}/12.2 = 2.46\ \text{ft}$

$$Q = \frac{4}{12.2} \times 6.56 \times 10^{-5}\ \text{ft/sec} \times 30\ \text{ft} = 6.45 \times 10^{-4}\ \text{cfs per ft}$$

Uplift pressures on the base are computed in points 1 to 6 and are represented graphically. The pressure distribution is of course symmetrical about the X–X axis. Since at the base $z = -5$ ft, it follows from the definition of h that $p = \gamma(h + 5)$. The final results are summarized in Table 5.1.

Table 5.1

Points	h in ft	p/γ in ft	p in lb/in.^{2}
1	$30 - 4.1 \times 2.46 = 19.92$	24.92	10.75
2	18.69	23.69	10.28
3	16.23	21.23	9.20
X	15.00	20.00	8.66
4	13.77	18.77	8.12
5	11.31	16.31	7.04
6	10.08	15.08	6.57

The effect of the uplift pressure may be felt near the downstream part of the dam. Because the specific weight of concrete is roughly 2.5 times that of water, every 2.5 ft of uplift pressure head may be neutralized by 1 ft of concrete. This condition is not satisfied in point 6, where there is only 5 ft of concrete slab, but 15.08 ft of uplift pressure head. Consequently, the thickness of the slab must be increased near the downstream end or holes must be drilled in that part to relieve the pressure.

5.6 *Dam on Infinitely Thick Stratum*

Figure 5.7 represents the cross section of a concrete dam of length L backing up water under a head difference H_t and resting on a homogeneous stratum of infinite thickness. Although such a case may seem to have only academic meaning, the flow net that results is very like that of a similar dam resting on a stratum of thickness above bedrock at least equal to $2L$. It is clear that the amount of seepage would be infinite if the stratum were infinitely thick. However, in the case where bedrock exists at great but finite depth, a few stream tubes are drawn and the seepage between the toe of the dam and a point C some distance downstream from this

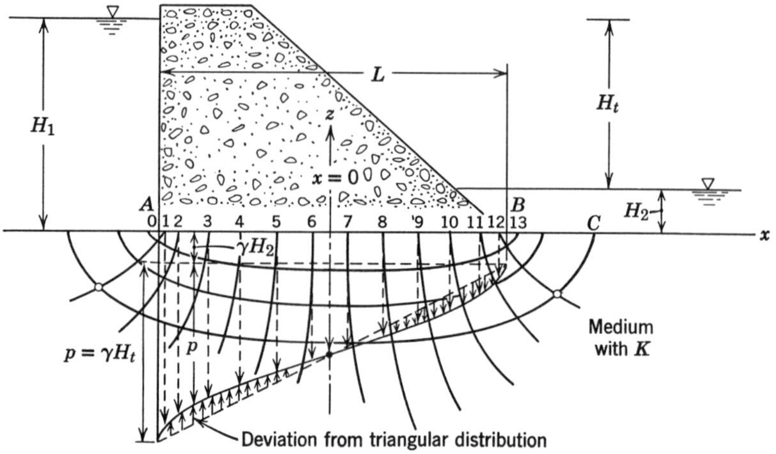

Fig. 5-7 Dam on infinitely thick stratum.

point is computed. The result which is graphically obtained may be compared with the analytical solution. Whereas the analytical treatment of the problem in the previous paragraph requires the use of elliptic functions, the present case involves only knowledge of conformal mapping by means of elementary functions. The streamlines and equipotential lines are respectively ellipses and hyperbolae, with common foci in *A* and *B*.

The equations of the equipotential lines have been found by the use of complex variables (see reference 8):

$$\frac{(2x/L)^2}{\cos^2 \varphi_1} - \frac{(2z/L)^2}{\sin^2 \varphi_1} = 1 \tag{5.13}$$

where

$$\varphi_1 = \frac{\pi\Phi}{KH_t} = \frac{\pi h}{H_t} \tag{5.14}$$

In Eq. 5.14,

$$h = \frac{p}{\gamma} + z - H_2 \tag{5.15}$$

as a result of Eq. 4.2* in which the constant $C = -H_2$ is so chosen as to have $h = 0$ along tailwater and $h = H_t$ along headwater. The equations of the streamlines are

$$\frac{(2x/L)^2}{\cosh^2 \psi_1} + \frac{(2z/L)^2}{\sinh^2 \psi_1} = 1 \tag{5.16}$$

where

$$\psi_1 = \frac{\pi\Psi}{KH_t} \tag{5.17}$$

Equations 5.13 and 5.16 have not been used to draw the flow net of Fig. 5.7. Instead, the flow net has been sketched graphically according to the procedure of section 5.4. Three streamlines pass between B and C and the number of equipotential drops between headwater and tailwater is 11.2. The seepage rate through the area downstream of the toe between B and C as determined by Eq. 5.12 is $Q = 0.268KH_t$. On the other hand, the analytical expression for the same seepage rate is given [8] by

$$Q = \frac{KH_t}{\pi} \log_e \frac{x_c + \sqrt{x_c^2 - (L/2)^2}}{L/2} \tag{5.18}$$

In the case of Fig. 5.7, in which $x_c = 0.735L$, Eq. 5.18 renders $Q = 0.306KH_t$, which shows that the graphical method has given rise to an error in Q slightly less than 13%. A comparison of the flow net that was obtained in a purely graphical way with the one that would result from plotting Eqs. 5.13 and 5.16 would show that the streamlines of Fig. 5.7 are drawn too close to the dam. This imperfection is also evident from the fact that streamlines and equipotential lines are not orthogonal at the intersections along the bottom streamline.

The pressure distribution along the bottom of the dam is obtained from Eqs. 5.13, 5.14, and 5.15 in which the value $z = 0$ for the elevation of the bottom of the dam has been inserted. Along the base of the dam it follows from Eq. 5.13 that

$$\frac{2x}{L} = \cos \varphi_1$$

and hence

$$\frac{\pi}{H_t}\left(\frac{p}{\gamma} - H_2\right) = \cos^{-1} \frac{2x}{L}$$

or

$$p = \gamma H_2 + \gamma \frac{H_t}{\pi} \cos^{-1} \frac{2x}{L} \tag{5.19}$$

This pressure distribution is plotted on Fig. 5.7 again by use of the flow net. Computations using Eq. 5.19 have shown that the errors in the pressures derived from the flow net are around 10%.

Figure 5.7 shows that the pressure diagram deviates from the trapezoidal distribution which is adopted as a first approximation.

The attention of the reader is drawn to the special value in this paragraph of the arbitrary constant of Eq. 4.2* which led to Eq. 5.15. Although this constant almost always assumes the value zero in this textbook, this is by no means always necessary. The value $(-H_2)$ used in Eq. 5.15 is compatible with the text of reference 8.

5.7 *Sheet Pile in Infinitely Deep Stratum*

A flow net depends on the geometry of the medium and on the nature of the boundary conditions, but not on the magnitude of the effective head or on the difference between head and tailwater. This property may be used to derive the flow net for the seepage underneath a sheet pile (Fig. 5.8) from that of the previous paragraph. Actually the right half of the flow net

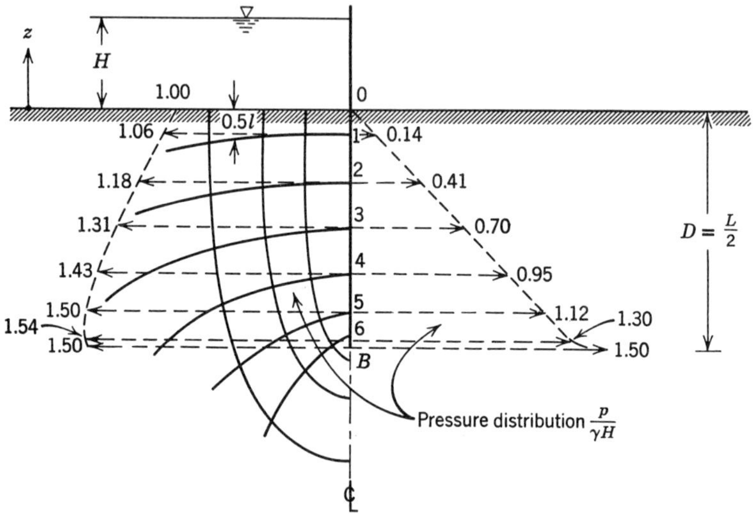

Fig. 5-8 Sheet pile in infinitely deep stratum.

of Fig. 5.7 may be thought of as taking place in an infinite quadrant bounded by the negative z-axis and the positive x-axis, which has an impervious segment OB (streamline). The rest of the x-axis and the entire z-axis are constant potential lines. The left half of the flow net of Fig. 5.8 is bounded by two mutually orthogonal lines, an horizontal line at constant potential and a vertical line, part of which acts as an impervious boundary (segment OB). The rest of this vertical line is also at constant potential. Therefore the quadrants of Figs. 5.7 and 5.8 are geometrically similar and the nature of the boundary conditions is the same in both cases so that the flow nets must be the same in both cases. To draw the flow net of Fig. 5.8, it is sufficient to assume a depth of penetration $D = OB = \dfrac{L}{2}$ and to rotate the entire net of Fig. 5.7 clockwise over an angle of 90° with preservation of streamlines and equipotential lines. In this rotation, however, the positions of the x-axis and z-axis remain unchanged; only the flow net is rotated. The datum plane for elevation is the soil surface as before. The dimensionless pressure distribution $p/\gamma H$ on both sides of the wall is computed from $h = z + p/\gamma$. For the points $n = 1, 2, 3, \cdots, 7$ it follows that

$$h_n = z_n + \frac{p_n}{\gamma}$$

with

$$h_n = H - \lambda_n H$$

and λ_n = ratio of the number of equipotential drops between datum plane and point n to the total number n_d of equipotential drops. The pressure distribution depends upon the ratio H/D and has to be computed in each different case. In Fig. 5.8 only the value one for this ratio has been considered.

5.8 *Singular Points in Flow Nets*

Difficulties arise in the construction of a flow net in points where both u and w of Eqs. 5.3 and 5.8 are either zero or infinite. In this case the orthogonality requirement of streamlines and equipotential lines breaks down. Such points are called singular points. It is possible to distinguish three categories of singular points, according to the following:

1. Streamlines and equipotential lines are not perpendicular. This occurs in general when the streamline runs along a boundary.
2. Streamlines or equipotential lines have a discontinuity in slope on the boundary.
3. The flow net has a source or sink. In this case, the velocity at the point source or sink becomes infinite as the squares around this point become vanishingly small. Examples of No. 3 will be described further in Chapter 6. Illustrations of 1 and 2 follow here.

Category 1 may be subdivided into two cases, depending on the magnitude of the angle α inside the porous medium between streamline and equipotential line (α smaller or larger than 90°).

$\alpha < 90°$. Inflow surface of dam (Fig. 5.9*a*).

It is easy to prove that the velocity in point A must tend to zero by continued subdivision of the flow element adjacent to A. This subdivision is done so that the flowrate is divided equally between the streamtubes resulting after the division of the element.

From the figure it follows that

$$a_1 < 2a_2, \quad \text{say}$$

$$\frac{a_1}{a_2} = \kappa 2, \quad \text{with } \kappa < 1$$

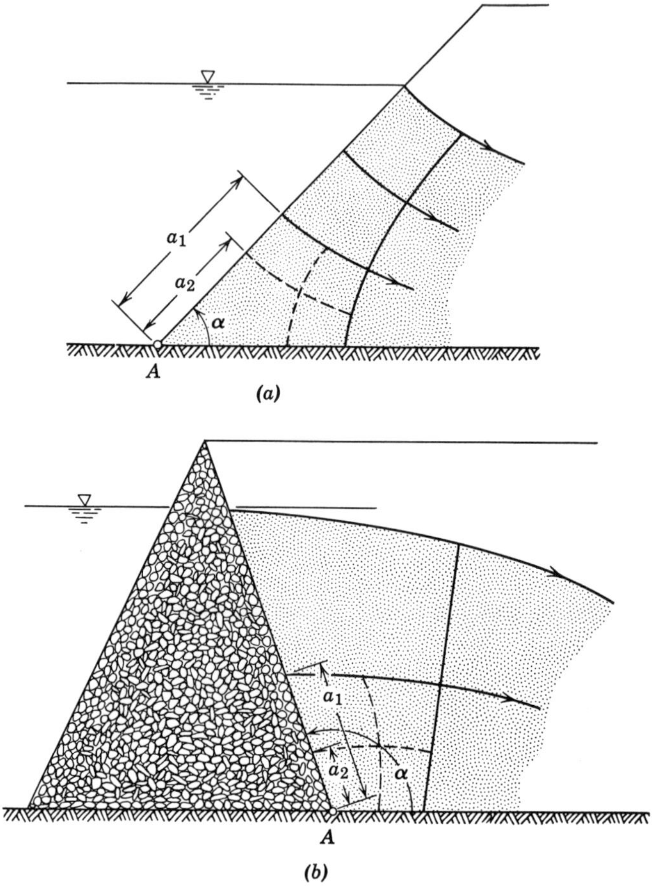

Fig. 5-9(*a*) Singular point *A*, velocity zero. (*b*) Singular point *A*, velocity infinite.

and by the next subdivision

$$\frac{a_2}{a_3} = \kappa 2, \quad \text{hence } \frac{a_1}{a_3} = \kappa^2 2^2$$

and so on until

$$\frac{a_1}{a_{n+1}} = \kappa^n 2^n \tag{5.20}$$

On the other hand, if $q_{\mathrm{I}} = V_1 a_1$ is the flowrate through the original element adjacent to *A*, then

$$V_2 = \frac{\frac{1}{2} q_{\mathrm{I}}}{a_2} = V_1 \frac{a_1}{2a_2}$$

Similarly
$$V_3 = V_2 \frac{a_2}{2a_3} = V_1 \frac{a_1}{2^2 a_3}$$

and
$$V_{n+1} = V_1 \frac{a_1}{2^n a_{n+1}} = V_1 \kappa^n \tag{5.21}$$

in view of Eq. 5.20. Equation 5.21 shows that

$$\lim_{n \to \infty} V_{n+1} = V_A = 0, \quad \text{because } \kappa < 1.$$

$\alpha < 90°$. Inflow surface of dam (Fig. 5.9*b*).

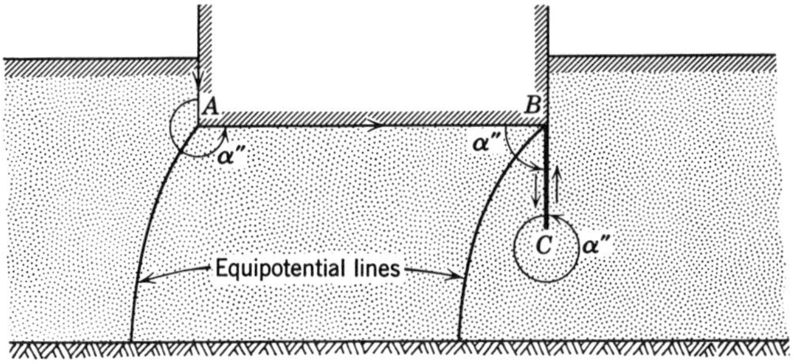

Fig. 5-10 Discontinuity in slope of streamline along boundary.

Inflow surface of dam protected by rockfill. By subdivision of the adjacent element, it follows that

$$\frac{a_1}{a_2} = \beta 2 \quad \text{with } \beta > 1$$

and by further subdivision exactly as for $\alpha < 90°$

$$\frac{a_1}{a_{n+1}} = \beta^n 2^n$$

The expression for V_{n+1} is derived in nearly the same way as is $\alpha < 90°$

$$V_{n+1} = V_1 \beta^n \tag{5.22}$$

This equation shows that $\lim_{n \to \infty} V_{n+1} = V_A = \infty$, because $\beta > 1$.

Category 2 is illustrated in Fig. 5.10. Here A, B, and C are singular points. The equipotential lines passing through A and B cannot be orthogonal to both branches of the streamline. The criterion now becomes

$$\alpha'' > 180°, \text{ velocity infinite}$$
$$\alpha'' < 180°, \text{ velocity zero}$$

Actually these examples are only special cases of those considered under category 1.

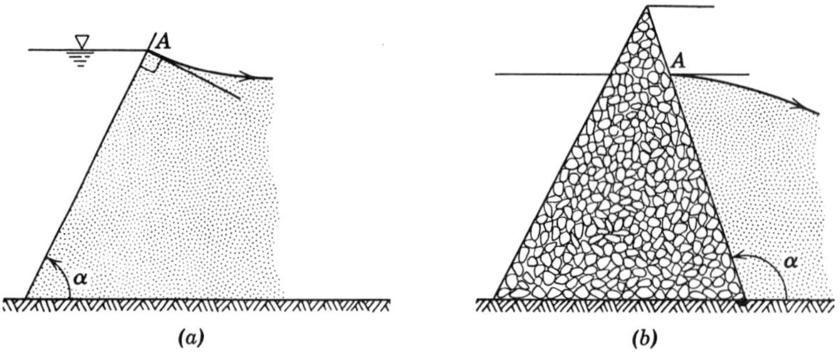

Fig. 5-11*a* and *b* Entrance conditions for free surface line.

5.9 *Flow Nets with a Free Surface*

In unconfined flow the boundaries are not completely defined and the location of the top flowline is found by suitable approximate methods. The top flowline is a streamline and at the same time a line of constant pressure. This means that equipotential lines intersect the top flowline orthogonally and in such a way that $\Delta h = \Delta z$ (Fig. 5.13).

The free surface line must also satisfy special entrance (Fig. 5.11*a*,*b*) and discharge conditions (Fig. 5.12*a*,*b*). In point *A* of Fig. 5.11*b* the free surface line must start horizontally because it cannot run uphill against gravity under constant atmospheric pressure. Therefore *A* is a singular point where the velocity is zero, as the angle between streamline and equipotential line is smaller than 90°.

In Fig. 5.12*a*, the free surface line is tangent to the sloping discharge face in point *B*. Suppose that this were not so, then a close-up of the net in the vicinity of *B* would require the same Δz for the intersection of a series of equidistant equipotential lines by two lines with different slopes,

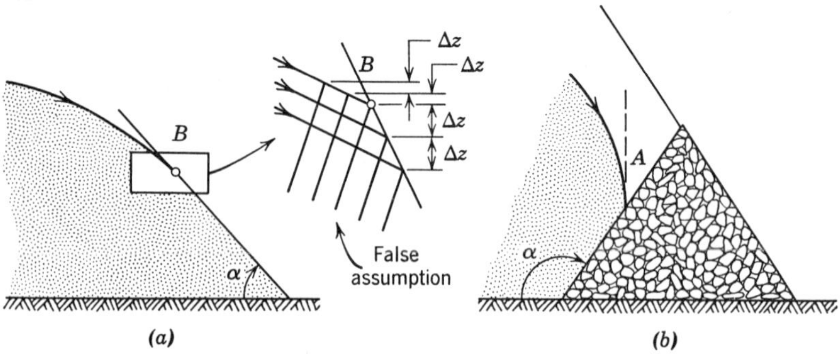

Fig. 5-12*a* and *b* Discharge conditions for free surface line.

which is impossible. Note that the equidistance of the equipotential lines stems from the fact that as a first approximation they are represented as straight lines and that an equilateral net is drawn.

In Fig. 5.12*b* the free surface seeps into the discharge surface or seepage surface in such a way that it is tangent to the vertical through point *A*. It also means that in the vicinity of *A* the equipotential lines are horizontal.

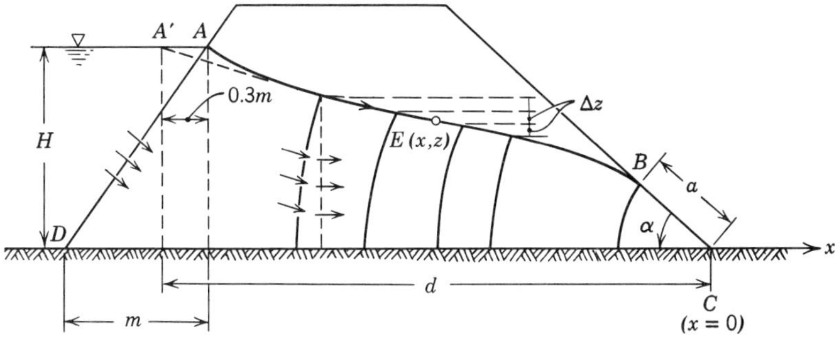

Fig. 5-13 Earthdam with seepage surface.

5.10 *Dam with Sloping Discharge Face—Dupuit's Assumptions*

A. Casagrande [5] applied Dupuit's assumptions to determine the equation of the free surface, the length *a* of the seepage segment, and the flow *Q* through a dam of the type illustrated in Fig. 5.13.

Dupuit's assumptions, which are also made in unconfined well flow, may be summarized as follows:

(a) In any vertical section, the flow is horizontal.

(b) The velocity is uniform over the depth of flow $z_f(x)$.

(c) The velocity at the free surface may be expressed as $v = -K(\partial h/\partial x)$ instead of as $v = -K(\partial h/\partial s)$. This assumption is reasonable for small slopes of the free surface.

In problems with a free surface it is very important to understand the nature of the variables in order to avoid serious errors as explained by Dicker [15]. In two-dimensional problems *h* is the dependent variable and x, and z are the independent variables. In the interior of the flow region

$$h(x, z) = z + \frac{p(x, z)}{\gamma}$$

When the head of a point at the free surface is considered, confusion may arise in the use of the symbol z. If atmospheric pressure is assumed to be constant, then at the free surface $h(x) = z_f(x) + \text{constant}$. The subscript

f is added to z to distinguish the elevation of the free surface, which is a function of x, from the elevation of an arbitrary point in the interior of the flow region, which is not a function of x.

In order to fit the Dupuit parabola (see section 5.14) to the free surface, an entrance correction is introduced which moves the real entrance point to the left in point A' over a distance $0.3m$ [5], where m is the horizontal projection of the inflow surface AD. The fictitious entrance point A' is at a distance d from the toe of the dam. The coordinates of A' and B are respectively $(-d, H)$ and $(-a \cos \alpha, a \sin \alpha)$.

By Dupuit's assumptions, the flowrate per unit width in any section is

$$Q = -Kz_f \frac{dz_f}{dx}$$

if in the first approximation at the free surface

$$\frac{dh}{dx} \cong \frac{dz_f}{dx}$$

This approximation is very good for mildly sloping water tables where

$$\frac{1}{\gamma} \frac{dp}{dx} \cong 0.$$

By continuity the same flowrate must pass through the section in point B, where, since the free surface is tangent to the discharge slope,

$$-\left(\frac{dz_f}{dx}\right)_B = \tan \alpha$$

Therefore

$$-Kz_f \frac{dz_f}{dx} = Ka \sin \alpha \tan \alpha \tag{5.23}$$

To find a, Eq. 5.23 may be integrated between A' and B

$$\int_{A'}^{B} z_f \, dz_f = -a \sin \alpha \tan \alpha \int_{A'}^{B} dx$$

or

$$\tfrac{1}{2}(a^2 \sin^2 \alpha - H^2) = -a \sin \alpha \tan \alpha(-a \cos \alpha + d)$$

The solution of this quadratic equation in a renders

$$a = \frac{d}{\cos \alpha} - \sqrt{\frac{d^2}{\cos^2 \alpha} - \frac{H^2}{\sin^2 \alpha}} \tag{5.24}$$

The equation of the free surface may be found by integration of Eq. 5.23 between A' and $E(x, z_f)$:

$$\tfrac{1}{2}(z_f^2 - H^2) = -a \sin \alpha \tan \alpha(x + d)$$

This is the equation of a parabola with vertex on the x-axis, at

$$x = -d + \frac{1}{2}\frac{H^2}{a \sin\alpha \tan\alpha} = -\frac{a}{2}\cos\alpha$$ [1]

The discharge Q is given by

$$Q = Ka \sin\alpha \tan\alpha \tag{5.25}$$

The above approximate method gives accurate results for small values of α, say below 30°. For $30° < \alpha < 60°$, L Casagrande [7, 9] has modified the above theory, using only the assumptions a and b made by Dupuit but keeping

$$v = -K\frac{\partial h}{\partial s} = -K\frac{dz}{ds}$$

as the velocity at the free surface, and hence as the uniform velocity. This leads to

$$Q = Ka \sin^2\alpha \tag{5.26}$$

where

$$a = s_0 - \sqrt{s_0^{\,2} - \frac{H^2}{\sin^2\alpha}} \tag{5.27}$$

and

$$s_0 = \sqrt{H^2 + d^2}$$

NUMERICAL EXAMPLE

Figure 5.14 illustrates the case of an earth dam with a sloping discharge face of 45°. The flow net has been constructed graphically starting from the discharge face after A' was located and H divided into fifteen equal Δz. A parabola was fitted through points A' and D after a was computed from Eq. 5.27.

Results:

$K = 2 \times 10^{-3}$ cm/sec $= 6.67 \times 10^{-5}$ ft/sec

$a = 22$ ft

$n_f = 3$

$n_d = 14.8$

$Q = 7.97 \times 10^{-4}$ cfs per unit length of dam (graphically)

$= 7.20 \times 10^{-4}$ cfs per unit length of dam (Eq. 5.26)

$= 10.2 \times 10^{-4}$ cfs per unit length of dam (Eq. 5.25)

[1] To prove the second equality, it suffices to prove that $\frac{1}{2}(H^2/a \sin\alpha \tan\alpha) = d - (a/2)\cos\alpha$. If $(a/2)\cos\alpha$ is subtracted from each member, and if the resulting equation is multiplied by a $\sin\alpha \tan\alpha$, the outcome is $\frac{1}{2}(a^2 \sin^2\alpha - H^2) = -a \sin\alpha \tan\alpha\,(-a\cos\alpha + d)$, as derived before.

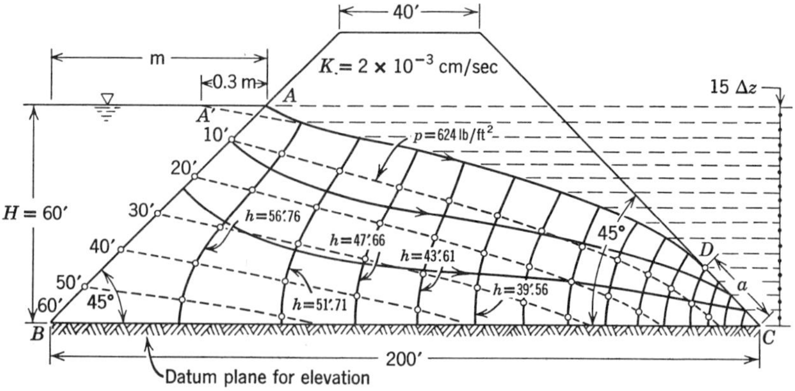

Fig. 5-14 Numerical example. Flow through earth embankment.

The lines of equal pressure corresponding to 10, 20, 30, 40, and 50 ft of water have been plotted on the net. The value of the total head h_n has been indicated on each equipotential line, and from the datum plane for elevation head heights equal to $h_n - 50$ ft, $h_n - 40$ ft, $h_n - 30$ ft, $\cdots$ have been measured so that the pressure head could be labeled at each equipotential line.

5.11 *Soils of Different Hydraulic Conductivity*

When streamlines from a medium of given hydraulic conductivity K_1 cross the boundary of a medium with different hydraulic conductivity K_2, they are refracted in a way similar to the optical refraction of light rays (Fig. 5.15).

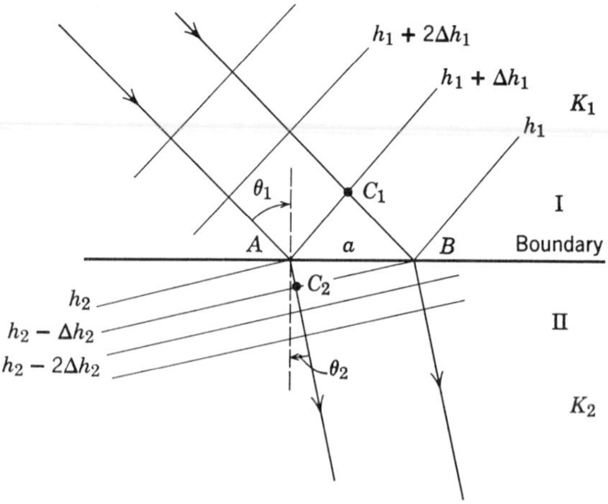

Fig. 5-15 Refraction of streamlines.

Consider a stream tube composed of streamlines passing through A and B. By continuity, all the fluid flowing between those streamlines in medium I also crosses into medium II and remains between the refracted streamlines; in other words the normal components of the average velocities in the streamtube on both sides of the boundary must be the same. Let θ_1 and θ_2 be the angles of the streamlines before and after refraction. The equality $V_{1,n} = V_{2,n}$ may be expressed as

$$K_1 \frac{(\Delta h_1)_{AB}}{C_1B} \cos \theta_1 = K_2 \frac{(\Delta h_2)_{AB}}{AC_2} \cos \theta_2 \tag{5.28}$$

The head difference between A and B is unique; therefore,

$$(\Delta h_1)_{AB} = (\Delta h_2)_{AB}$$

Furthermore, $C_1B = a \sin \theta_1$, and $AC_2 = a \sin \theta_2$. When these values of C_1B and AC_2 are inserted in Eq. 5.28, the result is

$$\frac{K_1}{\tan \theta_1} = \frac{K_2}{\tan \theta_2} \tag{5.29}$$

This relationship shows that the flow of Fig. 5.15 proceeds from a coarse to a fine-grained medium [10]. The flow elements in Fig. 5.15 are squares in medium I, but rectangles in medium II. If it is desired to draw a flow net with squares on both sides of the boundary and like flow quantities, then the required relationship between head drops is

$$\frac{\Delta h_2}{\Delta h_1} = \frac{K_1}{K_2} \tag{5.30}$$

A flow net in a medium consisting of two strata with different hydraulic conductivities is sketched in Fig. 5.16. It should be noted that the equipotential lines are refracted according to the law:

$$\frac{\tan \beta_1}{\tan \beta_2} = \frac{K_2}{K_1}$$

where $\alpha_1 + \beta_1 = \alpha_2 + \beta_2 = 90°$.

Only one streamline is refracted, and because $K_2 < K_1$, the streamline assumes a steeper descent when it crosses the boundary. This local behavior is very rapidly counteracted by the bedrock effect, so that the streamline tends to become parallel to the impervious foundation. The third streamtube originates entirely in medium I but towards the center of the dam foundation 80% of its flow takes place in medium I and 20%

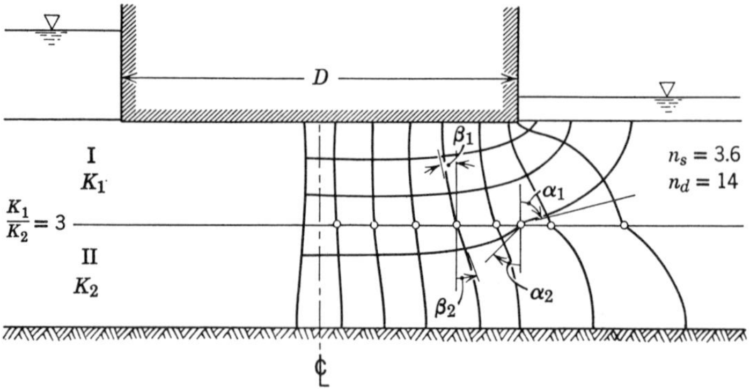

Fig. 5-16 Flow net in medium with different hydraulic conductivity.

in medium II. The lower flow channel is not a complete streamtube but only 60% of one. Since $K_2 = \frac{1}{3}K_1$, a full streamtube in medium II would have elements with a width equal to three times their length (i.e., in the direction of flow).

The problem for the same geometrical configuration but with $K_1 = \frac{1}{3}K_2$ is left as an exercise for the reader. The best procedure seems to consist in the construction of a "square" flow net in the region of the highest hydraulic conductivity and the extension of this net across the boundary. For the construction by means of an electrolytic tank, of nets under similar conditions and for anisotropic media, the reader is referred to reference 11.

5.12 *Anisotropic Soil*

Sediments are commonly deposited so that the hydraulic conductivity in one direction is greater than in another direction. This is the case in most stratified material where the flow proceeds more easily along the planes of deposition than across them. A condition in which the hydraulic conductivity in any point has preferential directions is called anisotropic. If this condition varies from point to point in a given stratum, the stratum is of heterogeneous composition. If, on the other hand, the condition of anisotropy is the same from point to point in the stratum, the stratum is still homogeneous. The simplest case of anisotropy is one in which the hydraulic conductivity is the same for all the directions of a given plane and is much smaller in the direction perpendicular to that plane. It will be shown in this case that the flow net is still a valuable tool.

In two-dimensional flow, Darcy's law becomes

$$u = -K_x \frac{\partial h}{\partial x}$$

$$w = -K_z \frac{\partial h}{\partial z}$$

If these values of u and w are inserted in Eq. 4.44 for ρ = constant and steady state flow, the following equation results

$$K_x \frac{\partial^2 h}{\partial x^2} + K_z \frac{\partial^2 h}{\partial z^2} = 0 \tag{5.31}$$

This equation differs from Laplace's equation but it may be reduced to Laplace's by a simple scale transformation.

Assume that $K_x > K_z$ and introduce a new variable

$$x_t = x\sqrt{K_z/K_x} \tag{5.32}$$

This means that the distances in the direction of greater hydraulic conductivity are reduced. To transform Eq. 5.31, consider

$$\frac{\partial h}{\partial x} = \frac{\partial h}{\partial x_t}\frac{\partial x_t}{\partial x} = \sqrt{K_z/K_x}\,\frac{\partial h}{\partial x_t}$$

$$\frac{\partial^2 h}{\partial x^2} = \sqrt{K_z/K_x}\,\frac{\partial}{\partial x}\left(\frac{\partial h}{\partial x_t}\right) = \sqrt{K_z/K_x}\,\frac{\partial}{\partial x_t}\left(\frac{\partial h}{\partial x_t}\right)\frac{\partial x_t}{\partial x} = \frac{K_z}{K_x}\frac{\partial^2 h}{\partial x_t^2}$$

Equation 5.30 may now be written as

$$\frac{\partial^2 h}{\partial x_t^2} + \frac{\partial^2 h}{\partial z^2} = 0 \tag{5.33}$$

This equation is Laplace's equation in the variables x_t and z. It is possible to apply a scale transformation as in Fig. 5.17 to draw a flow net and then to replot it to true scale. It should be noted that the outside flow elements in the deeper part of the flow region are singular. The velocities in this region become smaller as the inflow area becomes larger. The upper potential line must terminate orthogonally upon the impervious foundation. If K_{tr} is the unknown hydraulic conductivity of the transformed medium, the flowrate is given by

$$Q = \frac{n_s}{n_d} K_{tr} H \tag{5.34}$$

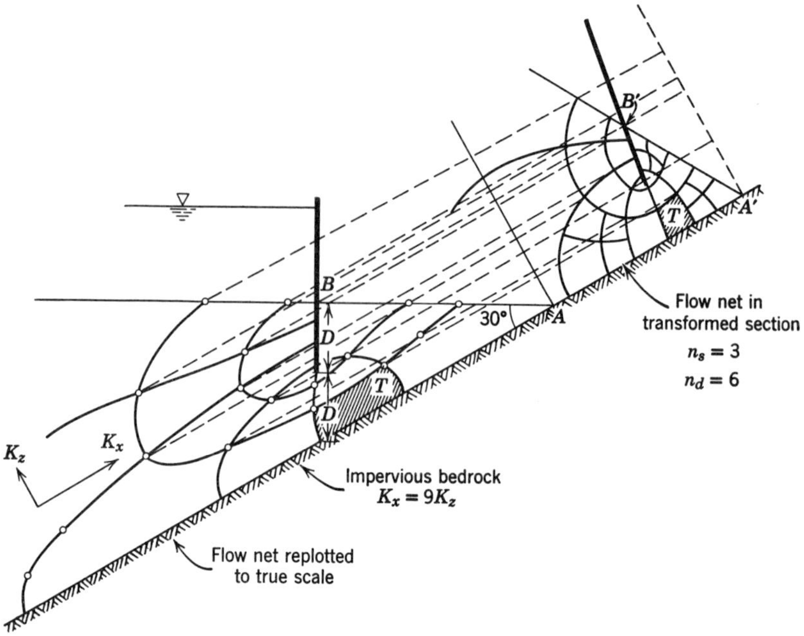

Fig. 5-17 Flow net in anisotropic medium.

To find K_{tr}, compare the two elements marked T in Fig. 5.17. The flowrate through the element in the transformed medium is

$$q = K_{tr} a \frac{\Delta h}{a} \tag{5.35}$$

where a is the average linear dimension of the square.

The same flowrate passes through the corresponding element in the original section in the direction of $K_{\max}$. The length of that element is clearly $a\sqrt{K_x/K_z}$. Hence

$$q = K_x \frac{a\,\Delta h}{a\sqrt{K_x/K_z}} \tag{5.36}$$

From Eqs. 5.35 and 5.36 the unknown K_{tr} is found

$$K_{tr} = \sqrt{K_x K_z}$$

and finally

$$Q = \frac{n_s}{n_d}\sqrt{K_x K_z}\,H \tag{5.37}$$

NUMERICAL EXAMPLE (Fig. 5.17)

$K_z = 10^{-5}$ cm/sec
$K_x = 9K_z$
$D =$ depth of penetration = 10 meters
$H =$ 5 meters
$n_s = 3;\ n_d = 6$
$Q = 7.5 \times 10^{-7} m^3$/sec per unit length of sheet pile
or 65 liters/day

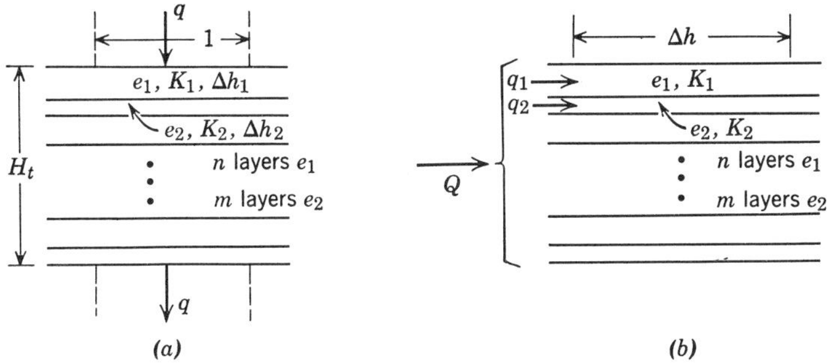

Fig. 5-18*a* and *b* Multilayered soils.

5.13 *Multilayered Soil*

Thin strata of variable thickness and of relatively high hydraulic conductivity, such as coarse sands, commonly alternate with less permeable lenses of clayey sands and it is necessary to have an idea of the maximum and minimum resulting or overall hydraulic conductivity of such a system.

Consider Fig. 5.18*a*. Let there be n layers of thickness e_1 and m layers of thickness e_2. Let Δh_1 be the head loss across a layer of thickness e_1, and Δh_2 the head loss across a layer of thickness e_2. The flowrate q perpendicular to the layers and across a unit area may be expressed as

$$q = K_1 \frac{\Delta h_1}{e_1} \cdot 1 = K_2 \frac{\Delta h_2}{e_2} \cdot 1 \tag{5.38}$$

On the other hand, the total head loss across the multilayered stratum is

$$H_t = n\,\Delta h_1 + m\,\Delta h_2$$

or

$$H_t = q\left(\frac{ne_1}{K_1} + \frac{me_2}{K_2}\right) \tag{5.39}$$

on account of Eq. 5.38.

The flowrate may also be expressed as

$$q = K_{\min} \frac{H_t}{ne_1 + me_2} \tag{5.40}$$

Elimination of H_t/q from Eqs. 5.39 and 5.40 renders

$$\frac{H_t}{q} = \frac{ne_1}{K_1} + \frac{me_2}{K_2} = \frac{ne_1 + me_2}{K_{\min}}$$

and

$$K_{\min} = \frac{ne_1 + me_2}{\dfrac{ne_1}{K_1} + \dfrac{me_2}{K_2}}$$

This formula may also be extended to a stratum composed of more than two different layers in which case

$$K_{\min} = \frac{\sum_i e_i}{\sum_i \dfrac{e_i}{K_i}} \tag{5.41}$$

in which $\sum_i$ means summation over all the layers.

To find the maximum hydraulic conductivity examine Fig. 5.18*b*. The flowrates q_1 and q_2 through the individual layers may be expressed as

$$q_1 = K_1 \frac{\Delta h}{\Delta l} e_1 \qquad q_2 = K_2 \frac{\Delta h}{\Delta l} e_2$$

and the total flowrate

$$Q = nq_1 + mq_2 = (nK_1e_1 + mK_2e_2)\frac{\Delta h}{\Delta l} \tag{5.42}$$

On the other hand, this flowrate may also be expressed as

$$Q = K_{\max}(ne_1 + me_2)\frac{\Delta h}{\Delta l} \tag{5.43}$$

From Eqs. 5.42 and 5.43 the value of $K_{\max}$ is found

$$K_{\max} = \frac{nK_1e_1 + mK_2e_2}{ne_1 + me_2}$$

This formula may also be extended to a stratum composed of more than two different layers, in which case

$$K_{\max} = \frac{\sum_i K_ie_i}{\sum_i e_i} \tag{5.44}$$

in which $\sum_i$ again means summation over all the layers.

NUMERICAL EXAMPLE

For $e_1 = 16.66$ ft
$e_2 = 10$ ft
$K_1 = 0.05 \times 10^{-4}$ cm/sec
$K_2 = 75 \times 10^{-9}$ cm/sec
and $n = m$, we obtain

$$K_{\min} = 2 \times 10^{-7} \text{ cm/sec}$$
$$K_{\max} = 32 \times 10^{-7} \text{ cm/sec}$$

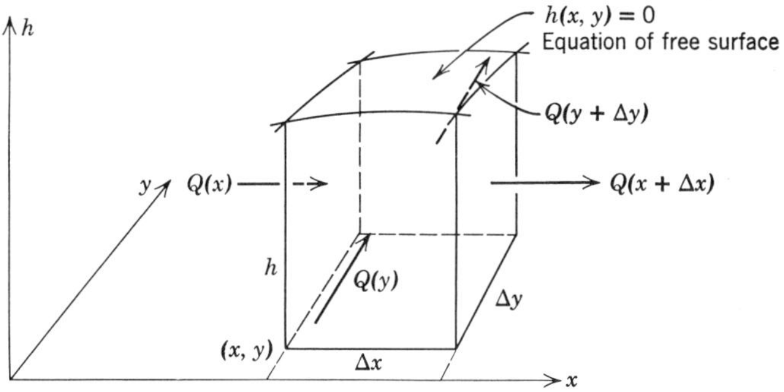

Fig. 5-19 Continuity for unconfined aquifer.

5.14 *Dupuit's Assumptions and* $\nabla^2 h^2 = 0$ *for Unconfined Flow*

In section 4.11 it was demonstrated that the piezometric head h satisfies Laplace's equation 4.52 on the basis of hydrodynamic considerations both in confined and unconfined aquifers. It was also stated that Eq. 4.52 could be modified for unconfined flow by the use of Dupuit's assumptions (see section 4.11) of the so-called hydraulic theory.

Consider the prism of Fig. 5.19 in which the bottom is parallel to the direction of flow and the top surface is the water table. The datum plane for h coincides with the upper surface of the impervious stratum underlying the aquifer. Under Dupuit's assumptions, $h(x, y)$ is the height of the free surface above the datum plane as well as the head in any point of the vertical dropped from a point of the free surface. Dupuit's assumptions are reasonable for mild curvatures of the free surface. In this case, if flow in the x-direction is considered, h in the section through x and parallel to y may also be taken as the average value of the head in that section. The flow rate through the elemental area at x, say $h\,\Delta y$, may be expressed as

$$Q(x) = -K \frac{\partial h}{\partial x} h\,\Delta y \tag{5.45}$$

The flowrate through the elemental area at $x + \Delta x$ may be expressed as

$$Q(x + \Delta x) = -K\frac{\partial h}{\partial x}h\,\Delta y + \Delta x\frac{\partial}{\partial x}\left(-K\frac{\partial h}{\partial x}h\,\Delta y\right) + \cdots \quad (5.46)$$

The difference of inflow rate and outflow rate, for constant K, is

$$\Delta x\,\Delta y K\frac{\partial}{\partial x}\left(h\frac{\partial h}{\partial x}\right) = \Delta x\,\Delta y K\frac{\partial}{\partial x}\left(\frac{1}{2}\frac{\partial h^2}{\partial x}\right)$$

Similarly, the difference between inflow rate and outflow rate in the y-direction may be found to be

$$K\,\Delta x\,\Delta y\frac{\partial}{\partial y}\left(\frac{1}{2}\frac{\partial h^2}{\partial y}\right)$$

By hypothesis there is no flow in the vertical direction. According to the principle of continuity the difference between inflow rate and outflow rate must be equal to the change of water volume inside the prism. This change is zero if there are neither sources nor sinks inside the prism, which will always be the case, although sources or sinks may occur at the boundaries of the domain. (See Chapter 6 on well flow.) Hence

$$\Delta x\,\Delta y\frac{K}{2}\left[\frac{\partial^2 h^2}{\partial y^2} + \frac{\partial^2 h^2}{\partial y^2}\right] = 0$$

and

$$\nabla^2 h^2 = 0 \quad (5.47)$$

In case recharge to the water table takes place through percolation of infiltrated water Eq. 5.47 may be modified without difficulty. If the rate of recharge per unit area is expressed as W (dimensions L/T), the rate of recharge to the prism of Fig. 5.19 is $W\,\Delta x\,\Delta y$ and conservation of mass requires that

$$\Delta x\,\Delta y\frac{K}{2}\left[\frac{\partial^2 h^2}{\partial x^2} + \frac{\partial^2 h^2}{\partial y^2}\right] + W\,\Delta x\,\Delta y = 0$$

or

$$\nabla^2 h^2 + \frac{2W}{K} = 0 \quad (5.47^*)$$

EXAMPLE: FREE SURFACE FLOW THROUGH RECTANGULAR EARTH EMBANKMENT (FIG. 5.20) WITHOUT REPLENISHMENT OF THE WATER TABLE

Equation 5.47 reduces to

$$\frac{d^2 h^2}{dx^2} = 0$$

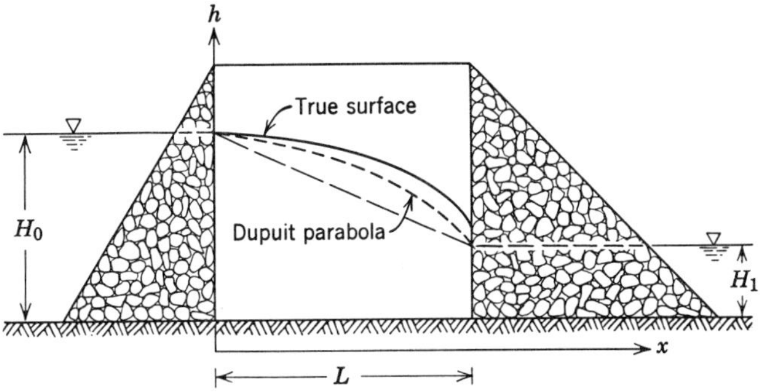

Fig. 5-20 Dupuit-flow through dam.

The integration of this ordinary differential equation renders

$$h^2 = Ax + B \tag{5.48}$$

in which A and B are constants of integration which remain to be determined. The constant A is extracted from Eq. 5.48 by differentiation

$$A = 2h\frac{dh}{dx}$$

Another expression of $h\frac{dh}{dx}$ is found in the flowrate Q through any vertical cross section, based on Dupuit's assumptions

$$Q = -Kh\frac{dh}{dx}$$

Therefore $$A = -\frac{2Q}{K}$$

The constant B is determined by the condition that for $x = 0$, h assumes the value H_0. Therefore

$$B = H_0^2$$

The equation of the free surface is

$$h^2 = -\frac{2Q}{K}x + H_0^2 \tag{5.49}$$

a parabola. If the existence of a seepage surface at the exit is ignored then Q is determined from Eq. 5.49 and from $h = H_1$ for $x = L$,

$$Q = \frac{K}{2L}(H_0^2 - H_1^2) \tag{5.50}$$

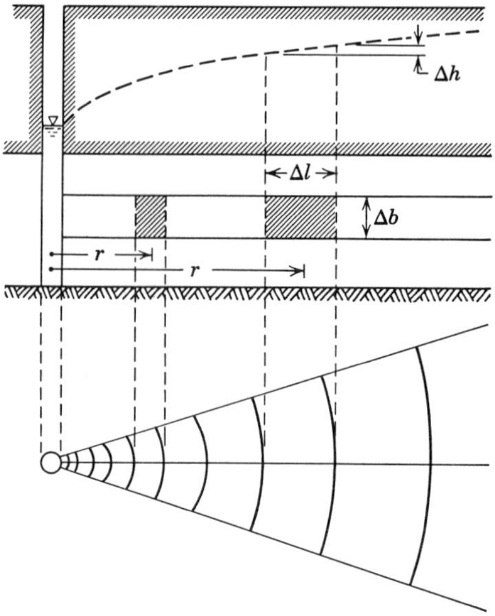

Fig. 5-21 Radial horizontal flow.

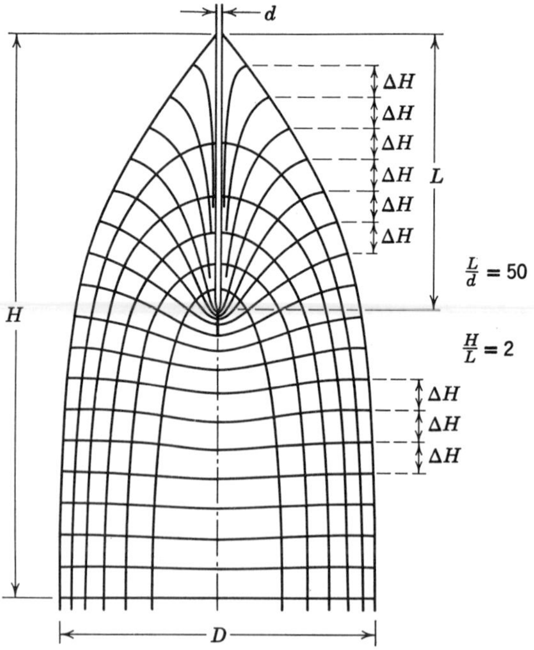

Fig. 5-22 Steady well recharge. (After Polubarinova.)

It may be proved that the flowrate expressed by Eq. 5.50 is exact, independent of the existence of the seepage surface. This proof was given by Charny [12]. The result of Eq. 5.50 was earlier arrived at by Muskat [13] without the use of Dupuit's assumption.

5.15 *Flow Nets for Wells*

The flow to or from a well under confined or unconfined conditions may be analyzed by means of flow nets in the case of radial symmetry where the flow picture is the same in every section through the axis of the well. Sometimes such flow is called axisymmetrical [14]. In this case it is possible to draw two flow nets, one in a vertical plane containing the axis of the well and the second in any plane perpendicular to the axis of the well. In the second type the streamlines are rays converging at the well while the equipotential lines are concentric circles. If the flow is confined, as in Fig. 5.21, then the horizontal flow net or net of the second type is the same for any plane perpendicular to the well. If the flow is unconfined, as in Fig. 5.22, then flow nets in planes perpendicular to the axis of the well are homothetic. It should be noted that they are not represented in Fig. 5.22.

The construction of the equilateral horizontal flow net is helpful in drawing the vertical net as is demonstrated in Fig. 5.21. Here the flow net is no longer equilateral, but the $(\Delta b/\Delta l)$ ratio for each element is inversely proportional to the radius. This follows from the consideration of the flowrate Δq in the inner stream tube of Fig. 5.21:

$$\Delta q = 2\pi r \, \Delta b \, K \frac{\Delta h}{\Delta l} \tag{5.51}$$

where Δh is taken in absolute value.

In order to have the same flowrate for the same Δh over all elements, it is evident that the ratio $\left(r \dfrac{\Delta b}{\Delta l}\right)$ must be constant in the construction of the net. The reader may complete the vertical flow net of Fig. 5.21 from the knowledge of the horizontal net and then check to see that the above ratio is constant.

The principle of keeping the ratio $\left(r \dfrac{\Delta b}{\Delta l}\right)$ constant has also been observed in the construction of the net illustrated in Fig. 5.22. The equipotential lines intersect the free surface at constant ΔH intervals.

The total flowrate in the case of Fig. 5.21 follows from $Q = n_s \, \Delta q$ and $H_t = n_d \, \Delta h$ so that

$$Q = 2\pi \, K \frac{n_s}{n_d} \left(r \frac{\Delta b}{\Delta l}\right) H_t \tag{5.52}$$

The flow net of Fig. 5.22 shows that the streamlines become almost parallel and vertical at a depth of twice the length of the well. In this case the influence of the existing water table on the recharge pattern has not been studied. Nasberg [14] who originally developed this flow net assumed that the influence of the water table was negligible at depths $2L$.

REFERENCES

1. Muskat, M., *The Flow of Homogeneous Fluids through Porous Media*, McGraw-Hill, New York, 1937; Second printing, Edwards Brothers, Ann Arbor, Mich., p. 301 (1946).
2. Polubarinova-Kochina, P. Ya., *Theory of Ground-Water Movement*, p. 32, Princeton, Princeton University Press, 1962. English translation by J. M. R. De Wiest.
3. Wayland, H., *Differential Equations Applied in Science and Engineering*, p. 67, Van Nostrand, Princeton, 1957.
4. Rouse, H., *Advanced Mechanics of Fluids*, p. 73, Wiley, New York, 1959.
5. Casagrande, A., "Seepage through Dams," *Journal of New England Water Works Association*, June 1937; see also *Journal of Boston Society of Civil Engineers*, pp. 295–337 (1940).
6. Polubarinova-Kochina, *Theory of Ground-Water Movement*, p. 432.
7. Taylor, D. W., *Fundamentals of Soil Mechanics*, p. 156, Wiley, New York, 1948.
8. Polubarinova-Kochina, *Theory of Ground-Water Movement*, pp. 56–58.
9. Casagrande, L., "Näherungsverfahren Zur Ermittlung der Sickerung in geschütteten Dämmen auf undurchlassiger Sohle," *Die Bautechnik*, Heft 15, 1934.
10. Hubbert, M. K., "The Theory of Ground-Water Motion," *Journal of Geology*, p. 846 (1940).
11. Todd, D. K. and J. C. Bear, "Seepage through Layered Anisotropic Porous Media," *ASCE Journal Hydraulics Division*, pp. 31–57 (May 1961).
12. De Wiest, J. M. R., "Russian Contributions to the Theory of Ground-Water Flow," *Ground Water Journal of the NWWA*, Vol. 1, No. 1, pp. 44–48 (January 1963).
13. Muskat, M., *The Flow of Homogeneous Fluids through Porous Media*, p. 377.
14. Polubarinova-Kochina, *Theory of Ground-Water Movement*, p. 439.
15. Dicker, D., "Discussions on Transient Development of the Free Surface in a Homogeneous Earth Dam," *Geotechnique*, Vol. 13, No. 3, pp. 260–262 (September 1963).
16. Forchheimer, P., *Hydraulik*, pp. 417–418, Teubner, Berlin, 1914.
17. Nelson, R. W., "Design of Interceptor Drains in Heterogeneous Soils," *ASCE Journal of the Irrigation and Drainage Division*, pp. 41–53 (December 1961).
18. Nelson, R. W., "Steady Darcian Transport of Fluids in Heterogenious Partially Saturated Porous Media, I. Mathematical and Numerical Formulation," *Report HW—72335 PTI, Hanford Atomic Products Operation, General Electric*, pp. 1–30, Richland, Washington, 1962.
19. Reisenauer, A. E., R. W. Nelson, and C. N. Knudsen, "Steady Darcian Transport of Fluids in Heterogeneous Partially Saturated Porous Media. II. The Computer Program," *Report HW-72335 PT2, Hanford Atomic Products Operation, General Electric*, pp. 1–83, Richland, Washington, 1963.

CHAPTER SIX

Mechanics of Well Flow

The principles of ground-water movement explained in Chapters 4 and 5 are applied in this chapter to problems of flow to and from wells. Simple boundary conditions make it possible to solve Eq. 4.52 and Eq. 5.47 without the use of mathematical techniques such as separation of variables, introduction of transforms, Green's functions, and conformal mapping. Results obtained by these methods which are beyond the scope of this elementary textbook are given only for student reference and are not developed analytically. On the other hand, potential and stream functions are constructed by means of the limited information that was made available in Chapters 4 and 5. The subject of well flow has been thoroughly treated in some of the existing textbooks and has been intensively investigated by research workers. There is some repetition of material that has become classic in the study of ground-water flow, also only the most idealized boundary conditions are treated. Nevertheless, practical problems have been solved satisfactorily with the equations which have been derived for these artificial boundary conditions even though the solutions are only a first approximation of the true situation in nature. The first part of this chapter (section 6.1–6.6) is devoted to steady flow with no special emphasis on either confined or unconfined flow, whereas sections 6.7–6.11 illustrate some problems of unsteady flow. Leaky aquifers are treated in the third part of the chapter.

◆ Steady Flow

6.1 *Steady Radial Flow to a Well*

(A) CONFINED FLOW

Assume that a well is pumped at a constant discharge Q (Fig. 6.1) and that the boundary conditions are such that the resulting lowering of the piezometric head h has the same distribution in any vertical section

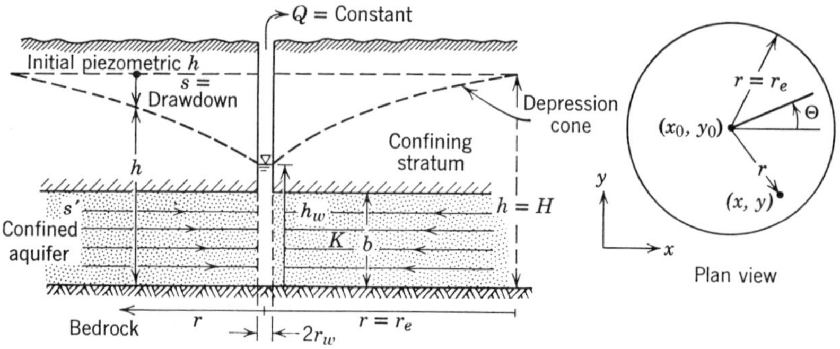

Fig. 6-1 Radial flow to a well completely penetrating a confined aquifer.

through the axis of the well. This means that the flow has radial symmetry (i.e., is independent of the angle θ in polar coordinates) and that the head must be constant along the perimeter of any circle concentric with the well. Such requirements are met only by a well centered on a circular island and penetrating a homogeneous and isotropic aquifer. If complete penetration of the aquifer by the well is assumed, the flow is everywhere parallel to the bedrock and follows streamlines s' in opposite direction to the rays r emanating from the center of the well, taken as the origin of plane polar coordinates. Hence Q equals the flowrate through a cylinder with radius r and height b, the thickness of the aquifer, so that

$$Q = -K\frac{\partial h}{\partial s'}2\pi rb = K\frac{dh}{dr}2\pi rb \tag{6.1}$$

Equation 6.1 may be integrated at once after separation of the variables. Since it is a first-order differential equation in one independent variable, it needs only one boundary condition, namely,

$$h = H \quad \text{for} \quad r = r_e \tag{6.2}$$

However, for didactic purposes it is better to consider Eqs. 6.1 and 6.2 as boundary conditions for the solution of Laplace's equation, Eq. 4.52, which in the present case may be written as

$$\frac{d^2h}{dr^2} + \frac{1}{r}\frac{dh}{dr} = \frac{1}{r}\frac{d}{dr}\left(r\frac{dh}{dr}\right) = 0 \tag{6.3}$$

Integration of Eq. 6.3 leads to

$$r\frac{dh}{dr} = \text{constant} = \frac{Q}{2\pi Kb} \tag{6.4}$$

because of condition 6.1. Equation 6.4 may now be integrated according to condition 6.2 as follows:

$$\int_h^H dh = \frac{Q}{2\pi Kb} \int_r^{r_e} \frac{dr}{r}$$

or

$$H - h = \frac{Q}{2\pi Kb} \log_e \frac{r_e}{r} \tag{6.5}$$

The difference H–h between the elevation of the initial piezometric surface and the elevation of this surface after Q was pumped is called the drawdown. It is commonly designated by the symbol s. Because in well problems we are more interested in the drawdown s than in the absolute value of h, and because the aquifer is confined, no special attention has been given to the choice of the datum plane for elevation. Note that in Fig. 6.1, to avoid confusion the symbol s' has been used to indicate the streamlines. The product $Kb = T$ is the transmissivity of the aquifer (see Eq. 4.50). It has the dimensions L^2/T and can be expressed in the United States in practical units of gallons per day and per foot (gpd/ft).

The integration of Eq. 6.4 may also be written as

$$h = \frac{Q}{2\pi Kb} \log_e r + \text{constant} \tag{6.6}$$

This formula avoids the concept of cylindrical boundaries at constant head, coaxial with the well and at a distance r_e from the center of the well. The radius r_e corresponds to the rather ill-defined (see reference 6, p. 361) radius of influence [40], which determines a so-called circle of influence. The circle of influence is nothing but the vertical projection of a cylinder at constant head, not affected by the pumping of the well.

From Eq. 6.6 it is clear that h increases indefinitely with an increase in r. For extensive natural aquifers, however, values of r can be very large but values of h are relatively small and are commonly limited by the elevation at which recharge enters the system. Thus Eq. 6.6 is applicable only within reasonable distances of a well. Stated in another way, steady radial flow to a well, which is an implied condition for Eq. 6.6, is achieved only near the well, and steady radial flow in large aquifers that approach infinite lateral dimensions is both theoretically and physically impossible.

Equations 6.5 and 6.6 still hold when the center of the well is located in the point (x_0, y_0) of the coordinate system, but of course

$$r^2 = (x - x_0)^2 + (y - y_0)^2$$

Equation 6.6 becomes

$$h(x, y) = \frac{Q}{4\pi Kb} \log_e [(x - x_0)^2 + (y - y_0)^2] + C \tag{6.7}$$

Equation 6.7 may be interpreted in the following way. It specifies the head h necessary in any point (x, y) to sustain a discharge Q in a well with coordinates (x_0, y_0).

Equation 6.5 is known as the equilibrium or Thiem [1] equation. It allows for the computation of Q or K, given one or the other, if the values of h or s are measured in two observation wells. One of these wells may be the well that is actually pumped. If h_w corresponds to r_w, the radius of the pumped well, and if h_1 is the head at an observation well a distance r_1 away from the pumped well, then

$$K = \frac{Q}{2\pi b(h_1 - h_w)} \log_e \frac{r_1}{r_w} \tag{6.8}$$

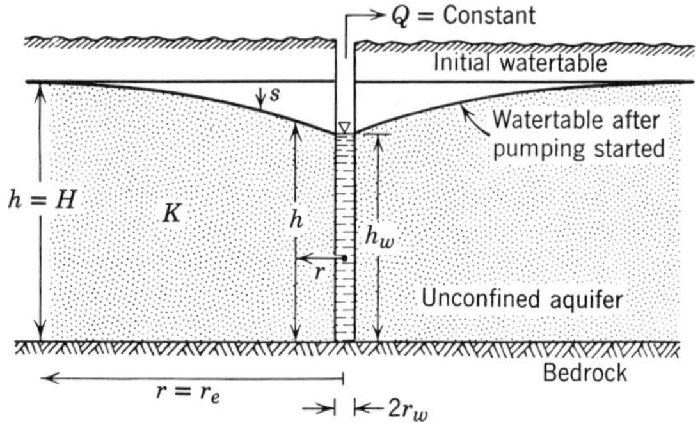

Fig. 6-2 Radial flow in water table aquifer.

Equation 6.8 is of limited use because local hydraulic conditions in and near the well strongly influence the value of h_w, r_w; r_w and h_w should be used with great caution and only when other methods of analysis are not available. Preferably two observation wells should be used, close enough to the pumped well to have significant drawdowns which are easy to measure. Although the nature of Eq. 6.8 is hypothetical, equations of this type have been widely used for field determinations of the hydraulic conductivity K of water-bearing strata.

(B) UNCONFINED FLOW (Fig. 6.2)

Under conditions similar to those just given, i.e., complete well penetration (to the horizontal base of an isotropic, homogeneous medium), laminar flow, and coaxial boundary of constant head, the equation for the drawdown of the water table is derived. Here h, as in section 5.14, measures the height of the water table above bedrock and satisfies Eq. 5.47, which may be written as

$$\frac{1}{r}\frac{d}{dr}\left(r\frac{dh^2}{dr}\right) = 0 \tag{6.9}$$

The boundary conditions are

$$(1)\quad Q = 2\pi rhK\frac{dh}{dr} \quad \text{or} \quad r\frac{dh^2}{dr} = \frac{Q}{\pi K} \tag{6.10}$$

which follows immediately from Fig. 6.2 and Dupuit's assumptions, again considering that $-\dfrac{dh}{ds'} = \dfrac{dh}{dr}$ as sub *a*.

$$(2)\quad h = H \quad \text{for} \quad r = r_e$$

Two successive integrations of Eq. 6.9 render

$$H^2 - h^2 = \frac{Q}{K\pi}\log_e\frac{r_e}{r} \tag{6.11}$$

or

$$h^2 = \frac{Q}{\pi K}\log_e r + \text{constant} \tag{6.12}$$

and

$$h^2(x, y) = \frac{Q}{2\pi K}\log_e[(x - x_0)^2 + (y - y_0)^2] + C \tag{6.12*}$$

Equations 6.11 and 6.12 do not consider the existence of a seepage surface above the water level in the well. Because of the underlying Dupuit assumptions they fail to describe accurately the drawdown curve near the well where the strong curvature of the water table contradicts these assumptions. Again K or Q may be determined as before and Eq. 6.8 may be replaced by

$$K = \frac{Q}{\pi(h_1^2 - h_w^2)}\log_e\frac{r_1}{r_w} \tag{6.13}$$

Remarks similar to those made for Eq. 6.8 also apply to Eq. 6.13. Equations 6.5 and 6.11 for radial flow may be generalized and extended to include the case where there is no radial symmetry if the concept of average head $\bar{h}_e$ on the circle of radius r_e is introduced. For confined flow,

$$\bar{h}_e - h = \frac{Q}{2\pi Kb}\log_e\frac{r_e}{r} \tag{6.5*}$$

with $\bar{h}_e = \dfrac{1}{2\pi}\displaystyle\int_0^{2\pi} h_e(\theta)\,d\theta$ = average head on the circle of radius r_e about the center of the well.

Similarly, for unconfined flow, we find

$$\bar{h}_e^2 - h^2 = \frac{Q}{\pi K}\log_e\frac{r_e}{r} \tag{6.11*}$$

These formulas are derived from solutions of Laplace's equation in polar coordinates r, θ.

6.2 *Several Wells*

Equations 6.7 and 6.12* may be extended without difficulty to include any number of wells ($i = 1, 2, \cdots, n$) located at (x_i, y_i). This follows from the linearity of Laplace's equation, allowing the superposition of solutions. In the case of confined flow, for example, h_i may be considered the head required in any point (x, y) to sustain a discharge Q_i of the ith well located in the point (x_i, y_i). Hence

$$h_i = \frac{Q_i}{4\pi Kb} \log_e [(x - x_i)^2 + (y - y_i)^2] + C_i.$$

To sustain simultaneous discharge $\sum_{i=1}^{n} Q_i$ of all the wells, a head $h(x, y) = \sum_{i=1}^{n} h_i$ will be needed. Therefore,

$$h(x, y) = \frac{1}{4\pi Kb} \sum_{i=1}^{n} Q_i \log_e [(x - x_i)^2 + (y - y_i)^2] + C \qquad (6.14)$$

where C is an arbitrary constant determined conveniently in each problem. The corresponding formula for unconfined flow is

$$h^2(x, y) = \frac{1}{2\pi K} \sum_{i=1}^{n} Q_i \log_e [(x - x_i)^2 + (y - y_i)^2] + C \qquad (6.15)$$

NUMERICAL EXAMPLE

Two wells are drilled 150 ft apart in a stratum of sand which has a hydraulic conductivity $K = 500 \times 10^{-4}$ cm/sec and is underlain by a horizontal impervious base. The original height H of the ground-water table above the impervious base was 40 ft. A discharge $Q = 300$ gal/min is pumped from each well, and it is estimated that a steady condition of flow is reached within a cylinder of $r_e = 2{,}000$ ft and that the original head of 40 ft is unaffected beyond this cylinder. Both wells have a radius of 1 ft. Assume that for each individual well the radius of the cylinder, beyond which the original head is unaffected by pumping, is also equal to 2,000 ft.

(a) Plot the drawdown curve along the vertical plane through the two wells.

(b) How much does the existence of the second well increase the pumping head for the first well?

(c) Prove in general that far away from a well field the drawdown surface approaches that which would exist for a single well at the center pumping the sum of the discharges of the individual wells. (This applies for either confined or unconfined steady flow.)

Solution. The coordinate system is chosen so that its origin $x = 0$, $y = 0$ coincides with the middle of the distance between the wells, and the vertical plane through the two wells has the equation $y = 0$.

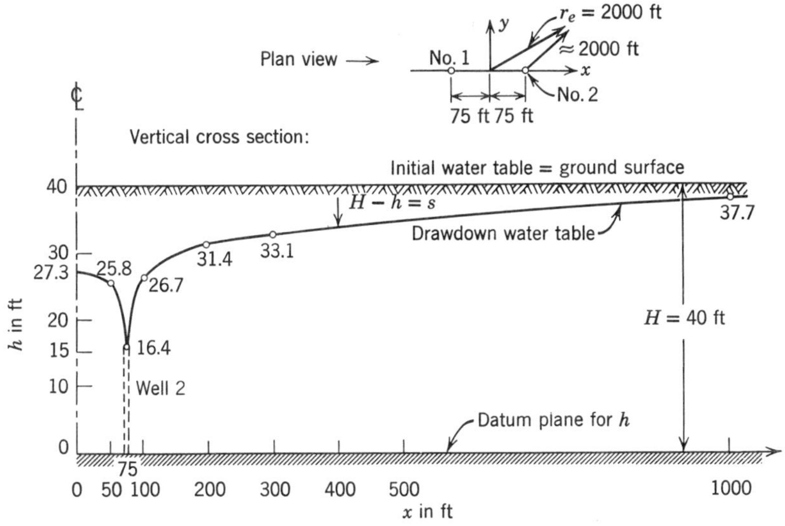

Fig. 6-3 Drawdown curve in vertical plane through two wells in water table conditions. (Well 1 is to the left of the diagram and produces a drawdown that is a mirror image of the drawdown shown above.)

Equation 6.15 renders

$$h^2 = \frac{1}{2\pi K}\,[Q_1 \log_e (x + 75)^2 + Q_2 \log_e (x - 75)^2] + C$$

and after insertion of the values for K and Q_1, Q_2

$$h^2 = 65\text{ ft}^2[\log_e (x + 75)^2 + \log_e (x - 75)^2] + C$$

C is determined from the boundary condition that $h = 40$ ft for $x = 2{,}000$ ft. Its value is $C = -380\text{ ft}^2$, and the final equation of the drawdown curve in the plane through the wells is

$$h^2 = 65\text{ ft}^2[\log_e (x + 75)^2 + \log_e (x - 75)^2] - 380\text{ ft}^2$$

The drawdown curve is represented in Fig. 6.3. Because of the symmetry of the total drawdown curve, Well 1 is not sketched in this figure. This

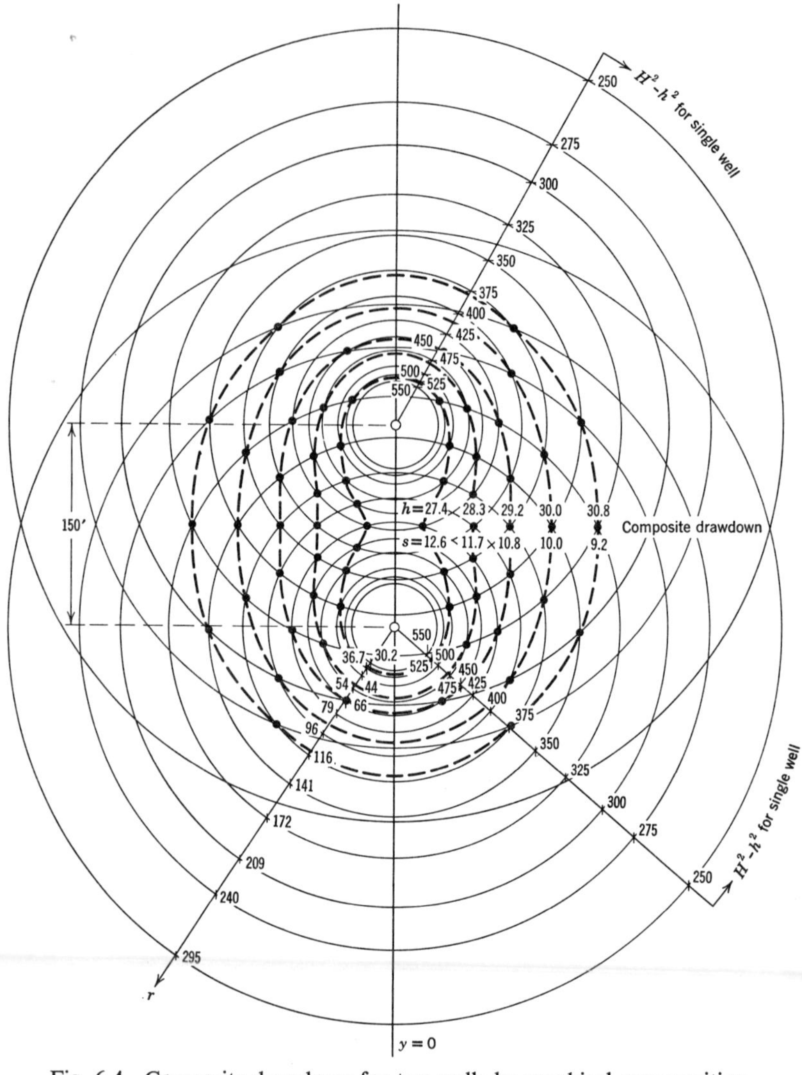

Fig. 6-4 Composite drawdown for two wells by graphical superposition.

well produces a drawdown curve that is the mirror image of the drawdown curve shown in Fig. 6.3. The drawdown at both wells is found to be 40 ft − 16.4 ft = 23.6 ft.

If one well existed alone, Eq. 6.11 for $r_w = 1$ ft would render

$$H^2 - h_w^{\,2} = \frac{Q}{\pi K} \log_e r_e = 995 \text{ ft}^2$$

after insertion of known numerical values. This would lead to $h_w =$ 24.6 ft, and therefore the increase in pumping head would be 24.6 ft − 16.4 ft = 8.2 ft.

To prove part c, let the coordinates of the center of the well field be

$$\sum_{i=1}^{n} \frac{x_i}{n}, \qquad \sum_{i=1}^{n} \frac{y_i}{n}$$

Then we have to prove that the expression

$$Q \log_e \left[\left(x - \frac{\Sigma x_i}{n}\right)^2 + \left(y - \frac{\Sigma y_i}{n}\right)^2\right] \tag{I}$$

which is proportional to the head or drawdown in any point (x, y) for a single well drilled at the center of the well field, and discharging $Q = \sum_{i=1}^{n} Q_i$ in the limit is equal to the expression

$$\sum_{i=1}^{n} Q_i \log_e [(x - x_i)^2 + (y - y_i)^2] \tag{II}$$

which is proportional to the head in any point (x, y) in the case of n wells, and where the proportionality constants are the same in I and II. This is true because x and y become very large if the wells are scattered around the origin of coordinates and if we go far away from the well field. Therefore x_i, y_i and $\frac{\Sigma x_i}{n}$, $\Sigma \frac{y_i}{n}$ may be neglected compared to x and y, coordinates of the observer at a great distance from the well field. Both expressions tend to a common limit $Q \log r^2$.

Finally, in Fig. 6.4 for each well contour maps are drawn of $(H^2–h^2)$ equal to 275, 300, 325, $\cdots$, 550 ft², where h is the elevation of the free surface above the impervious base. By graphical superposition the contours for $H^2–h^2$ equal to 650, 700, 750, 800, and 850 ft² for the combined effect of both wells are constructed. For each contour the value of h has been labeled. The exact intersection of the composed drawdown curves of Fig. 6.4 with $y = 0$ could be determined from the curve of Fig. 6.3.

6.3 *Flow Between a Well and a Recharge Well*

The center of the pumped well is on the x-axis at a distance x_0 from the origin, that of the recharge well at $(-x_0)$, and the flow rate of the recharge well is assumed to be $(-Q)$, where Q is the discharge of the pumped well. The recharge takes place into a confined aquifer and Eq. 6.14 reduces to

$$h = \frac{Q}{4\pi Kb} \log_e \frac{(x - x_0)^2 + y^2}{(x + x_0)^2 + y^2} + C \tag{6.16}$$

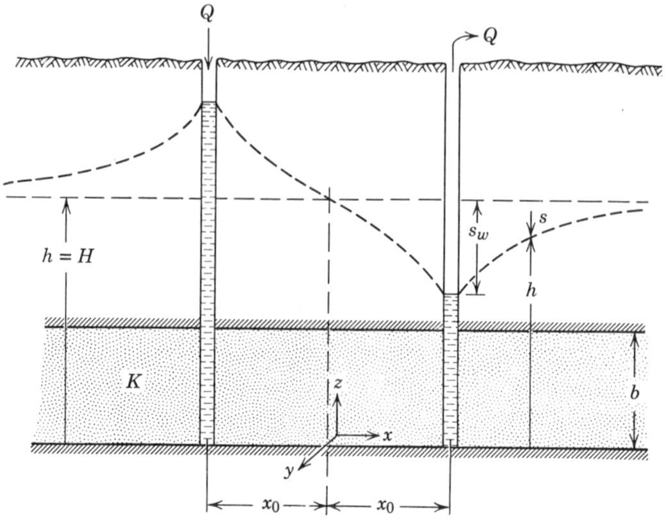

(a)

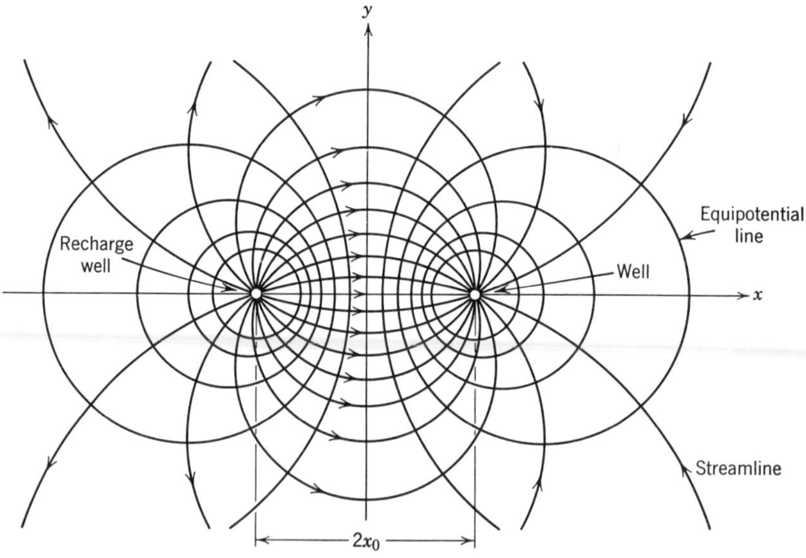

(b)

Fig. 6-5(*a*) Well and recharge well (confined aquifer). (*b*) Streamlines and equipotential lines for system of well and recharge well.

The constant C is determined by the condition that for very large $r(r \to \infty)$ the head H is not affected by the wells. Therefore $C = H$, which also implies that $h = H$ for $x = 0$, $y = 0$. The final equation for the head h becomes

$$h = \frac{Q}{4\pi Kb} \log_e \frac{(x - x_0)^2 + y^2}{(x + x_0)^2 + y^2} + H \tag{6.17}$$

Equipotential lines h = constant follow from Eq. 6.17 as

$$e^{\frac{4\pi Kb(h-H)}{Q}} = \frac{(x - x_0)^2 + y^2}{(x + x_0)^2 + y^2} = \text{constant, say } m, \tag{6.18}$$

or, after some rearrangement,

$$x^2 + y^2 - 2xx_0 \frac{1 + m}{1 - m} + x_0^{\,2} = 0 \tag{6.19}$$

Equation 6.19 represents a family of circles with the center at $\left(x_0 \dfrac{1 + m}{1 - m}, 0\right)$ and radii $\dfrac{2x_0\sqrt{m}}{1 - m}$. Since the flowlines have to be perpendicular to the equipotential lines, it follows immediately that they are circles with their center on the y-axis [2, 3]. (See Fig. 6.5*b*.)

The drawdown s_w at the discharging well may be found from Eq. 6.17 for $x = x_0 - r_w$, $y = 0$:

$$s_w = H - h_w = \frac{Q}{4\pi Kb} \log_e \frac{(2x_0 - r_w)^2}{r_w^{\,2}} \approx \frac{Q}{2\pi Kb} \log_e \frac{2x_0}{r_w}$$

A comparison with Eq. 6.5 shows that in first approximation this drawdown is equal to the drawdown at the face of a well in a circular island aquifer of radius $2x_0$.

6.4 *Cylindrical Sink and Source Flows—Uniform Flow*

At this point it is particularly helpful to introduce the hydrodynamical concepts of plane parallel flow and source and sink flow and to compare them with their counterparts in ground-water flow, say uniform flow, and flow to and from a well. It is worthwhile to derive briefly the expressions for Φ and Ψ of these particular flows, as they will be applied in the method of images. The reader is reminded of the Cauchy-Riemann equations (5.9), which may be written in polar coordinates, or by mere inspection, considering that the cartesian components dx, dz of a path of length ds are replaced by the components dr and $r\,d\theta$ in polar coordinates. Therefore Eqs. 5.9 in polar coordinates become

$$\frac{\partial \Phi}{\partial r} = \frac{1}{r}\frac{\partial \Psi}{\partial \theta}$$
$$\frac{1}{r}\frac{\partial \Phi}{\partial \theta} = -\frac{\partial \Psi}{\partial r} \qquad (6.20)$$

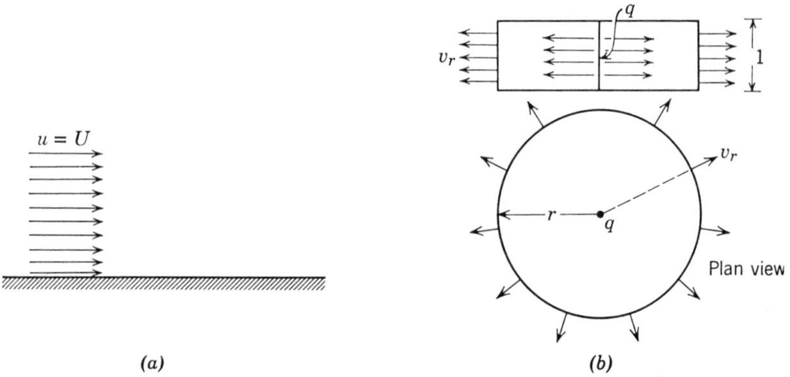

Fig. 6-6(*a*) Parallel flow of ideal fluid. (*b*) Line source in ideal flow.

(A) PARALLEL FLOW

This type of flow is visualized (Fig. 6.6*a*) by an ideal fluid flowing along an infinite wall at a constant velocity U. Therefore $u = U$, i.e.,

$$-\frac{\partial \Phi}{\partial x} = U \quad \text{or} \quad \Phi = -Ux + c$$

In view of Eqs. 5.9 (6.21)

$$-\frac{\partial \Psi}{\partial z} = U \quad \text{or} \quad \Psi = -Uz + c$$

The analogous flow in ground water is uniform flow where the slope of the hydraulic grade line or hydraulic gradient S^* is constant. Here

$$-\frac{\partial \Phi}{\partial x} = KS^* \quad \text{or} \quad \Phi = -KS^*x + c$$

and

$$-\frac{\partial \Psi}{\partial z} = KS^* \quad \text{or} \quad \Psi = -KS^*z + c \qquad (6.22)$$

(B) CYLINDRICAL SOURCE AND SINK FLOW

Consider the flow from a line source (Fig. 6.6*b*) coinciding with the axis of a cylinder. If q is the strength of the source per unit length, then by definition

$$q = 2\pi r \cdot 1 \cdot v_r \tag{6.23}$$

where $v_r = -\partial\Phi/\partial r$, radial velocity.

Integration of Eq. 6.23 leads to

$$\Phi = -\frac{q}{2\pi}\log_e r + c$$

and

$$\Psi = -\frac{q}{2\pi}\theta + c \tag{6.24}$$

after use of the first Eq. 6.20.

Equations 6.24 are those of a line source which has its analogy in a recharge well in ground-water flow. If q is replaced by $-q$, the equations for a line sink are found, say,

$$\Phi = \frac{q}{2\pi}\log_e r + c$$

$$\Psi = \frac{q}{2\pi}\theta + c \tag{6.25}$$

The analogy with flow to a well follows immediately from Eq. 6.6 if $q = Q/b$, discharge per unit thickness of aquifer, and because $\Phi = Kh$ (Eq. 4.26). This analogy emphasizes the need for a continued and fully understood use of the concept of velocity potential.

The foregoing derived equations for the potential and stream functions may be used in superposition of several ground-water flows.

EXAMPLE: WELL AND RECHARGE WELL OF EQUAL STRENGTH (FIG. 6.7)

The stream function Ψ for a flow per unit thickness of aquifer follows immediately from Eqs. 6.24 and 6.25:

$$\Psi = \frac{q}{2\pi}\theta_1 + \frac{-q}{2\pi}\theta_2 + c = \frac{q}{2\pi}(\theta_1 - \theta_2) + c \tag{6.26}$$

The streamlines follow from Eq. 6.27 for Ψ = constant or $\Psi/(q/2\pi)$ = constant, i.e., $\theta_1 - \theta_2$ = constant. The values of θ_1 and θ_2 may be expressed simply from the geometry of Fig. 6.7. It is more convenient to use $\tan(\theta_1 - \theta_2)$ = constant to derive the equations of the streamlines. Hence

$$\tan(\theta_1 - \theta_2) = \frac{\tan\theta_1 - \tan\theta_2}{1 + \tan\theta_1 \tan\theta_2} = \frac{\dfrac{y}{x - x_0} - \dfrac{y}{x + x_0}}{1 + \dfrac{y^2}{x^2 - {x_0}^2}} = \text{constant, say } n, \tag{6.27}$$

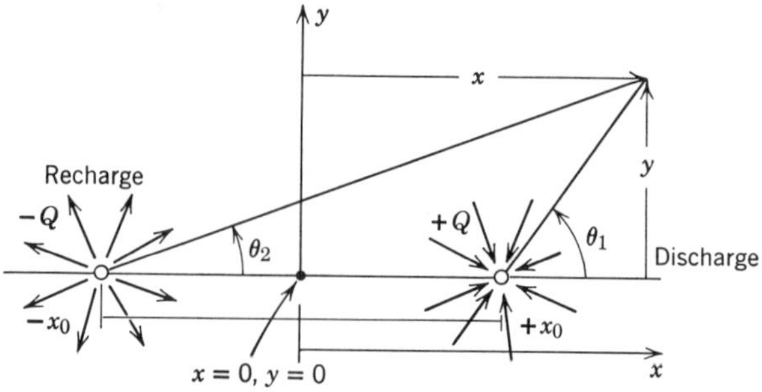

Fig. 6-7 Well and recharge well. Plan view.

or, after some rearrangement,

$$x^2 + y^2 - 2y\frac{x_0}{n} - x_0^{\ 2} = 0 \tag{6.28}$$

This equation is that of the circles represented on Fig. 6.5*b*, with their center at $\left(0, \dfrac{x_0}{n}\right)$ and radii $x_0\sqrt{1 + (1/n^2)}$.

6.5 *A Well in a Uniform Flow*

Consider a confined aquifer with thickness $b = 100$ ft and having a uniform flow for which $KS^* = 6.9$ ft/day $= 8 \times 10^{-5}$ ft/sec. A completely penetrating well is pumped at $Q = 107.5$ gal/min, i.e., $q = Q/b = 24 \times 10^{-3}$ ft²/sec, and it is necessary to determine the ground-water divide for the water entering the well.

The flowlines of the composite flow Ψ = constant are determined by adding the values of Ψ_u (uniform flow) and Ψ_w (well flow) at points of intersection as indicated on Fig. 6.8.

$$\Psi_u = -KS^*y + c = -8 \times 10^{-5}y \text{ ft}^2/\text{sec}$$

if $c = 0$ by assumption. The lines Ψ_u = constant are represented on Fig. 6.9 as parallel lines for $y = 0, \pm 25, \pm 50, \cdots, \pm 200$ ft.

$$\Psi_w = \frac{q}{2\pi}\theta + C = 24\frac{\theta}{2\pi} \times 10^{-3} \text{ ft}^2/\text{sec},$$

if $C = 0$ by assumption.

In Fig. 6.9, θ varies from 0 to π counterclockwise and from 0 to $-\pi$ clockwise measured from the x-axis. The well is placed at the origin of the

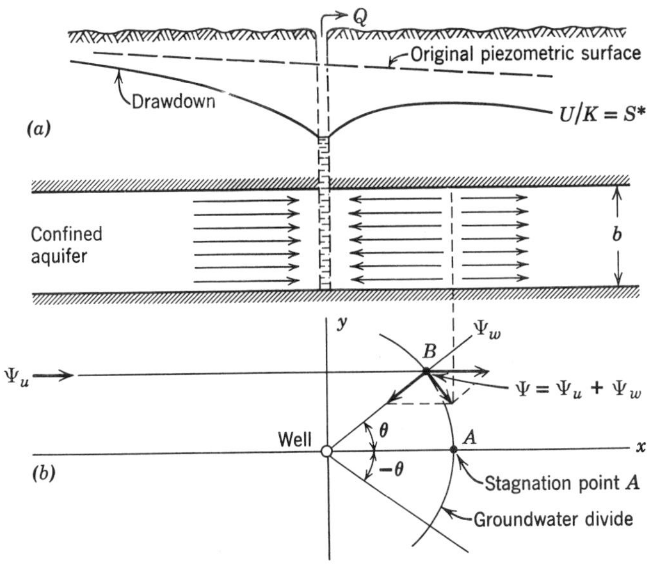

Fig. 6-8 Well in uniform flow. (*a*) Vertical cross-section. (*b*) Plan view for superposition of stream-functions.

coordinate axis. The rays Ψ_w = constant are labeled for $C = 0$; for example, for

$$\theta = \frac{\pi}{2}, \Psi_w = \frac{q}{2\pi} \cdot \frac{\pi}{2} = 6 \times 10^{-3}\ \text{ft}^2/\text{sec}$$

To construct a streamline, points where the sum of the added stream functions assumes the same value are joined by a smooth line. With the particular choice $C = 0$ in the expressions of Ψ_u and Ψ_w, the ground-water divide happens to be the streamline $\Psi = 0$ up to the stagnation point A. For another choice of the constant C a different value of Ψ along the ground-water divide would be found, emphasizing once more that Ψ is determined only up to an arbitrary constant. It would also be possible to construct in any point B (Fig. 6.8*b*) of the aquifer two-velocity vectors: one horizontal and of constant length, the other inversely proportional to r according to Eq. 6.23 and pointed towards the origin. Their resultant would be tangent to the streamline passing through B. In point A the vectors are in the opposite direction. If these vectors are equal, A is a stagnation point. Its location is determined by the equality

$$\frac{q}{2\pi x_s} = KS^* \tag{6.29}$$

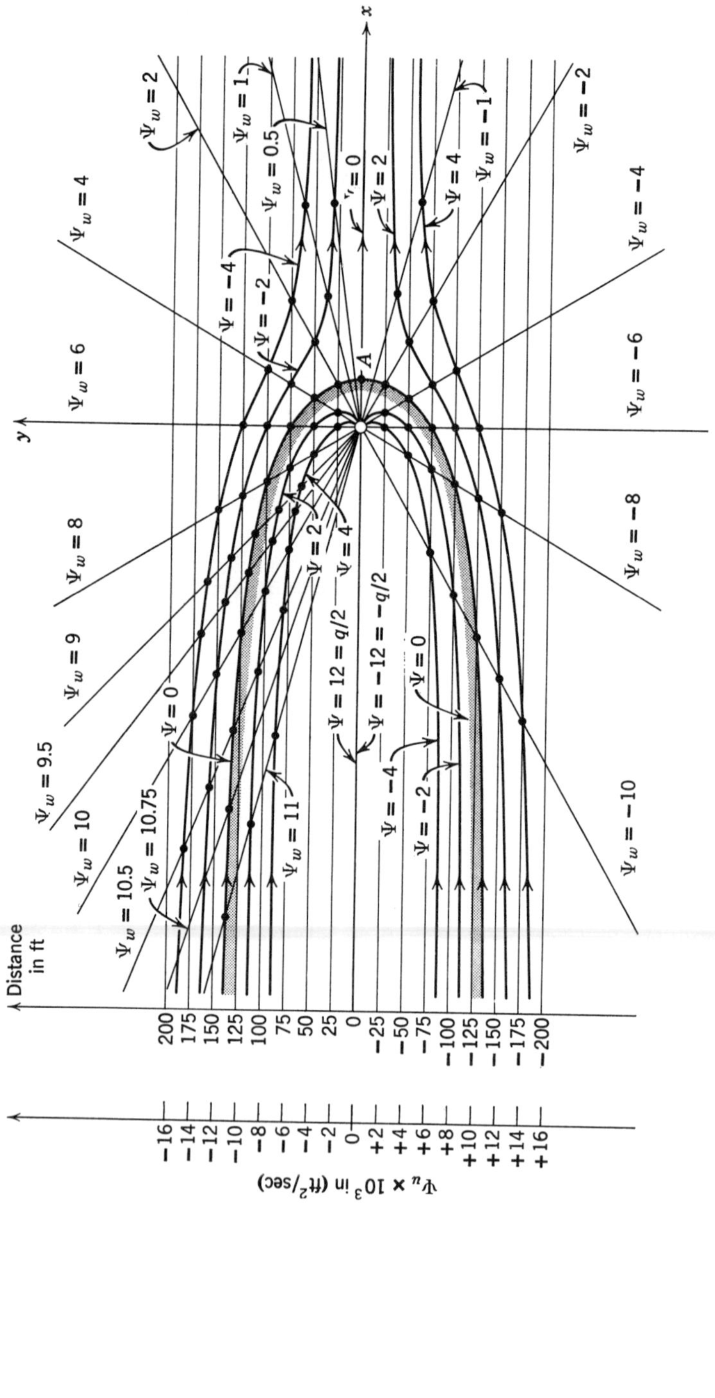

Fig. 6-9 A well in uniform flow. Graphical construction of flow lines. Note: All Ψ'_w and Ψ' are multiplied by 10^3.

i.e., in the present case

$$x_s = \frac{24 \times 10^{-3}\ \text{ft}^2/\text{sec}}{2\pi \times 8 \times 10^{-5}\ \text{ft/sec}} = 47.7\ \text{ft}$$

The ground-water divide defines the boundary of the aquifer region contributing to the well discharge. From Fig. 6.9 it would appear that the width $2y$ of the contributing region infinitely far to the left would approach $2 \times 150 = 300$ ft. This is true from

$$\Psi = \frac{q}{2\pi}\theta - KS^*y \tag{6.30}$$

At the boundary, $\Psi = 0$ and $\theta = \pi$ infinitely far to the left. From Eq. 6.30 it follows that

$$2y = \frac{q}{KS^*} = \frac{24 \times 10^{-3}\ \text{ft}^2/\text{sec}}{8 \times 10^{-5}\ \text{ft/sec}} = 300\ \text{ft}$$

From the foregoing it is evident that the relative strengths of uniform flow and well flow determine the location of the ground-water divide. The ratio of these characteristics is very important in the study of the recirculation between a recharge and discharge well when uniform ground-water flow is present. For each of these ratios there is a critical distance between the well and recharge well above which recirculation between the wells ceases [4].

6.6 *Method of Images*

The right half of Fig. 6.5*a* also represents the solution to the problem of a single well near a stream. The recharge well may be considered as the negative image of the discharging well reflected in the line $x = 0$ which is coincident with the axis of the stream. The system "well-stream" is replaced by the system "well-image well with negative discharge" and both systems observe the same boundary conditions along the line $x = 0$. This artificial device called "method of images" makes it possible to combine solutions of the type given by Eqs. 6.5 and 6.6 in the case of noncircular aquifers. If Eq. 6.6 is applied to both the well and the image well the result is, according to Fig. 6.10*a*,

$$h = \frac{Q}{2\pi Kb}\log_e r_1 + \text{constant}$$

and

$$h = \frac{-Q}{2\pi Kb}\log_e r_2 + \text{constant}$$

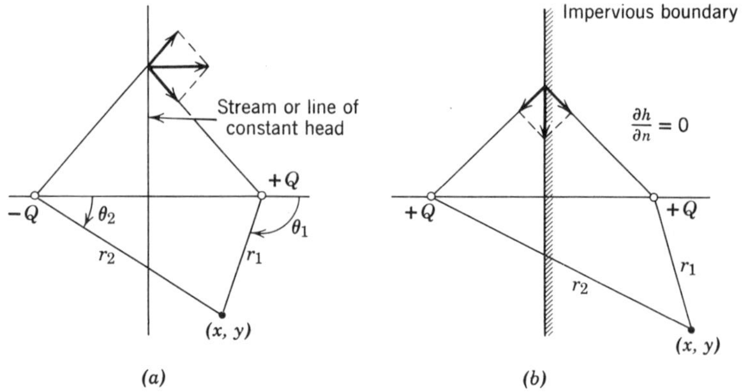

Fig. 6-10 Method of images. (*a*) Well near a stream. (*b*) Well near impervious boundary.

The head for the composite system is

$$h = \frac{Q}{2\pi K b} \log_e \frac{r_1}{r_2} + C$$

in which C again is determined by the boundary condition at a far distance from the wells. For very large r_1 and r_2 (where $\log(r_1/r_2)$ tends to zero), the head h is not affected and preserves its original value H. Hence $C = H$ and the final result, with the distances squared for convenience, becomes

$$H - h = \frac{Q}{4\pi K b} \log_e \frac{r_2^2}{r_1^2} \tag{6.31}$$

This equation may be identified with Eq. 6.17. It is possible to prove that Eq. 6.31 is correct by computing the flowrate through a cylinder of radius r_2 and height b concentric with the image well:

$$\text{Flowrate} = \int_0^{2\pi} bK\left(-\frac{\partial h}{\partial r_2}\right) r_2 \, d\theta_2$$

The result is Q if h in this integral is replaced by its expression from Eq. 6.31. For a well pumped in a semi-infinite aquifer near an impervious boundary, the condition that this boundary is a streamline is observed by putting a positive image well opposite the real well (Fig. 6.10*b*). By superposition,

$$H - h = \frac{Q}{2\pi K b} \log_e \frac{r_e^2}{r_1 r_2}$$

or

$$h - h_w = \frac{Q}{2\pi K b} \log_e \frac{r_1 r_2}{r_w^2} \tag{6.32}$$

if the pair (r_w, h_w) is used instead of (r_e, H).

The method may be extended to aquifers with boundaries intersecting at right angles to form quadrants or even rectangles and strips [5]. For strips and rectangles infinite series of images are used as the insertion of one image to satisfy the conditions along one boundary disturbs the flow conditions along the other boundaries. This requires a correction which is

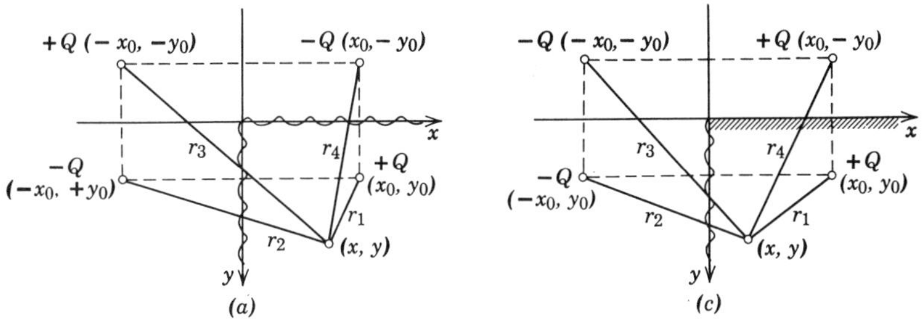

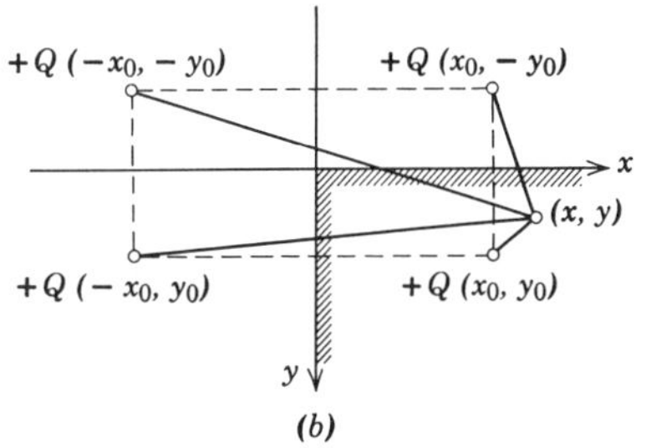

Fig. 6-11 Well in infinite quadrant. (*a*) Two streams intersecting at right angle. (*b*) Two impervious boundaries intersecting at right angles. (*c*) Stream and impervious boundary intersecting.

accomplished by the insertion of new images, and in this way infinite series of image wells arise. The reader is referred to reference 3 for a detailed review of this technique. Figure 6.11 illustrates systems of images with a well in an infinite quadrant:

(a) A well near the intersection of two rectilinear canals.

(b) A well near the intersection of two rectilinear impervious boundaries.

(c) A well near the intersection of a canal and an impervious boundary.

In case a the head h may be found by superposition as

$$H - h = \frac{Q}{4\pi T}\left(\log_e \frac{r_e^{\,2}}{r_1^{\,2}} - \log_e \frac{r_e^{\,2}}{r_2^{\,2}} + \log_e \frac{r_e^{\,2}}{r_3^{\,2}} - \log_e \frac{r_e^{\,2}}{r_4^{\,2}}\right)$$

$$H - h = \frac{Q}{4\pi T} \log_e \frac{r_2^{\,2} r_4^{\,2}}{r_1^{\,2} r_3^{\,2}}$$

or

$$H - h = \frac{Q}{4\pi T} \log_e \frac{[(x + x_0)^2 + (y - y_0)^2][(x - x_0)^2 + (y + y_0)^2]}{[(x - x_0)^2 + (y - y_0)^2][(x + x_0)^2 + (y + y_0)^2]} \tag{6.33}$$

It is left as an exercise for the reader to derive similar expressions for cases b and c.

++

As another example, consider the flow to an eccentric well in a circular acquifer with constant head H along its boundary (Fig. 6.12). Here a negative

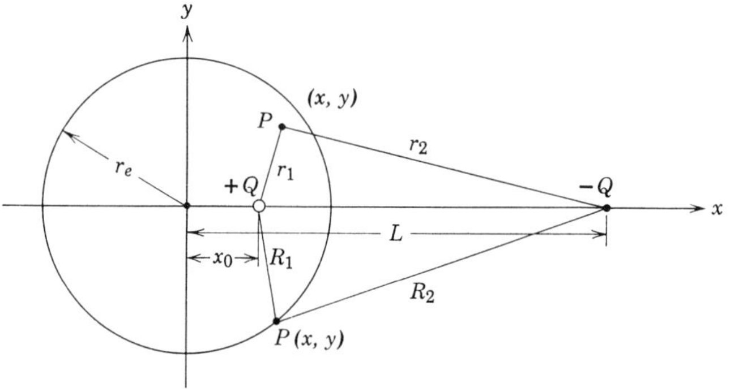

Fig. 6-12 Flow to an eccentric well.

image is put at a distance L from the origin. From the theory of inverses (see reference 6, p. 149) it is known that $L = r_e^2/x_0$, and this results also from the following considerations. Let r_e^* be the unknown radius of influence of the image well, and $r_e^* = \lambda r_e$, where λ remains to be determined. By superposition

$$H - h = \frac{Q}{4\pi T} \log_e \frac{r_e^2}{r_1^2} - \frac{Q}{4\pi T} \log_e \frac{r_e^{*2}}{r_2^2} = \frac{Q}{4\pi T} \log_e \frac{r_2^2}{r_1^2} \frac{1}{\lambda^2}$$

The coordinates of any point P inside the circle with radius r_e of Fig. 6.12 are indicated by (x, y). When a point is on that circle the boundary condition $h = H$ must be satisfied, the coordinates of that point are indicated by (X, Y), and its position vectors becomes $r_1 = R_1$ and $r_2 = R_2$. The boundary condition requires

$$\log_e \frac{R_2^2}{R_1^2} \frac{1}{\lambda^2} = 0, \quad \text{or} \quad R_2^2 = \lambda^2 R_1^2$$

This becomes

$$(L - X)^2 + Y^2 = \lambda^2[(X - x_0)^2 + Y^2]$$

or, after rearrangement,

$$X^2 + Y^2 - 2\frac{(L - x_0\lambda^2)X}{1 - \lambda^2} = \frac{x_0^2\lambda^2 - L^2}{1 - \lambda^2} \tag{6.34}$$

Equation 6.34 is that of a circle fulfilling the requirement that the head h along its circumference equals H. On the other hand the circular boundary of the aquifer has the same property $h = H$. Both circles therefore must be the same, or Eq. 6.34 must become $X^2 + Y^2 = r_e^2$, and this requires

$$L = x_0 \lambda^2$$

$$x_0^2\lambda^2 - L^2 = (1 - \lambda^2)r_e^2$$

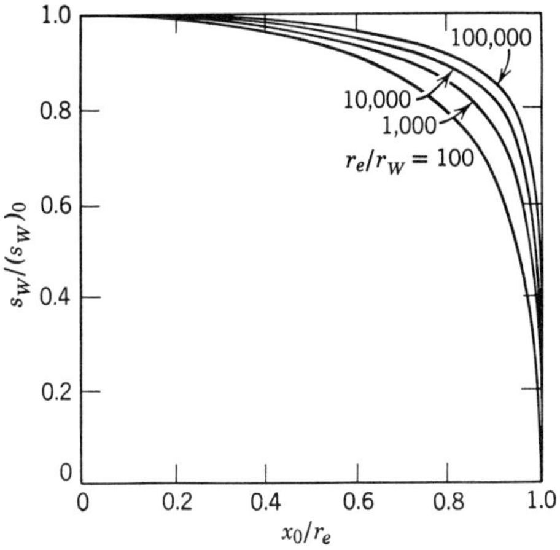

Fig. 6-13 Ratio of drawdowns of eccentric well and concentric well of same capacity versus eccentricity. (After Jacob.)

These two equations in λ, L have the solution

$$L = \frac{r_e^2}{x_0}$$

$$\lambda = \frac{r_e}{x_0}$$

The drawdown $s = H - h$ finally becomes

$$s = \frac{Q}{4\pi T} \log_e \frac{r_2^2}{r_1^2} \frac{x_0^2}{r_e^2} \tag{6.35}$$

The drawdown at the face of an eccentric well may be compared to the drawdown of a concentric well for the same aquifer and discharge. The result is given in Fig. 6.13 (see reference 2, p. 357). It follows from the fact that the drawdown $(s_w)_0$ at the concentric well may be written from Eq. 6.5 as

$$(s_w)_0 = \frac{Q}{2\pi T} \log_e \frac{r_e}{r_w}$$

Equation 6.35 may be transformed by multiplying the term under the logarithm by $(r_e/r_w)(r_w/r_e)$ and remarking that at the well $r_1 = r_w$ and $r_2 \approx L - x_0$.

Therefore,

$$s_w = \frac{Q}{2\pi T}\log_e \frac{(L - x_0)}{r_w}\cdot\frac{x_0}{r_e}\cdot\frac{r_e}{r_w}\cdot\frac{r_w}{r_e}$$

$$= \frac{Q}{2\pi T}\left[\log_e \frac{r_e}{r_w} + \log_e \frac{x_0(L - x_0)}{r_e^2}\right] = \frac{Q}{2\pi T}\left[\log_e \left(1 - \frac{x_0^2}{r_e^2}\right) + \log_e \frac{r_e}{r_w}\right]$$

Hence

$$\frac{s_w}{(s_w)_0} = \frac{\log_e \dfrac{r_e}{r_w} + \log_e \left(1 - \dfrac{x_0^2}{r_e^2}\right)}{\log_e \dfrac{r_e}{r_w}} \tag{6.36}$$

Polubarinova-Kochina [6] computes the ratio of the discharges of an eccentric and a concentric well for the same drawdown as a function of the eccentricity. She finds

$$\frac{Q_e}{Q_c} = \frac{\log_e \dfrac{r_e}{r_w}}{\log \dfrac{r_e}{r_w} + \log_e \left(1 - \dfrac{x_0^2}{r_e^2}\right)}$$

This result, of course, agrees with Eq. 6.36. The method of images is not limited to confined flow. As an example, compute the discharge for a well at a distance x_0 from a canal cutting completely through a water table aquifer. The answer is

$$Q = \frac{(H^2 - h_w^2)\pi K}{\log_e \dfrac{2x_0}{r_w}}$$

◆ Unsteady Flow [44]

6.7 *Radial Flow to a Well in an Extensive Confined Aquifer*

This problem was investigated by C. V. Theis [7], who made an analogy with the conductive flow of heat to a sink in a plate. The solution given is that for an aquifer of infinite extent having no lateral inflow from surrounding water bodies. The entire discharge must be provided by the release of stored water. According to the definition of storage coefficient (see section 4.11), the discharge must equal the product of the storage coefficient and the rate of decline in head integrated over the area affected by pumping. According to Fig. 6.14,

$$dQ = -Sr\,d\theta\,dr\,\frac{\partial h}{\partial t}$$

and in general

$$Q(t) = -S\int_{r_w}^{\infty}\int_0^{2\pi} r\,\frac{\partial h(r, \theta, t)}{\partial t}\,d\theta\,dr$$

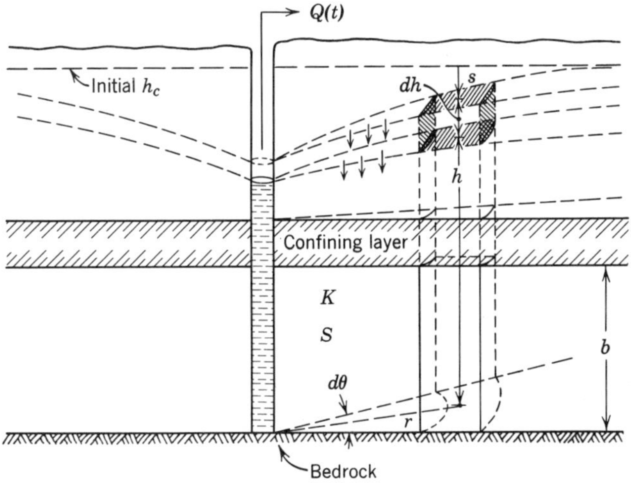

Fig. 6-14 Unsteady flow in an extensive aquifer.

For an aquifer of infinite extent tapped by a single well, the distribution of head is radially symmetric and therefore

$$Q = -2\pi S \int_{r_w}^{\infty} r \frac{\partial h(r, t)}{\partial t} \, dr \tag{6.37}$$

From Eq. 6.37 it becomes evident that the rate of decline of head must decrease continuously as the area affected by pumping spreads out (as time goes on) in order to make the integral a constant. This is required if the well is pumped at a constant flow rate Q and if it is assumed that S remains constant as a first approximation, during pumping. Consequently, there can be no steady flow in a confined aquifer of infinite extent.

Theis replaced the well by a mathematical sink of constant strength and solved Eq. 4.51, which may be written as

$$\frac{\partial^2 h}{\partial r^2} + \frac{1}{r}\frac{\partial h}{\partial r} = \frac{S}{T}\frac{\partial h}{\partial t} \tag{6.38}$$

for the boundary conditions

$$\left.\begin{aligned} &h \to h_0 \quad \text{as} \quad r \to \infty \\ &\lim_{r \to 0}\left(r \frac{\partial h}{\partial r}\right) = \frac{Q}{2\pi T} \end{aligned}\right\} \quad \text{for } t > 0$$

and the initial condition

$$h(r, 0) = h_0 \qquad \text{for } t \leqslant 0$$

The initial condition means that the head in the aquifer was uniform up to the time pumping started. The solution is

$$h = h_0 - \frac{Q}{4\pi T}\int_{r^2S/4Tt}^{\infty} \frac{e^{-x}\,dx}{x} \tag{6.39}$$

The foregoing integral is a function of the lower limit

$$u = \frac{r^2 S}{4Tt} \tag{6.40}$$

and is tabulated as the exponential integral [8] under the symbol $-Ei(-u)$ [see Table B.1, Appendix B]. It can be verified that Eq. 6.39 is a solution to Eq. 6.38 under the foregoing conditions by substitution of the expression for h in the differential equation. The exponential integral can be expanded in a convergent series so that the drawdown $s = h_0 - h$ from Eq. 6.39 may be written as

$$\begin{aligned} s = h_0 - h &= \frac{Q}{4\pi T}[-Ei(-u)] \\ &= \frac{Q}{4\pi T}\left[-0.5772 - \log_e u + u - \frac{u^2}{2 \cdot 2!} + \frac{u^3}{3 \cdot 3!} - \frac{u^4}{4 \cdot 4!} + \cdots\right] \end{aligned} \tag{6.41}$$

The drawdown at the well face is found from Eq. 6.41 for $r = r_w$:

$$s_w = \frac{Q}{4\pi T}\left[-Ei\left(-\frac{r_w^2 S}{4Tt}\right)\right] \tag{6.42}$$

If Q is not constant, s_w may be computed by summing the increments of drawdown produced by increments of Q.

Equation 6.39 is known as the nonequilibrium equation or Theis equation. Although its derivation has been based on several assumptions which are seldom justified in field tests, the nonequilibrium formula has been applied with reasonable success in determining the formation constants T and S of an aquifer. Among the limiting assumptions, complete penetration and the infinite areal extent of the aquifer should be emphasized. Other conditions require a homogeneous and isotropic medium, an infinitesimally small well diameter, and an instantaneous removal of water with decline in head.

Wenzel [9] gave a tabulation of the exponential integral written symbolically as $W(u)$, or well function of u, for values of u from 10^{-15} to 9.9. A condensed version of his table is given in Table 6.1. This table is helpful in the application of a graphical method devised by Theis [7] to compute the formation constants T and S from pumping test data.

THEIS' METHOD

With $-Ei(-u) = W(u)$ and for American practical hydrology units, Eq. 6.41 may be written as

$$s = \frac{114.6Q}{T} W(u) \tag{6.43}$$

$$u = \frac{1.87S}{T} \frac{r^2}{t} \tag{6.44}$$

where s = drawdown, in ft, measured in observation well due to constant discharge of pumped well
Q = discharge of pumped well, in gal/min
T = transmissivity, in gal/day and /ft
r = distance, in ft, from pumped well to observation well
S = coefficient of storage, dimensionless
t = time in days since pumping started

From Eqs. 6.43 and 6.44 it follows that

$$\log s = \log W(u) + \log \frac{114.6Q}{T} \tag{6.45}$$

and

$$\log \frac{r^2}{t} = \log u + \log \frac{T}{1.87S} \tag{6.46}$$

This rearrangement is useful in separating the constants $114.6Q/T$ and $T/1.87S$ from the variables. The logarithm is taken for convenient plotting of the variables which have a wide range of values. Theis used the following graphical method of superposition. First a plot of $W(u)$ versus u is made on logarithmic paper (Fig. 6.15). This plot is known as the type curve and may be prepared by means of Table 6.1.

Next, values of the drawdown s are plotted against values of r^2/t on logarithmic paper of the same size as was used for the type curve. If the discharge Q is constant, Eqs. 6.45 and 6.46 show that $W(u)$ is a function of u in the same way that s is a function of r^2/t. Therefore it will be possible to superimpose the data curve on the type curve, holding the coorindate axes of the two curves parallel in such a way that the data best fit the type curve. A common point, the match point, arbitrarily chosen on the overlapping part of the curves or even anywhere on the overlapping portion of the sheets, determines mutual values of s, $W(u)$, r^2/t, and u which may be inserted in Eqs. 6.43 and 6.44 so that these equations may be solved for S and T. Once the values of S and T are determined, it is possible to make a check on the trial by computing the values of $T/1.87S$ and $114.6Q/T$. According to Eqs. 6.43 and 6.44, these values must correspond to r^2/t and s respectively for $u = 1$ and $W(u) = 1$. This uniquely determines the translation of the curves as is illustrated in Fig. 6.15.

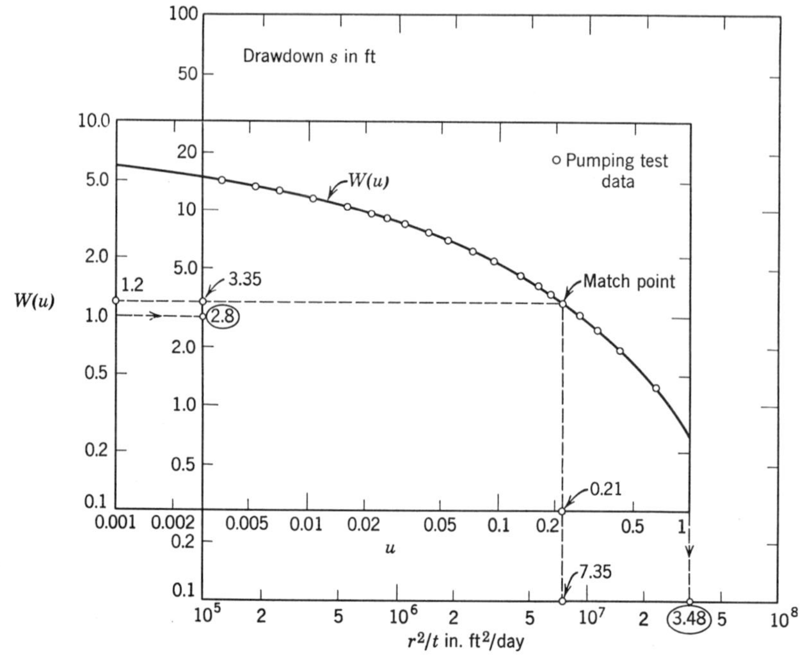

Fig. 6-15 Theis' graphical method of superposition. Nonequilibrium equation.

NUMERICAL EXAMPLE

A completely penetrating well is pumped at a constant rate of 500 gpm. Drawdowns during the pumping period are measured in an observation well 150 ft from the pumped well, at times varying from 2 min to 6 hr.

Table 6.1 Values of $W(u)$ for Values of u

(After Wenzel)

u	1.0	2.0	3.0	4.0	5.0	6.0	7.0	8.0	9.0
	0.219	0.049	0.013	0.0038	0.0011	0.00036	0.00012	0.000038	0.00
$\times 10^{-1}$	1.82	1.22	0.91	0.70	0.56	0.45	0.37	0.31	0.26
$\times 10^{-2}$	4.04	3.35	2.96	2.68	2.47	2.30	2.15	2.03	1.92
$\times 10^{-3}$	6.33	5.64	5.23	4.95	4.73	4.54	4.39	4.26	4.14
$\times 10^{-4}$	8.63	7.94	7.53	7.25	7.02	6.84	6.69	6.55	6.44
$\times 10^{-5}$	10.94	10.24	9.84	9.55	9.33	9.14	8.99	8.86	8.74
$\times 10^{-6}$	13.24	12.55	12.14	11.85	11.63	11.45	11.29	11.16	11.04
$\times 10^{-7}$	15.54	14.85	14.44	14.15	13.93	13.75	13.60	13.46	13.34
$\times 10^{-8}$	17.84	17.15	16.74	16.46	16.23	16.05	15.90	15.76	15.65
$\times 10^{-9}$	20.15	19.45	19.05	18.76	18.54	18.35	18.20	18.07	17.95
$\times 10^{-10}$	22.45	21.76	21.35	21.06	20.84	20.66	20.50	20.37	20.25
$\times 10^{-11}$	24.75	24.06	23.65	23.36	23.14	22.96	22.81	22.67	22.55
$\times 10^{-12}$	27.05	26.36	25.96	25.67	25.44	25.26	25.11	24.97	24.86
$\times 10^{-13}$	29.36	28.66	28.26	27.97	27.75	27.56	27.41	27.28	27.16
$\times 10^{-14}$	31.66	30.97	30.56	30.27	30.05	29.87	29.71	29.58	29.46
$\times 10^{-15}$	33.96	33.27	32.86	32.58	32.35	32.17	32.02	31.88	31.76

They have been recorded in Table 6.2. The type curve is first drawn and then the data curve is superimposed on it. The match point gives values of $u = 0.21$, $r^2/t = 7.35 \times 10^6$ ft²/day, $s = 3.35$ ft and $W(u) = 1.2$. From Eq. 6.43, we find

$$T = \frac{114.6 \times 500 \times 1.2}{3.35} = 20{,}500 \text{ gal/day ft}$$

and from Eq. 6.44

$$S = \frac{0.21 \times 20{,}500}{1.87 \times 7.35 \times 10^6} = 0.000315$$

The same results would have been obtained if the choice had been the arbitrary point $s = 1$ ft, $r^2/t = 10^5$ ft²/day for which $u = 2.9 \times 10^{-3}$ and $W(u) = 0.355$.

Table 6.2 Aquifer Test Data: Observation Well $r = 175$ ft

Time Elapsed since Beginning of Pumping, t		Drawndown s in Observation Well	r^2/t
Minutes	Days	Feet	Feet²/Day
0	0	0	∞
2	1.39×10^{-3}	1.2	2.20×10^7
3	2.09×10^{-3}	1.9	1.47×10^7
4	2.78×10^{-3}	2.45	1.10×10^7
5	3.48×10^{-3}	2.90	8.80×10^6
6	4.17×10^{-3}	3.35	7.35×10^6
7	4.86×10^{-3}	3.65	6.31×10^6
8	5.57×10^{-3}	4.10	5.51×10^6
10	6.96×10^{-3}	4.60	4.4×10^6
14	9.72×10^{-3}	5.50	3.16×10^6
18	1.25×10^{-2}	6.15	2.45×10^6
24	1.67×10^{-2}	7.00	1.84×10^6
30	2.09×10^{-2}	7.75	1.47×10^6
40	2.78×10^{-2}	8.50	1.10×10^6
50	3.48×10^{-2}	9.00	8.8×10^5
60	4.17×10^{-2}	9.50	7.36×10^5
80	5.57×10^{-2}	10.05	5.50×10^5
120	8.33×10^{-2}	10.30	3.68×10^5
180	1.25×10^{-1}	10.50	2.45×10^5
240	1.67×10^{-1}	10.65	1.83×10^5
360	2.50×10^{-1}	10.80	1.22×10^5

To check the translation of the graphs, the values of

$$\frac{T}{1.87S} = \frac{20{,}500}{1.87 \times 0.000315} = 3.48 \times 10^7 \text{ ft}^2/\text{day}$$

and
$$\frac{114.6Q}{T} = \frac{114.6 \times 500}{20{,}500} = 2.80 \text{ ft}$$

are computed. From Fig. 6.15 it is evident that these respective values for r^2/t and s are the opposites of the values $u = 1$ and $W(u) = 1$.

Figure 6.15 indicates that the test has been run under almost ideal conditions. Sometimes there are boundary effects, either because of recharge from a nearby stream or because of the presence of impervious boundaries [10]. In the case of a recharging boundary, the drawdowns would be smaller than in the absence of such a boundary and the data would fall below the type curve with more pronounced deviations as time goes on after pumping has been started. Finally, Fig. 6.15 could have rendered s as a function of $1/t$ because only one observation well was used.

JACOB'S METHOD

For small values of u (i.e., for small r and/or large t) compared to $\log_e u$, series 6.41 may be terminated after the first two terms. Considering that $0.5772 = \log_e 1.78$, Eq. 6.41 becomes

$$s = \frac{Q}{4\pi T}\left(\log_e \frac{1}{u} - \log_e 1.78\right)$$

or in decimal logarithms

$$s = \frac{2.30Q}{4\pi T} \log \frac{2.25Tt}{r^2S} \qquad (6.47)$$

Jacob [2, 11] first introduced this simplification, which should be restricted to values of u less than about 0.01. The data of Table 6.2 therefore would not be very suitable for the approximation. Equation 6.47 may be applied 1° to observations in one well at different times, 2° to observations in different wells at same time, and 3° to observations in various wells at various times.

EXAMPLES

In the first case, where observations are made in a single well, only t varies in Eq. 6.47, which may be rewritten as

$$s = \frac{2.30Q}{4\pi T} \log \frac{2.25T}{r^2S} + \frac{2.30Q}{4\pi T} \log t \qquad (6.48)$$

This is the equation of a straight line on semilogarithmic paper s versus $\log t$. The slope of the straight line is equal to $2.30Q/4\pi T$ and is found as the vertical projection of the intercept of the straight line between two numbers on the time scale that have logarithms one unit apart, say 1,000 and 100 on Fig. 6.16. Thus T may be found from the slope. S may be found from any point on the straight line and Eq. 6.47; T is known. Also,

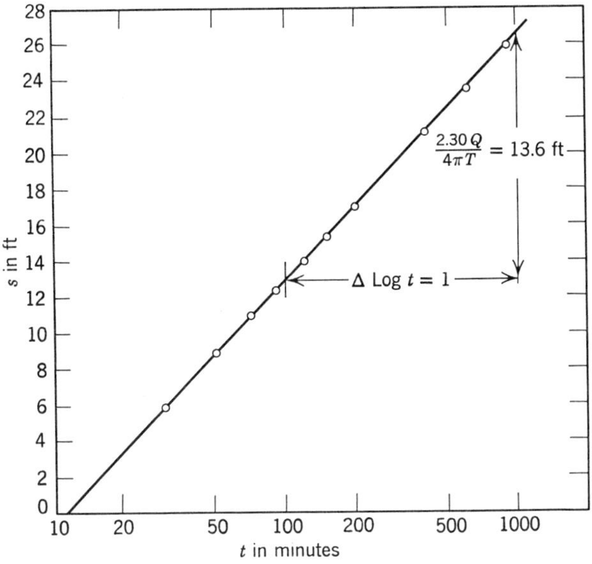

Fig. 6-16 Jacob method for solution of nonequilibrium equation. One observation well.

it may be determined from the intercept t_0 of the straight line on the $\log t$ axis. In Eq. 6.47 s becomes zero for

$$\frac{2.25Tt_0}{r^2S} = 1 \tag{6.49}$$

NUMERICAL EXAMPLE

Tabulated below are data on an observation well 50 ft from a well which is pumped for a test at 250 gpm. Find the transmissivity T and storage coefficient S for the aquifer.

Time, min	30	50	70	90	120	150	200	400	600	900
Drawdown, ft	6.5	9.0	11.0	12.4	14.1	15.4	17.0	21.2	23.6	26.0

From the slope of the plot on Fig. 6.16

$$\frac{2.30Q}{4\pi T} = 13.6 \text{ ft}$$

$$\text{or} \quad T = \frac{2.30}{\pi} \times \frac{250\ ft^3}{449\ \text{sec}} \times \frac{1}{13.6\ \text{ft}} = 7.48 \times 10^{-3}\,\frac{\text{ft}^2}{\text{sec}} = 4{,}860\ \text{gpd/ft}$$

From Eq. 6.49

$$S = \frac{2.25 \times 7.48 \times 10^{-3}\ \text{ft}^2/\text{sec} \times 672\ \text{sec}}{2{,}500\ \text{ft}^2} = 0.00045$$

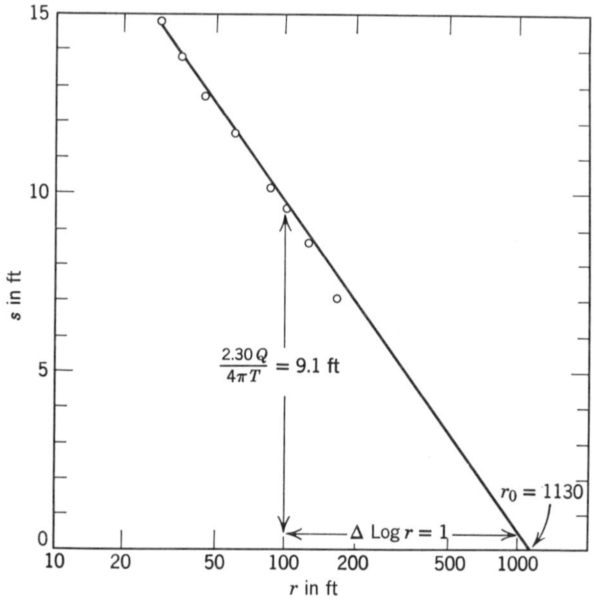

Fig. 6-17 Jacob method for solution of nonequilibrium equation. Simultaneous observations.

Once the values of T and S are computed, the assumption that u is sufficiently small for all data is checked. Here it suffices to examine the point for $t = 30$ min.

In the second case, only r varies in Eq. 6.47, which may be rewritten as

$$s = \frac{2.30Q}{4\pi T} \log \frac{2.25Tt}{S} - \frac{2.30Q}{2\pi T} \log r \tag{6.50}$$

This is the equation of a straight line on semilogarithmic paper, s versus $\log r$. The slope of the straight line is equal to $(-2.30Q/2\pi T)$, and is found in nearly the same way as before (see Fig. 6.17). T may be found from the slope of the straight line. S may be found as it was before. Here it may be determined from the intercept r_0 of the straight line on the log r axis. For $s = 0$, from Eq. 6.47

$$\frac{2.25Tt}{r_0^2 S} = 1 \tag{6.51}$$

NUMERICAL EXAMPLE

Tabulated below are data of a 2-hour aquifer test at 350 gpm. The drawdowns are measured in a number of nearby observation wells. Determine T and S:

Well	1	2	3	4	5	6	7	8
Distance, ft	29	35	44	60	85	100	125	163
Drawdown, ft	14.9	13.8	12.7	11.7	10.1	9.6	8.6	7.0

From the slope of the plot on Fig. 6.17,

$$\frac{2.30Q}{2\pi T} = 9.1 \text{ ft}$$

or

$$T = \frac{2.30}{2\pi} \times \frac{350}{449} \text{ ft}^3/\text{sec} \times \frac{1}{9.1 \text{ ft}} = 3.14 \times 10^{-2} \text{ ft}^2/\text{sec} = 20{,}400 \text{ gpd/ft}$$

From Eq. 6.51,

$$S = \frac{2.25 \times 3.14 \times 10^{-2} \text{ ft}^2/\text{sec} \times 2 \times 3600 \text{ sec}}{(1030)^2 \text{ ft}^2} = 0.00048$$

With the computed values of S and T, the values of u are sufficiently small for wells 1 to 7.

Finally, as shown by Jacob [2], a composite drawdown graph may be obtained by having values of drawdown measured at various times in several wells plotted against the logarithms of the respective values of t/r^2. Equation 6.47 should now be rewritten as

$$s = \frac{2.30Q}{4\pi T} \log \frac{2.25T}{S} + \frac{2.30Q}{4\pi T} \log \frac{t}{r^2}$$

On semilogarithmic paper this equation is plotted as a straight line from which slope T can be determined. S is determined as before.

THEIS' RECOVERY METHOD

Theis [7] and Wenzel [9] pointed out that the nonequilibrium formula could be used to determine the formation constants from the analysis of the recovery of a shutdown well. If a well is pumped at a constant rate and then shut down, the head will recover from its lowest value $h_{T'}$ at time T'

when pumping stopped to attain a value $h' > h_{T'}$ at times t' counted from the time of shutdown ($t' = 0$ corresponds to $t = T'$; Fig. 6.18). If H is the initial value of the head before pumping started, then $H - h' = s'$ is called the residual drawdown. For the computation of the residual drawdown it is assumed that the discharge goes on at the same Q for $t > T'$ but that at $t = T'$ a recharge well of strength $-Q$ is superimposed on the discharging well so that the net discharge is zero from $t = T'$ on. Therefore,

$$s' = (H - h) + (h - h') = \frac{Q}{4\pi T}\left[\int_{\frac{r^2S}{4Tt}}^{\infty} \frac{e^{-x}}{x}\,dx - \int_{\frac{r^2S}{4Tt'}}^{\infty} \frac{e^{-x}}{x}\,dx\right] \qquad (6.52)$$

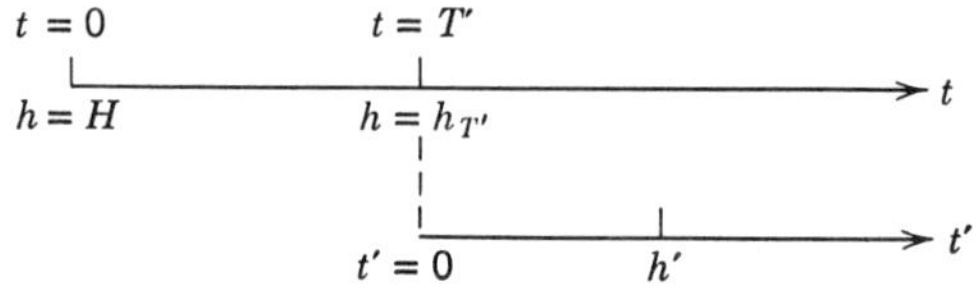

Fig. 6-18 Time axes for Theis' Recovery method.

For $r^2S/4Tt'$ sufficiently small, the approximation made in the derivation of Eq. 6.47 is also valid here, and therefore

$$s' = \frac{2.30Q}{4\pi T}\left[\log \frac{2.25Tt}{r^2S} - \log \frac{2.25Tt'}{r^2S}\right] = \frac{2.30Q}{4\pi T} \log \frac{t}{t'} \qquad (6.53)$$

This equation is represented by a straight line on semi-logarithmic paper, and its slope allows for the determination of T if measured residual drawdowns are plotted against $\log (t/t')$. S may be determined from the value of the drawdown at the time of shutdown.

6.8 *Boundary Effects on Unsteady Well Flow*

METHOD OF IMAGES

The equations derived in section 6.7 of this chapter were all valid in the case of aquifers of infinite extent. Such idealized aquifers are not the rule and it should be possible to account for the presence of nearby rivers and impervious boundaries. The principles explained in section 6.6 are also applicable here. If, for example, nonsteady flow takes place to a well near a river, positioned as in Fig. 6.10*a*, an image well with strength $(-Q)$ is placed across the river and the drawdown is

$$s = \frac{Q}{4\pi T}\left[-Ei\left(-\frac{r_1{}^2S}{4Tt}\right)\right] - \frac{Q}{4\pi T}\left[-Ei\left(-\frac{r_2{}^2S}{4Tt}\right)\right] \qquad (6.54)$$

Unsteady flow toward a well is expected in the early stages after pumping has just started and before the flow pattern is fully established. To have an idea of when the flow is established, a comparison is made between u and $\log_e u$ in both terms of Eq. 6.54. When these terms may be replaced by their logarithmic approximations, time drops out of the equation and the flow becomes steady. Indeed at that time

$$s = \frac{Q}{4\pi T}\left[\log_e \frac{2.25Tt}{r_1^2 S} - \log_e \frac{2.25Tt}{r_2^2 S}\right] = \frac{Q}{4\pi T}\log_e \frac{r_2^2}{r_1^2}$$

and this is exactly the steady state of Eq. 6.31. This result is significant and shows that the presence of a recharge boundary in the aquifer tends to offset the unsteady character of the flow. The method of unsteady images may further be applied to other configurations analogous to those of section 6.6.

◆ Leaky Aquifers [43]

Aquifers in nature are not always perfectly confined between completely impervious strata. This became more evident when the recharge conditions were studied for aquifers that apparently had larger yields than were available from the replensihment in the outcrop area. It was found that those aquifers were overlain by poorly pervious yet water-transmitting strata which, over large contact areas, contributed significantly to the recharge of the aquifer. The semiconfining beds in turn were overlain by ponded water, as in the case of the Dutch polder areas, or by other more pervious strata which again were confined or semiconfined. Sometimes water from the main aquifer seeped through underlying semipervious beds and then again, for other head conditions in adjacent water bodies, the aquifer was recharged from below. The percolation of water into and away from the main aquifer through the semiconfining strata is called leakage. The phenomenon was subjected to mathematical analysis first by Dutch hydrologists and engineers [12, 13] and later in this country by Jacob [14] and Hantush [15]. From 1954 to 1964, first in collaboration with Jacob [16, 17, 18, 19] and later by himself, Hantush [20, 21, 22, 23, 24, 25, 26, 27, 28] developed the theory of the leaky aquifer. The first important Russian contribution on this subject seems to come from Mjatiev [29], who analyzed the interaction of pervious strata separated by semiconfining beds [30]. Although this research has been highly analytical, there is still room for relaxation of some of the boundary conditions considered in the problems or for extension to more practical cases [De Wiest 31, 32]. Walton [33] has applied the theory of the leaky aquifer to some regional conditions in Illinois. Finally, it is worthwhile to mention that the state of semiconfinement may be brought about on a large scale by the presence of lenticular zones in which clays predominate, as in many coastal plain deposits, rather than by a single well-defined layer as pointed out by Jacob [14] and De Wiest [41].

6.9 *Modification of Laplace's Equation, Taking Leakage into Account*

Consider Fig. 6.19 where for reasons of simplicity only leakage from above takes place. The semiconfining stratum is overlain by a sand and capacity for lateral inflow is sufficient to maintain essentially constant head

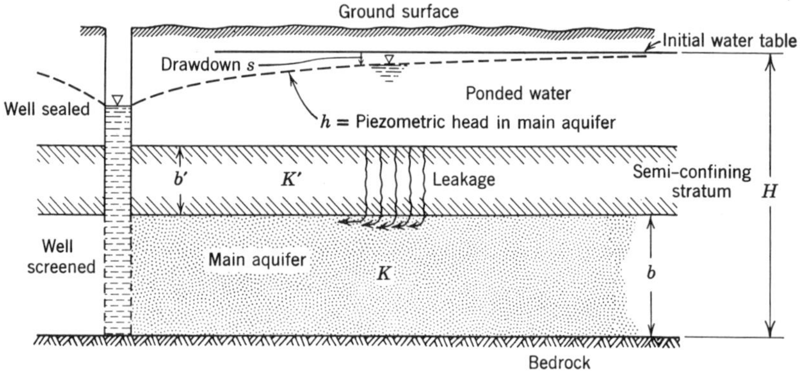

Fig. 6-19 Leakage into aquifer.

in spite of the downward leakage into the main aquifer [14]. The x-y plane coincides with the completely impervious bedrock and is also the datum plane for the head. Before water was withdrawn from the main aquifer the head h in the main aquifer was uniform and equal to the height H of the water table above bedrock. After water is withdrawn from the main aquifer the head h is represented by a curved line as indicated on Fig. 6.19, and because the ponded water remains under head H a head difference is established between top and bottom of the semiconfining stratum which will induce leakage through this stratum. The hydraulic conductivity K of the main aquifer is so large compared to that of the semiconfining stratum K' that it is safe to assume that water seeps vertically through the semiconfining layer and is refracted over 90° to proceed horizontally in the main aquifer, according to the principles of section 5.11. For steady state flow and water, continuity equation 4.44 becomes

$$\frac{\partial u}{\partial x} + \frac{\partial v}{\partial y} + \frac{\partial w}{\partial z} = 0 \tag{6.55}$$

The horizontal velocity components u and v vary according to their distance above bedrock, but if these changes are minor, average values of u and v, say $\bar{u}$ and $\bar{v}$, may be introduced to make the problem hydraulic [30], i.e.:

$$\bar{u}(x, y) = \frac{1}{b}\int_0^b u(x, y, z)\,dz$$
$$\bar{v}(x, y) = \frac{1}{b}\int_0^b v(x, y, z)\,dz \tag{6.56}$$

The adjective "hydraulic" is used to indicate the averaging process in the z-direction, which eliminates the independent variable z. Velocity components that are function of three independent variables x, y, z, are called hydrodynamic. In this simplified framework Eq. 6.55 may be easily integrated along the z-axis over the height of the aquifer. We have

$$\int_0^b \frac{\partial u}{\partial x}\,dz = \frac{\partial}{\partial x}\int_0^b u\,dz = b\frac{\partial \bar{u}}{\partial x}$$
$$\int_0^b \frac{\partial v}{\partial y}\,dz = \frac{\partial}{\partial y}\int_0^b v\,dz = b\frac{\partial \bar{v}}{\partial y} \tag{6.57}$$

and also

$$\int_0^b \frac{\partial w}{\partial z}\,dz = w(b) - w(0) \tag{6.58}$$

where $w(b)$ is the value of the vertical velocity at the top of the main aquifer and $w(0)$ that value at the bottom of the aquifer. In the present case, $w(0) = 0$ because the foundation of the main aquifer is completely impervious by assumption. To find $w(b)$ the head h is averaged over the thickness of the main aquifer so that

$$\bar{h}(x, y) = \frac{1}{b}\int_0^b h(x, y, z)\,dz \tag{6.59}$$

The head H of the ponded water is essentially constant by assumption so that $w(b)$ may be expressed by means of Darcy's law:

$$w(b) = +\frac{K'}{b'}(\bar{h} - H) \tag{6.60}$$

in which the plus sign accounts for the fact that $w(b)$ is measured against the z-direction. The integration of Eq. 1.55 gives

$$b\left(\frac{\partial \bar{u}}{\partial x} + \frac{\partial \bar{v}}{\partial y}\right) + w(b) = 0 \tag{6.61}$$

Now let

$$\bar{u} = -K\frac{\partial \bar{h}}{\partial x}$$
$$\bar{v} = -K\frac{\partial \bar{h}}{\partial y} \tag{6.62}$$

in analogy with Eq. 4.24 of Chapter 4, and insert the value of $w(b)$ from Eq. 6.60, to rewrite Eq. 6.61 in final form:

$$-Kb\left(\frac{\partial^2 \bar{h}}{\partial x^2} + \frac{\partial^2 \bar{h}}{\partial y^2}\right) + \frac{K'}{b'}(\bar{h} - H) = 0$$

or

$$\nabla^2 \bar{h} - \frac{K'}{Kbb'}(\bar{h} - H) = 0 \tag{6.63}$$

This equation, when compared with Eq. 5.1, shows that h is replaced by its average value $\bar{h}$ and that there is an additional linear term in $\bar{h}$ (linear leakage). In the rest of this derivation h will be written for $\bar{h}$ with the understanding that it represents average values. Let $s = H - h$ as usual be the drawdown; then $\nabla^2 h = -\nabla^2 s$ and Eq. 6.63 becomes

$$\nabla^2 s - \frac{s}{B^2} = 0 \tag{6.64}$$

in which

$$B = \sqrt{Kbb'/K'} \tag{6.65}$$

is the leakage factor [16].

By analogy with Eq. 4.51, Eq. 6.63 may be extended to include unsteady flow as

$$\nabla^2 \bar{h} - \frac{K'}{Kbb'}(\bar{h} - H) = \frac{S}{T}\frac{\partial \bar{h}}{\partial t}$$

or with the above conventions

$$\nabla^2 s - \frac{s}{B^2} = \frac{S}{T}\frac{\partial s}{\partial t} \tag{6.66}$$

The leakage factor B has been introduced only because of convenience in the computations. In fact a large leakage factor means that little leakage takes place and vice versa. To characterize the amount of leakage, values of another coefficient, the leakance or leakage coefficient K'/b', are usually given numerically. This coefficient may be defined as the quantity of water that flows across a unit area of the boundary between the main aquifer and its semiconfining bed, if the difference between the head in the main aquifer and that of the ponded water supplying leakage is unity. Values of the leakage coefficient for the Roswell artesian basin in New Mexico as given by Hantush [20] vary from 4.8×10^{-8} sec^{-1} to 10^{-10} sec^{-1}, as compared to values of $3.5 \times 10^{-7}\ \text{sec}^{-1}$ to $8 \times 10^{-9}\ \text{sec}^{-1}$ reported by Walton [33] for glacial drift deposits in the southern half of Illinois.

6.10 *Determination of the Formation Constants of Leaky Aquifers* [42]

Considering that leakage may account for a significant percentage of water pumped from deep wells (up to 11% for the Maquoketa formation in northeastern Illinois, or about 8,400,000 gpd [33]), it seems hardly necessary to justify the attention paid in this chapter to the methods available to determine leakance and formation constants. Two of these methods as described hereafter are completely analogous to those for nonleaky conditions and their application has become a routine operation. The third method, devised by Hantush [20], requires the proof of some properties of the unsteady drawdown curve given in Appendix A.

(A) TYPE CURVE METHOD FOR NONSTEADY-STATE TIME-DRAWDOWN

Hantush and Jacob [17] gave a solution to Eq. 6.66 involving a uniform aquifer of infinite extent completely penetrated by a well of infinitesimal diameter, under the assumptions made in the derivation of Eq. 6.66. It should be noted that these assumptions are the same as those made by Theis (section 6.7) supplemented by those of linear leakage, constant head of the ponded water supplying the leakage, and horizontal refraction of the leakage.

This solution is

$$s = \frac{Q}{4\pi T} W\left(u, \frac{r}{b}\right) \tag{6.67}$$

in which

$$W\left(u, \frac{r}{b}\right) = \int_u^\infty \frac{1}{x} e^{-x-\frac{r^2}{4B^2 x}}\, dx \tag{6.68}$$

is the well function for leaky artesian aquifers and in which

$$u = \frac{r^2 S}{4Tt} \tag{6.40}$$

All symbols are as defined before. For $B \to \infty$, i.e., when leakage $\to 0$, Eq. 6.68 reduces to Eq. 6.39, implying that a graphical superposition method similar to that devised by Theis would be possible if $W(u, r/B)$ were tabulated. Hantush [20] suggested the following method (Table B.2, Appendix B) in which a number of observations are made in one well. Let t be expressed in min instead of in days, Q in gal/min, T in gal/day/ft,

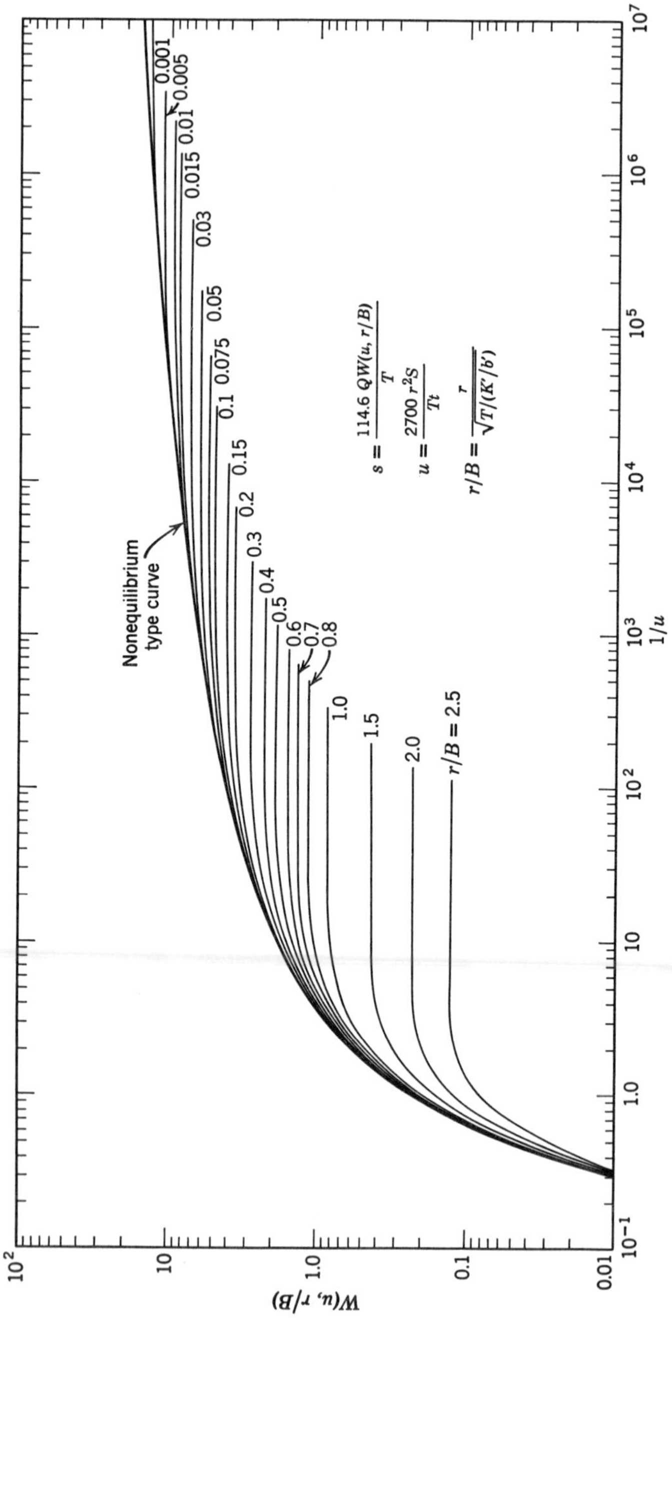

Fig. 6-20 Nonsteady-state leaky artesian type curves. (After Walton.)

s in ft, and r be the distance from the pumped well to the observation well, in ft. The formulas used in the method become

$$s = \frac{114.6Q}{T} W\left(u, \frac{r}{B}\right) \tag{6.69}$$

$$u = \frac{2700r^2S}{Tt} \tag{6.70}$$

$$\frac{r}{B} = \frac{r}{\sqrt{T/(K'/b')}} \tag{6.71}$$

From Eqs. 6.69 and 6.70 the following equations are derived:

$$\log s = \log W(u, r/B) + \log \frac{114.6Q}{T} \tag{6.72}$$

$$\log t = \log \frac{1}{u} + \log \frac{2700r^2S}{T} \tag{6.73}$$

The analogy with the Theis method becomes more obvious when Eq. 6.44 or Eq. 6.46 is rewritten as

$$\log t = \log \frac{1}{u} + \log \frac{1.87r^2S}{T} \tag{6.74}$$

In the present case log $W(u, r/B)$ is plotted versus log $1/u$ for a series of values of r/B. Figure 6.20 shows curves for r/B varying from 2.5 to zero; zero, of course, corresponds to the nonequilibrium type curve. Next, values of the drawdown s are plotted against values of t on logarithmic paper of the same size as was used for the series of type curves. For a constant discharge Q Eqs. 6.72 and 6.73 show that $W(u, r/B)$ is a function of $1/u$ in the same way s is a function of t. Therefore, it will be possible to superimpose the data curve on the series of type curves, holding the coordinate axes of the plots parallel, and in such a way that the data curve fits one of the type curves for a given r/B or may be interpolated between two such curves. A match point is chosen anywhere on the overlapping portion of the sheets and its coordinates $W(u, r/B)$, $1/u$, s, and t, together with the specific value of r/B that caused the fitting of the curves are inserted in Eqs. 6.69, 6.70, and 6.71 S and T follow from the first two equations while the knowledge of r/B and T allows for the solution of the leakance K'/b' from Eq. 6.71. Walton [33] has applied this method in an aquifer test near Dieterich, Illinois (Fig. 6.21). This figure shows that during the first thirty minutes the effect of leakage was not measurable as the data coincide with the nonequilibrium (nonleaky) type curve and that subsequently the field data deviated to follow the curve for $r/B = 0.22$. As time goes on,

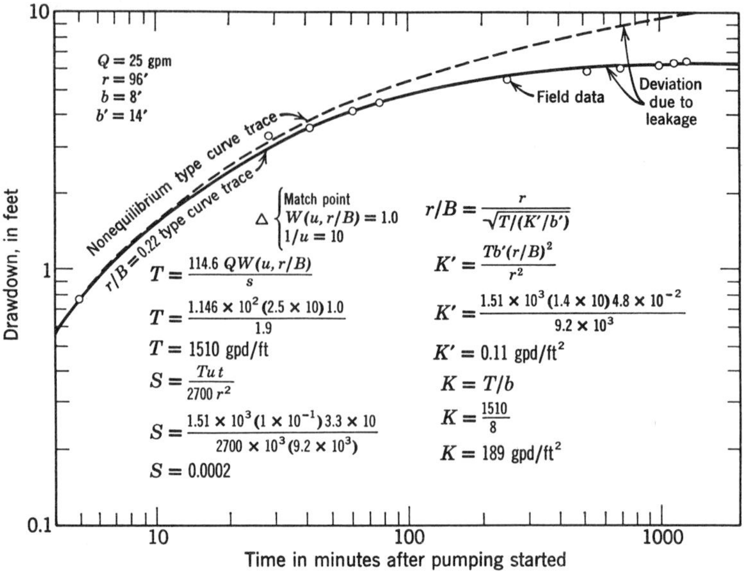

Fig. 6-21 Time-drawdown graph for Well 19 near Dieterich, Ill. (After Walton.)

more and more well discharge is derived from leakage and ultimately, as the flow becomes steady, the entire yield of the well is due to leakage. This was first observed by Jacob [16] who devised the following graphical method of superposition.

(B) JACOB'S METHOD FOR STEADY STATE DRAWDOWN

Jacob [16] and Hantush [20] pointed out that the steady state solution for the drawdown is proportional to $K_0(r/B)$, where K_0 is the modified Bessel function of the second kind and of zero order. They gave the solution

$$s = \frac{Q}{2\pi T} K_0(r/B) \tag{6.75}$$

as a limit case of Eq. 6.69 when $t \to \infty$. When Q is expressed in gallons per minute, T in gallons per day per foot, and s in feet, the drawdown may be computed by

$$s = 229 \frac{Q}{T} K_0 (r/B) \tag{6.76}$$

This equation is combined with

$$\frac{r}{B} = \frac{r}{\sqrt{T/(K'/b')}} \tag{6.71}$$

in the following graphical way. Values of $K_0(r/B)$, prepared for example from Table B.1, from reference 20, or from reference 8, are plotted as a type curve on logarithmic paper versus values of r/B as shown in Fig. 6.22. Aquifer tests data from several observation wells collected under steady state conditions are plotted on logarithmic paper of the same size as was

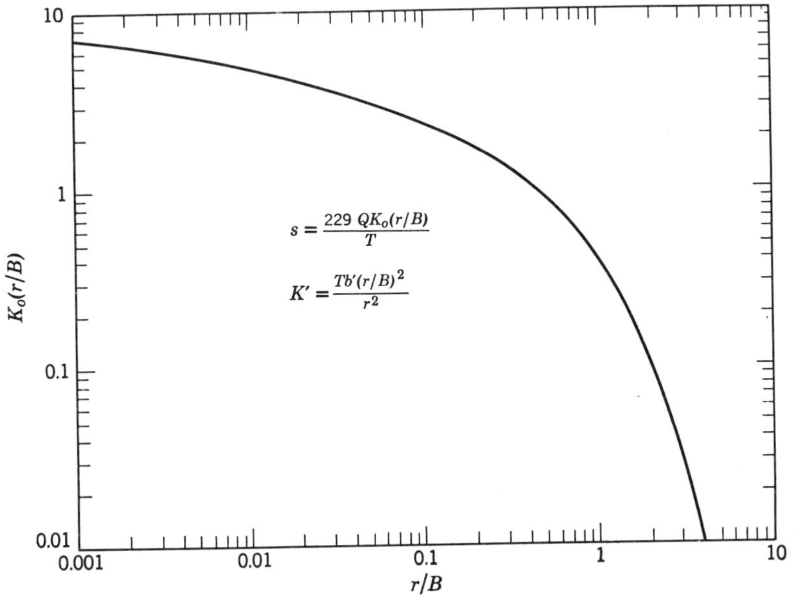

Fig. 6-22 Steady-state leaky artesian type curve. (After Walton.)

used for the type curve, s versus r. The two curves are then superimposed in the usual way, keeping the axes of the two graphs parallel. A match point is taken anywhere in the overlapping portions of the two sheets, and the values of $K_0(r/B)$, r/B, r and s of this point are inserted in Eqs. 6.76 and 6.71 to render T and K'/b'. It should be emphasized that this method is reliable only if enough time has passed since pumping started to be sure of steady state flow and that the drawdowns of the observation wells should not vary during the time when all of them are recorded. Figure 6.23 gives an example of a test run by Walton [33] near Dieterich, Illinois. As noted before, S cannot be determined by this test since the entire yield of the pumped well is derived from leakage. S may be determined by the method explained sub 6.10(A) applied to drawdowns observed during the early phase of the transient state. Simultaneously the value of T may be checked. For Table B.1, see Appendix B.

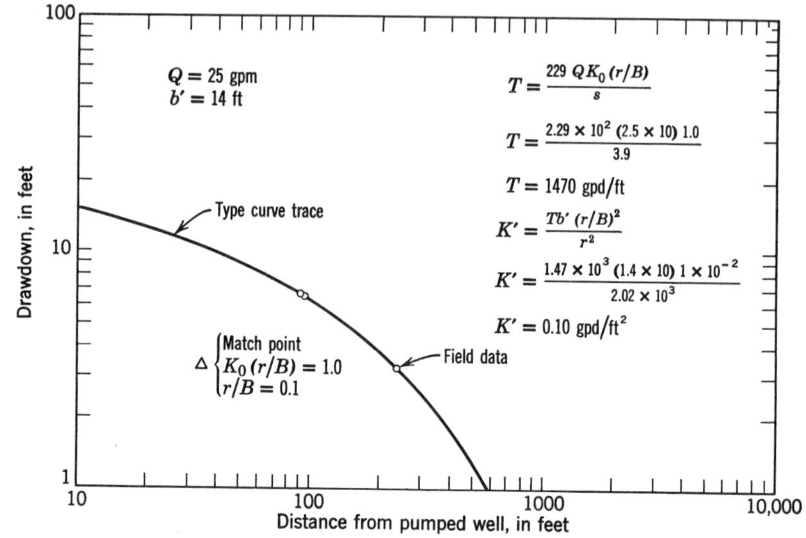

Fig. 6-23 Distance-drawdown graph for test near Dieterich, Ill. (After Walton.)

(c) HANTUSH'S METHOD

Hantush [20] devised another elegant method to determine the formation constants of and the leakance for a leaky aquifer. The proof of the properties underlying the method is given in Appendix A. Step by step the procedure is as follows:

1. Plot the drawdown (in feet) measured in an observation well versus time in minutes on semilogarithmic paper and extrapolate the data till the maximum drawdown s_{max} is found (Fig. 6.24).
2. Locate the inflection point on the drawdown curve by taking $s_i = \frac{1}{2}s_{max}$, where s_i is the drawdown at the inflection point.
3. Determine graphically the slope m_i of the drawdown curve at the inflection point and read the time t_i corresponding to the inflection point.
4. As is proved in Appendix A,

$$e^{r/B}K_0\left(\frac{r}{B}\right) = \frac{2.3s_i}{m_i}$$

Values of the function $e^x K_0(x)$ have been tabulated by Hantush [20]. Here the function is determined by the ratio $2.3s_i/m_i$. The value of the argument, say r/B, may be found from the available table (see Table B.2, Appendix B). This determines B.

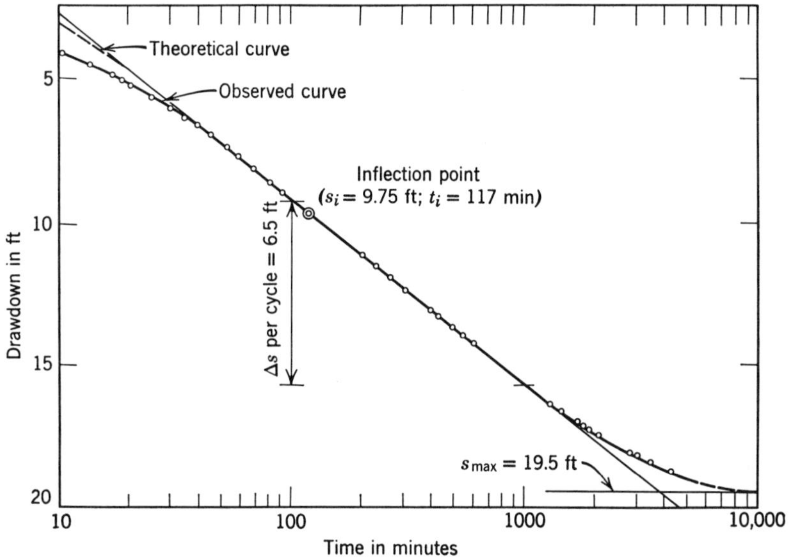

Fig. 6-24 Time-drawdown curve, Roswell Artesian Basin, New Mexico. (After Hantush.)

5. Compute T from the fact that the slope m_i of the curve at the inflection point is given by

$$m_i = (2.3Q/4\pi T)e^{-r/B}$$

T is found in ft²/sec if Q is in cfs and m_i in ft.

6. At the inflection point

$$u_i = \frac{r^2 S}{4 \times 60 T t_i} = \frac{r}{2B}$$

where r and B are in ft, T in ft²/sec, and t_i in min. From this relationship, S may be determined.

7. Finally the leakance K'/b' follows from

$$B = \sqrt{\frac{T}{K'/b'}}$$

with B and T determined.

NUMERICAL EXAMPLE

From the drawdown curve of Fig. 6.24, we find the values of T, S, and K'/b'. Check that $T = 0.09$ ft²/sec, $S = 0.000046$, and $K'/b' = 1.35 \times 10^{-10}$ sec⁻¹.

Hantush [20] also prepared a method to obviate the difficulty of extrapolating the value of s_{max} provided that at least two wells are observed and that the straight portions of the semilogarithmic drawdown-time curves are fully developed. He also examined the case where leakage is very high and affects the flow immediately so that more than 50% of the maximum drawdown is developed within a few minutes of pumping.

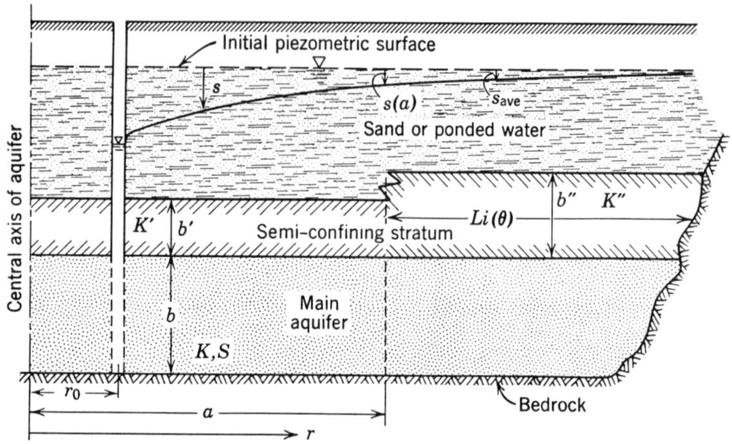

Fig. 6-25 Aquifer with vertical leakage and varied lateral replenishment. (After De Wiest.)

6.11 *Nonidealized Boundary Conditions*

The value of the foregoing methods for leaky and nonleaky aquifers has often been questioned in view of simpler existing methods [34] and in view of the simplifying assumptions regarding infinite extent or idealized boundary conditions of uniform head or zero lateral influx. This attitude is shortsighted in the first place because the methods of graphical superposition both for leaky and nonleaky aquifers are now easily accessible to hydrologists with limited mathematical backgrounds. Furthermore, it will be possible in the future, owing to the development of machine computers, to solve problems with natural boundary conditions as indicated in Figs. 6.25 and 6.26. Here we examine the flow to an eccentric well in a leaky aquifer with varied lateral replenishment, i.e., where the drawdown at the boundary is proportional to its radial derivative. However, the proportionality factor α which for the idealized boundary conditions treated so far was either zero or infinity now is a function of θ mainly through the variable distance $L_i(\theta)$ from the circular boundary to the impervious geological boundary of the artesian basin. But variations of

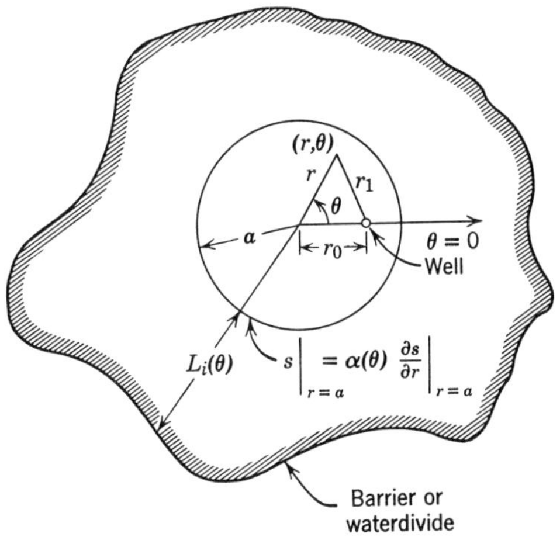

Fig. 6-26 Well eccentrically located in circular leaky aquifer with varied lateral replenishment. (After De Wiest.)

b'' and K'' (see Fig. 6.25) in function of θ as well may be accounted for. The problem is tractable in its present form if α is given a constant value differing from the trivial (i.e., zero or infinite) [32]. This value could be selected carefully upon examination of the geology of the basin. In some cases the average value $\bar{\alpha}(\theta) = \dfrac{1}{2\pi}\displaystyle\int_0^{2\pi} \alpha(\theta)\,d\theta$ may be the best one.

◆ Partial Penetration of Wells

In this chapter complete penetration of the aquifer by the well has always been assumed, although this condition is seldom realized in the field. The question arises as to how much error is made when formulas for complete penetration are applied to partially penetrated aquifers. Flow into a partially penetrating well has been investigated by De Glee [12], Kozeny [35], Muskat [36], and, more recently, by Boreli [37], Hantush, and Jacob [18], [21], and Kirkham [38]. Hantush's study [21] leads to the conclusion that three-dimensional flow towards a steady well just tapping (i.e., almost zero penetration) an infinite leaky aquifer rapidly changes to a radial type of flow. At a distance from the well equal to 1.5 times the thickness of the aquifer, both types of flow can barely be distinguished from each other. When the depth of penetration increases the radial flow pattern becomes more dominant. Hence it is safe to apply the formulas for complete penetration to compute drawdowns in points at least 1.5 times the thickness of the aquifer away from the partially penetrating well. For distances closer to the well, formulas derived in the foregoing references must be used.

Often the flow to a partially penetrating well has axisymmetrical properties. In this case, as explained in section 5.15, a flow net may be drawn without too much difficulty. An example of the flow to a well with impervious sidewall and with protecting filter at the bottom is given in Fig. 6.27 [39].

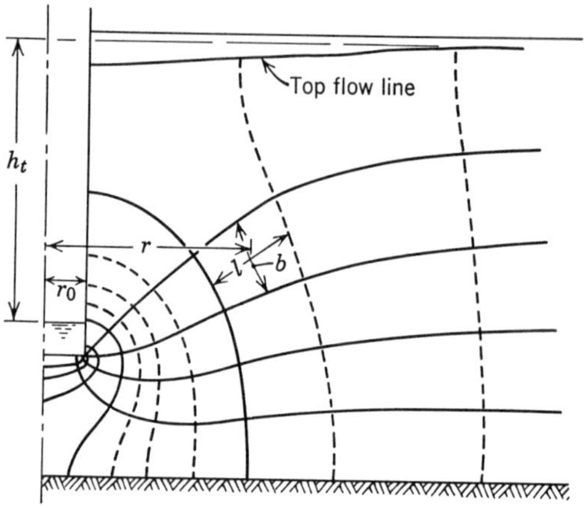

Fig. 6-27 Flow net for partially penetrating well. (After Taylor.)

REFERENCES

1. Thiem, G., *Hydrologische Methoden*, Gebhardt, Leipzig, 1906 (56 pp.).
2. Jacob, C. E., "Flow of Groundwater," *Engineering Hydraulics*, p. 346, H. Rouse (Ed.), Wiley, New York, 1950.
3. Ferris, J. G., D. B. Knowles, R. H. Brown, and R. W. Stallman, "Theory of Aquifer Tests," *U. S. Geological Survey Water Supply Paper* 1536-E, p. 148, Washington, D.C., 1962.
4. Jacob, C. E., "Flow of Groundwater" in *Engineering Hydraulics*, pp. 348–349.
5. Hantush, M. S. and C. E. Jacob, "Plane Potential Flow of Ground Water with Linear Leakage," *Transactions American Geophysical Union*, Vol. 35, No. 6, pp. 917–936 (December 1954).
6. Polubarinova-Kochina, P. Ya., *Theory of Ground-water Movement*, p. 365, Princeton University Press, Princeton, 1962.
7. Theis, C. V., "The Relation between the Lowering of the Piezometric Surface and the Rate and Duration of Discharge of a Well Using Groundwater Storage," *Transactions American Geophysical Union*, Vol. 16, pp. 519–524 (1935).
8. Jahnke, E. and F. Emde, *Tables of Functions*, pp. 6–7, Dover Publications, New York, 1945.
9. Wenzel, L. K., "Methods for Determining Permeability of Water-Bearing Materials with Special Reference to Discharging-Well Methods," *U.S. Geological Survey Water-Supply Paper* 887, Washington, D.C., 1942 (192 pp.).

10. Lang, S. M., "Interpretation of Boundary Effects from Pumping Test Data," *Journal of American Water Works Association*, Vol. 52, No. 3, pp. 356–364 (March 1960).
11. Cooper, H. H., Jr. and C. E. Jacob, "A Generalized Graphical Method for Evaluating Formation Constants and Summarizing Well-Field History," *Transactions American Geophysical Union*, Vol. 27, pp. 526–534 (1946).
12. De Glee, G. J., *Over Grondwaterstromingen by Wateronttrekking by middel van Putten*, T. Waltman, Jr., Delft, 1930 (175 pp.).
13. Steggewentz, J. H. and B. A. Van Nes, "Calculating the Yield of a Well Taking Account of Replenishment of the Groundwater from Above," *Water and Water Engineering*, Vol. 41, pp. 561–563 (1939).
14. Jacob, C. E., "Radial Flow in a Leaky Artesian Aquifer," *Transactions American Geophysical Union*, Vol. 27, pp. 198–205 (1946).
15. Hantush, M. S., *Plain Potential Flow of Groundwater with Linear Leakage*, Ph.D. dissertation, University of Utah, 1949 (86 pp.).
16. Hantush, M. S. and C. E. Jacob, "Plane Potential Flow of Ground Water with Linear Leakage," *Transactions American Geophysical Union*, Vol. 35, pp. 917–936 (1954).
17. Hantush, M. S. and C. E. Jacob, "Nonsteady Radial Flow in an Infinite Leaky Aquifer and Nonsteady Green's Functions for an Infinite Strip of Leaky Aquifer," *Transactions American Geophysical Union*, Vol. 36, pp. 95–112 (1955).
18. Hantush, M. S. and C. E. Jacob, "Steady Three-Dimensional Flow to a Well in a Two-Layered Aquifer," *Transactions American Geophysical Union*, Vol. 36, pp. 286–292 (1955).
19. Hantush, M. S. and C. E. Jacob, "Flow to an Eccentric Well in a Leaky Circular Aquifer," *Journal of Geophysical Research*, Vol. 65, pp. 3425–3431 (1960).
20. Hantush, M. S., "Analysis of Data from Pumping Tests in Leaky Aquifers," *Transactions American Geophysical Union*, Vol. 37, pp. 702–714 (1956).
21. Hantush, M. S., "Nonsteady Flow to a Well Partially Penetrating an Infinite Leaky Aquifer," *Proceedings Iraqi Scientific Societies*, Vol. 1, pp. 10–19 (1957).
22. Hantush, M. S., "Nonsteady Flow to Flowing Wells in Leaky Aquifer," *Journal of Geophysical Research*, Vol. 64, pp. 1043–1052 (1959).
23. Hantush, M. S., "Analysis of Data from Pumping Wells near a River," *Journal of Geophysical Research*, Vol. 64, pp. 1921–1932 (1959).
24. Hantush, M. S., "Modification of the Theory of Leaky Aquifers," *Journal of Geophysical Research*, Vol. 65, pp. 3713–3725 (1960).
25. Hantush, M. S., "Drawdown around a Partially Penetrating Well," *ASCE Journal of Hydraulics Division*, pp. 83–98, July 1961.
26. Hantush, M. S., "Economical Spacing of Interfering Wells. Groundwater in Arid Zones," *IASH Publication No.* 57, pp. 350–364 (1961).
27. Hantush, M. S., "Flow of Groundwater in Sands of Nonuniform Thickness," *Journal of Geophysical Research*, Vol. 67, pp. 703–720 and 1527–1535 (1962).
28. Hantush, M. S., "Drainage Wells in Leaky Water-Table Aquifers," *ASCE Journal of Hydraulics Division*, pp. 123–137, March 1962.
29. Mjatiev, A. N., "Pressure Complex of Underground Water and Wells," *Isvestiya Akademiya Nauk*, USSR Division Technical Sciences, No. 9, 1947.
30. Polubarinova-Kochina, P. Ya., *Theory of Ground-Water Movement*, pp. 377–383.
31. De Wiest, R. J. M., "On the Theory of Leaky Aquifers," *Journal of Geophysical Research*, Vol. 66, pp. 4257–4262 (1961).
32. De Wiest, R. J. M., "Flow to an Eccentric Well in a Leaky Circular Aquifer with Varied Lateral Replenishment," *Geofisica Pura e Aplicata*, Vol. 54, pp. 87–102 (1963).

33. Walton, W. C., *Leaky Artesian Aquifer Conditions in Illinois, Report of Investigation No.* 39, Illinois State Water Survey, 1960.
34. Kazmann, R. G., "Discussion of Paper by M. S. Hantush, Analysis of Data from Pumping Wells near a River," *Journal of Geophysical Research*, Vol. 65, pp. 1625–1626 (1960).
35. Kozeny, J., "Theorie und Berechnung der Brunnen," *Wasserkraft und Wasserwirtschaft*, Vol. 28, pp. 101–105 (1953).
36. Muskat, M., *The Flow of Homogeneous Fluids through Porous Media*, Chapter V, McGraw-Hill, New York, 1937.
37. Boreli, M., "Free Surface Flow toward Partially Penetrating Wells," *Transactions American Geophysical Union*, Vol. 36, pp. 664–672 (1955).
38. Kirkham, D., "Exact Theory of Flow into a Partially Penetrating Well," *Journal of Geophysical Research*, Vol. 64, pp. 1317–1327 (1959).
39. Taylor, D., *Fundamentals of Soil Mechanics*, p. 193, Wiley, New York, 1948.
40. Todd, D. K., *Ground Water Hydrology*, p. 82, Wiley, New York, 1959.
41. De Wiest, R. J. M., "Replenishment of Aquifers Intersected by Streams," *Journal of the Hydraulics Division*, pp. 165–191, American Society Civil Engineers, November, 1963.
42. Walton, W. C., "Estimating the Infiltration Rate of a Streambed by Aquifer-Test Analysis," *International Association of Scientific Hydrology, Publication No.* 63, pp. 409–420 (1963).
43. Hantush, M. S., "Depletion of Storage, Leakage, and River Flow by Gravity Wells in Sloping Sands," *Journal of Geophysical Research*, Vol. 69, pp. 2551–60 (1964).
44. Maasland, D. E. and M. Bittinger. *Proceedings of the Symposium of Transient Ground-Water Hydraulics*, Colorado State University, July 25–27, 1963 (223 pp.).

CHAPTER SEVEN

Multiple-Phase Flow. Dispersion

7.1 *Movement of Oil and Gas under Hydrodynamical Conditions*

The first attempt to establish the laws of oil and gas accumulation in subterranean media, the so-called anticlinal theory, conceived of a static system of gas, oil, and water with horizontal interfaces, resting on the crests of anticlines. This concept prevailed in the early days after the discovery of oil at Titusville, Pennsylvania, in 1859, and except for some sporadic thinking along fluid dynamics lines in the 1920's and 1930's that made the flow of water an essential requirement for oil and gas movement, the anticlinal theory with its hydrostatic connotation was by and large the most accepted in the first half of the twentieth century. In 1953, Hubbert [1] put a new emphasis on the hydrodynamical aspect of the problem, and at present there remains little doubt as to the basic validity of this approach. The salient features of his paper, which are outlined in this paragraph, are the derivation of force potentials for the fluids in question and the immiscibility of the fluids. This leads to the determination of distinct fluid-fluid interfaces and the study of the surface tension effects along these interfaces.

Petroleum and gas, formed by the decomposition of organic matter deposited in sedimentary rocks, are normally found in a water-saturated environment. For this reason it is convenient to express the potential energy of these fluids in terms of that of the ambient ground water. In the study of the movement of oil and gas in porous rocks, a distinction should be made between the primary migration from the source rocks in which the fluids are originated in a dispersed state, to the reservoir rocks, and the movement in the reservoir rocks themselves.

(A) PRIMARY MIGRATION

The force potential per unit mass for a given liquid petroleum at a given point may be derived by extension from Eq. 4.83 to include, besides the terms expressing the work against gravity and pressure, a term to account for the interfacial energy between petroleum water and rock. Thus

$$\Phi_l{}^* = gz + \frac{p}{\rho_l} + \frac{p_c}{\rho_l} + \text{constant} \tag{7.1}$$

where p is the pressure of the ambient water at the point, ρ_l the density of the oil, and p_c the difference of pressure across the interface between oil and water. The value of p_c depends on the wettability of the rocks. In general, rocks are preferentially wet by water with respect to oil; then p_c is positive, which means that the pressure inside the oil exceeds that inside the water by the amount p_c. If petroleum occurs in the gaseous phase, Eq. 7.1 should be replaced by

$$\Phi_g{}^* = gz + \int_0^p \frac{dp}{\rho} + \int_p^{p+p_c} \frac{dp}{\rho} + \text{constant} \tag{7.2}$$

where the variable density ρ is a function of the pressure p only.

To find the forces acting on the mass of fluid in a given point, the gradient ∇ of Eq. 7.1 must be taken as was demonstrated in the derivation of Eq. 4.85. Here, in the primary migration of oil from source rocks to reservoir rocks, p_c/ρ_l is the important term of Eq. 7.1 to consider in the region of the boundary. This term alone depends on the geometry of the porous medium, and therefore its gradient taken across the boundary of the different rocks shows the discontinuity in permeability as paramount in the driving force. This follows from the expression of p_c:

$$p_c = \frac{C\sigma \cos \theta}{2r} \tag{7.3}$$

where C is a dimensionless factor of proportionality, $2r$ the mean grain diameter of the rock, σ the surface tension, and θ the contact angle in the water phase which the oil-water interface makes with the rock (see section 4.15 for $h_c = p_c/\gamma$). The force due to capillarity may be derived from Eq. 7.3 as

$$-\frac{1}{\rho_l} \text{grad } p_c = \frac{C\sigma \cos \theta}{2\rho_l} \frac{\text{grad } r}{r^2} \quad \text{(see footnote 1)} \tag{7.4}$$

[1] $\text{grad} \frac{u}{v} = \frac{v \text{ grad } u - u \text{ grad } v}{v^2}$. Here $u = 1$, $v = r$, grad $1 = 0$. Rule is analogous to the rule for differentiation $d\frac{u}{v} = \frac{v\,du - u\,dv}{v^2}$.

Hubbert [1] assumes that grad $r = \beta r$, where β is a proportionality constant, so that Eq. 7.4 may be written as

$$-\frac{1}{\rho_l}\,\text{grad}\, p_c = \frac{C'}{r} \tag{7.5}$$

where C' lumps all constants together.

Equations 7.4 and 7.5 show that the force acting on the liquid petroleum is directed along the steepest rate of increase of the grain size of the rock

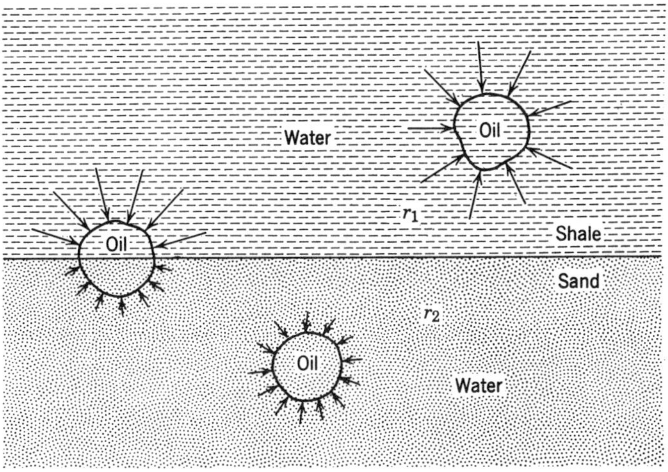

Fig. 7-1 Passage of oil globules due to unbalanced capillary pressures. [After Hubbert.]

(meaning of grad r) and also that this force, for a given grad $r = \beta r$, is inversely proportional to the grain size of the rock. The steepest rate of change in grain size normally occurs perpendicularly to the planes of sedimentary deposition so that oil globules have a tendency to move in that direction under influence of capillarity. Once a globule reaches the boundary between rocks of different texture, as in Fig. 7.1, part of its mass is subjected to a capillary force proportional to $1/r_1$ in the medium with grain size $2r_1$ and part of its mass is subjected to a force proportional to $1/r_2$ in the medium with grain size $2r_2$, the net resultant force being proportional to the difference $(1/r_1 - 1/r_2)$, with $r_1 < r_2$. The magnitude of the forces is indicated by the lengths of the arrows in Fig. 7.1. It follows that an oil globule in a liquid or gaseous state in a water-saturated environment can only pass from fine- to coarse-textured rocks and not in the opposite way. Once more it should be emphasized that capillary action would not be present if oil were not surrounded by water. If only single-phase flow

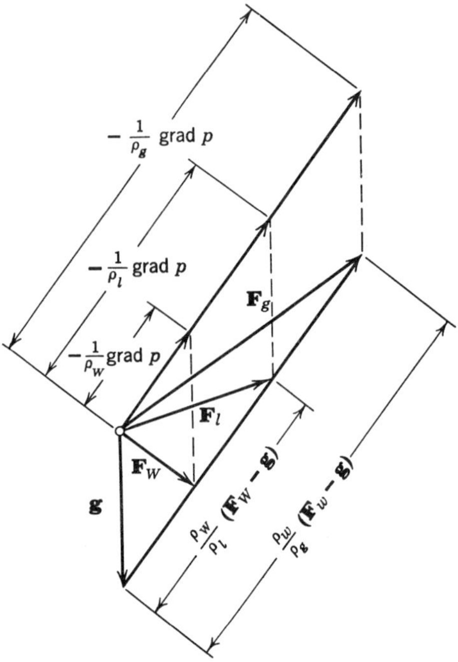

Fig. 7-2 Impelling forces on water, oil, and gas in hydrodynamic environment. [After Hubbert.]

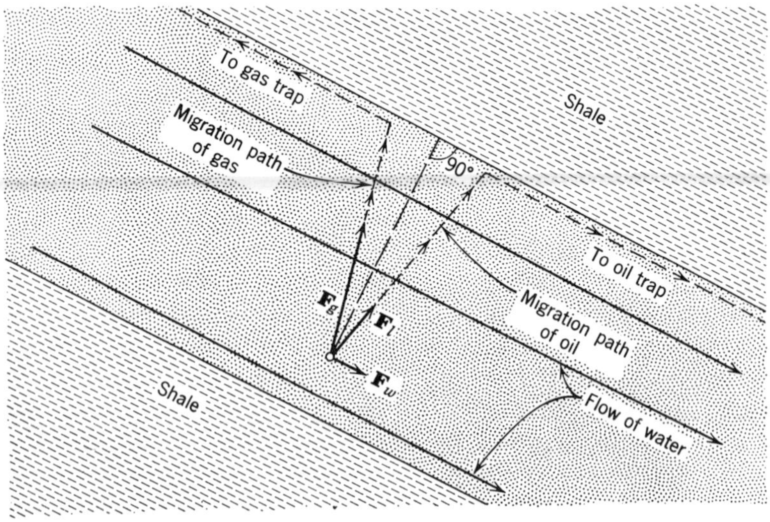

Fig. 7-3 Divergent migration of oil and gas in hydrodynamic environment. [After Hubbert.]

were to be considered, the flow would proceed equally well in both directions and obey the law of section 5.11. This is the case for water constituting the environment in which oil is dispersed. The sand-shale boundary of Fig. 7.1 may then be considered as an impermeable barrier for oil present in the sand, but not for water. In the source rock, oil is also subjected to gravity and pressure forces derived from the two remaining terms in Eq. 7.1. They have to be superimposed on the capillary forces which for oil in water-saturated shales are of the order of tens of atmospheres.

(B) MOVEMENT IN RESERVOIR ROCKS

In the flow of oil in reservoir rocks, the role of the capillary force expressed by Eq. 7.4 may be neglected because of the large value of the grain size. The capillary pressure of oil in sandstone is only of the order of tenths of an atmosphere. To find the impelling force $\mathbf{F}_l$ per unit mass of liquid petroleum, the gradient is taken of Eq. 7.1 with the negative sign as explained in the derivation of Eq. 4.85.

$$\mathbf{F}_l = -\text{grad}\,\Phi_l{}^* = \mathbf{g} - \frac{1}{\rho_l}\,\text{grad}\,p \tag{7.6}$$

as compared to

$$\mathbf{F}_w = -\text{grad}\,\Phi_w{}^* = \mathbf{g} - \frac{1}{\rho_w}\,\text{grad}\,p \tag{7.7}$$

which is the equivalent of Eq. 4.85 for the force on a unit mass of water. The pressure p is the same in both equations and may therefore be eliminated between these equations, and the result is

$$\mathbf{F}_l = \mathbf{g} + \frac{\rho_w}{\rho_l}(\mathbf{F}_w - \mathbf{g}) \tag{7.8}$$

The force $\mathbf{F}_g$ per unit mass of petroleum in the gaseous state is derived in a similar way:

$$\mathbf{F}_g = \mathbf{g} + \frac{\rho_w}{\rho_g}(\mathbf{F}_w - \mathbf{g}) \tag{7.9}$$

Equations 7.6 and 7.8 may be represented by a vector diagram as shown in Fig. 7.2. From this vector diagram it is evident that for a given flow of the ambient water the variation in density of the hydrocarbon is responsible for the direction of the total force $\mathbf{F}_l$ or $\mathbf{F}_g$ and for the separation of liquid petroleum and gas under given geological conditions. Such a case is represented in Fig. 7.3, where the angle of the homoclinal dip is such that the gas would be deflected up the dip and the liquid petroleum down the dip since the boundary sand-shale constitutes an impermeable barrier for the hydrocarbons because of the surface tension effects. A change in the magnitude of $\mathbf{F}_w$ could change this picture so that both fluids would migrate in the same direction. From the foregoing derivations, where the force-potential is used, it is clear that elements of petroleum will move from regions of higher energy to regions of lower energy and will

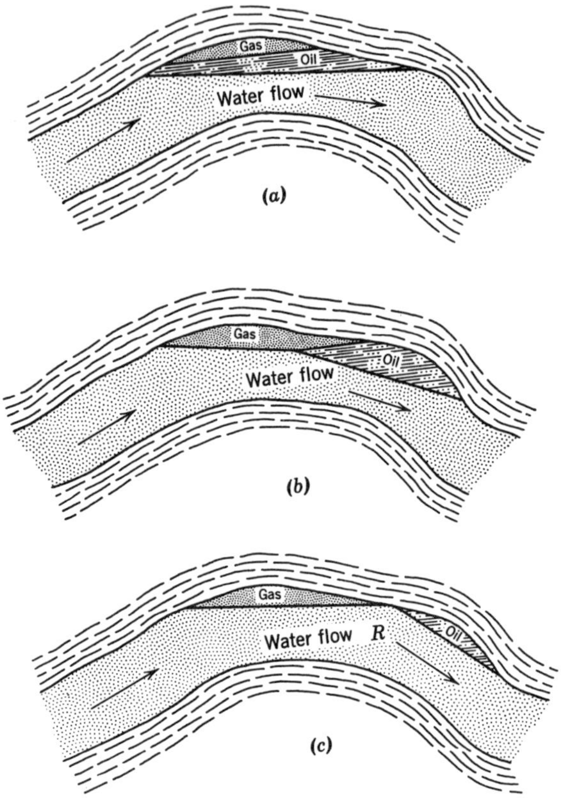

Fig. 7-4 Hydrodynamic oil and gas accumulation in gently folded thick sand. (*a*) Gas entirely underlain by oil. (*b*) Gas partly underlain by oil. (*c*) Gas and oil traps separated. [After Hubbert.]

come to rest when the energy of their surroundings is higher than their own energy or when they are trapped between regions of higher energy and impermeable barriers. Types of such traps of oil and gas accumulations are given in Fig. 7.4. In Fig. 7.4*a*, oil and gas form a static horizontal interface; the oil-water interface is sloping because water is flowing underneath the oil. Figure 7.3 may be considered a close-up of region *R* of Fig. 7.4*c*. In all three cases the hydrocarbons and the water occupy different regions separated by sharp interfaces because the fluids have been assumed to be immiscible. In the next paragraph the slope of the interface between two regions will be examined.

(C) SLOPE OF THE HYDROCARBON-WATER INTERFACE

According to the concept of potential, pressure p and elevation z which are unique for any point in the two regions may be expressed in either $\Phi_w{}^*$ or $\Phi_l{}^*$, by the use of Eq. 4.83:

$$\Phi_w{}^* = gz + \frac{p}{\rho_w}$$
$$\Phi_l{}^* = gz + \frac{p}{\rho_l} \tag{7.10}$$

By elimination of p from Eq. 7.10, the elevation z of any point in terms of its potentials in the two regions is found:

$$z = \frac{1}{g}\left(\frac{\rho_w \Phi_w{}^*}{\rho_w - \rho_l} - \frac{\rho_l \Phi_l{}^*}{\rho_w - \rho_l}\right) \tag{7.11}$$

Similar computations may be made for a system of water and gaseous hydrocarbon. In particular, Eq. 7.11 is valid for points along the interface. This equation, however, does not make it possible to compute the elevation of the interface, for the values of $\Phi_w{}^*$ and $\Phi_l{}^*$ are in general not known along the interface. Equation 7.11 becomes useful, however, if the potentials are related to the velocities in each region by means of Darcy's law. This is done by computing

$$\sin\alpha = \frac{\partial z}{\partial s} = \frac{1}{g}\left[\frac{\rho_w}{\rho_w - \rho_l}\frac{\partial \Phi_w{}^*}{\partial s} - \frac{\rho_l}{\rho_w - \rho_l}\frac{\partial \Phi_l{}^*}{\partial s}\right] \tag{7.12}$$

where s is the trace of the interface in a vertical plane and α the angle of the trace with the horizontal (Fig. 7.5).

In view of Eq. 4.86

$$\frac{\partial \Phi_w{}^*}{\partial s} = -\frac{g}{K_w} V_{w,s}$$
$$\frac{\partial \Phi_l{}^*}{\partial s} = -\frac{g}{K_l} V_{l,s}$$

where $V_{w,s}$ and $V_{l,s}$ are the components of the specific discharge respectively of the water and the liquid hydrocarbon along the trace of the interface s so that

$$\sin\alpha = \frac{\partial z}{\partial s} = -\left[\frac{\rho_w}{\rho_w - \rho_l}\frac{1}{K_w} V_{w,s} - \frac{1}{K_l}\frac{\rho_l}{\rho_w - \rho_l} V_{l,s}\right] \tag{7.13}$$

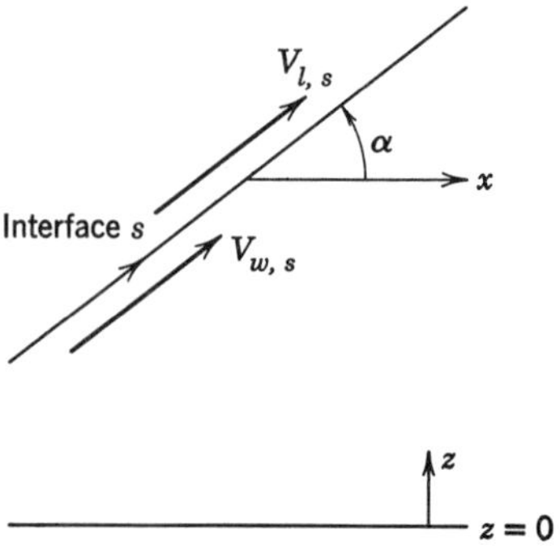

Fig. 7-5 Slope α of the interface in immissible flow.

From this equation it is easy to predict the trend in the slope of the interface in general. In particular, when one of the fluids is stagnant, α assumes a definite sign. When the hydrocarbon is trapped as in Fig. 7.4, $V_{l,s} = 0$ and $\sin \alpha < 0$ or the interface slopes downward when $\rho_w > \rho_l$ and the slope will be steeper with increasing $V_{w,s}$. On the other hand, this slope is milder when ρ_l is replaced by ρ_g, i.e., when the hydrocarbon is in the gaseous state. Where one fluid is stagnant it is convenient to replace Eq. 7.13 by

$$\tan \alpha = \frac{dz}{dx} = -\frac{1}{K}\left[\frac{\rho_w}{\rho_w - \rho_l}\right] V_{w,x} = \frac{\rho_w}{\rho_w - \rho_l} \frac{dh_w}{dx} \tag{7.14}$$

and the slope of the interface may be visualized by the reading of two standpipes introduced as sketched in Fig. 7.6.

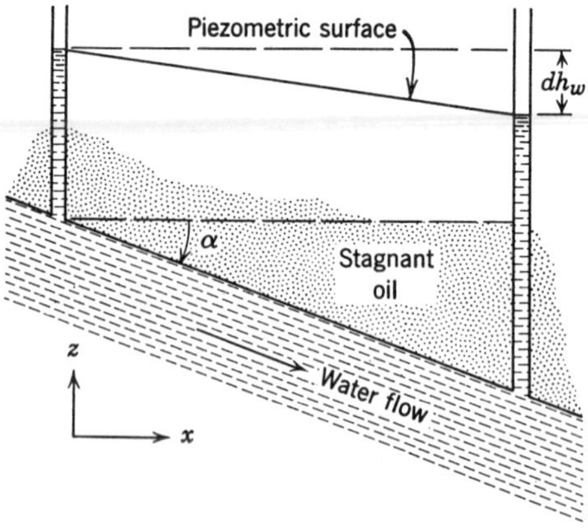

Fig. 7-6 Slope α of interface and standpipe reading.

The influence of capillarity on the slope of the interface is zero when the medium is homogeneous, even where the fluids are immiscible [2].

For further details on the different types of hydrodynamic traps and on their shifts in position with changes in the state of flow, the reader is referred to the original papers by Hubbert.

7.2 *Salt Water Encroachment*

Salt water encroachment or intrusion is the shoreward movement of water from a sea or ocean into confined or unconfined coastal aquifers and the subsequent displacement of fresh water from these aquifers. If, as a first approximation, fresh water and salt water are treated as two immiscible fluids, they are separated by a sharp interface with a slope that may be computed by formulas similar to those of section 7.1*c*. Because of the slope of the interface the front of the salt water is sometimes compared to a tongue progressing into the land as a result of overdraft of the overlying fresh water and pushed back seaward when fresh water is replenished through precipitation. Actually, fresh water and sea water mix in a region of dispersed water, and it is sometimes necessary to consider salt water dispersion in order to obtain a more accurate picture of the nature of ground water in a coastal aquifer. Salt water encroachment that may reach several thousand feet inland, is by its very definition of interest because of islands and coastal regions which rely on ground water for their water supply. Wells may become contaminated with salt water and may have to be abandoned, or fresh water may have to be injected in order to stop the inland movement of the salt water and to establish a fresh water barrier. Furthermore, a study of the location of even the idealized interface is useful to obtain an idea of the fresh water losses to the ocean through discharge under sea level.

(A) GHYBEN-HERZBERG HYDROSTATIC CONDITIONS

The hydrostatic equilibrium between immiscible fresh water and salt water bodies in contact with each other along a certain interface was studied first by Badon Ghyben [3] and Herzberg [4]. The equation for the depth of the interface was in good agreement with measurements made in the field indicating that for every foot of fresh water above mean sea level the thickness of the fresh water lens resting on the salt water was about 40 ft. Near the shore the interface is sloping and hence narrows the fresh water lens [37]. In this elementary interpretation, as indicated in Fig. 7.7, the cross sections of coastal line, interface, ocean level, and water table meet all in one point. The depth z_s of the interface below sea level may be

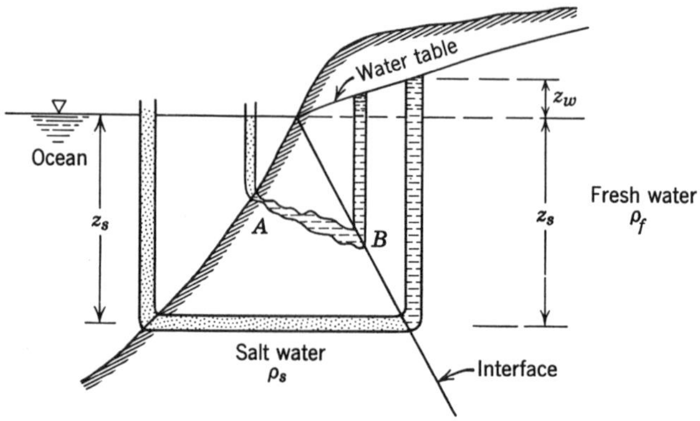

Fig. 7-7 Escape mechanism for fresh water suggested by Wentworth [5].

determined from the hydrostatic law in the U tube of Fig. 7.7. Hence if the datum plane[2] for z_s is taken at the bottom of the U tube,

$$z_s \rho_s g = g \rho_f z_s + g \rho_f z_w$$

or

$$z_s = \frac{\rho_f}{\rho_s - \rho_f} z_w \tag{7.15}$$

where ρ_s is the density of salt water, ρ_f the density of fresh water, and z_w the height of the water table above mean sea level.

For $\rho_s = 1.026$ gram-mass/cm³ and $\rho_f = 1.000$ gram-mass/cm³,

$$z_s = 38 z_w \tag{7.16}$$

The limitations of this theory are obvious. If both fluids were truly in static condition, the water table would have zero slope and the interface would become horizontal with fresh water overlying salt water by mere density difference. Furthermore, fresh water is in a continuous state of motion due to changes in the water table through replenishment, evaporation, and discharge. Seepage surfaces above sea level have been found along many shores and have been tapped in earlier times as potable water for use on sea-going vessels. The existence of such surfaces is not considered in the Ghyben-Herzberg theory, nor is there any provision for the escape of fresh water below sea level. Wentworth [5] in a discussion of salt water intrusion near Honolulu suggested an escape mechanism whereby

[2] If the datum plane for z is taken at mean sea level, Eq. 7.15 should be written as

$$z_s = -\frac{\rho_f}{\rho_s - \rho_f} z_w \tag{7.15'}$$

with all the z's above the datum plane measured as positive.

fresh water would escape below sea level through the interface via highly permeable avenues in the porous medium due to interconnecting tubes or cracks of larger than average size. He proposed the existence of local disturbances as indicated in Fig. 7.7, where there is an unbalance between the salt water column in *A* and the fresh water column in *B*. Essentially this would mean that the interface in a crack is displaced from *B* to *A*.

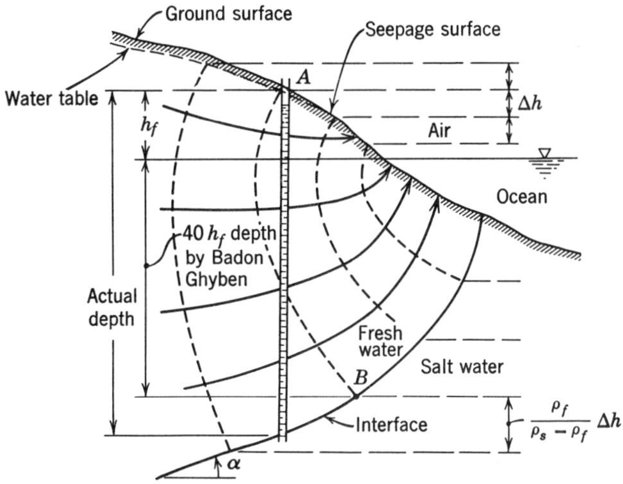

Fig. 7-8 Discrepancy between actual depth to salt water and depth calculated by Ghyben-Herzberg relation. [After Hubbert.]

The foregoing relationship, derived for water-table conditions, is also valid for confined aquifers if the height of the water table is replaced by the piezometric head.

(B) HUBBERT'S HYDRODYNAMICAL APPROACH

Hubbert [6], pointing at the dynamic rather than at the hydrostatic equilibrium of the fresh water–salt water interface, showed the discrepancy between the actual depth of salt water and the depth as calculated by the Ghyben-Herzberg formula for flow conditions near the shore line (Fig. 7.8).

The slope of the interface may be derived from Eq. 7.15 by direct analogy in which the subscripts *f* and *s* replace the subscripts *w* and *l* respectively to indicate fresh and salt water. Hence

$$\sin \alpha = \frac{\partial z}{\partial s} = -\left[\frac{1}{K_f}\frac{\rho_f}{\rho_f - \rho_s} V_{f,s} - \frac{1}{K_s}\frac{\rho_s}{\rho_f - \rho_s} V_{s,s}\right] \qquad (7.17)$$

If we assume the salt water to be stagnant and the fresh water to be flowing over it, $V_{s,s} = 0$ and $V_{f,s} \neq 0$. It follows that $\sin \alpha > 0$, because $\rho_f < \rho_s$, and α will increase with $V_{f,s}$. This explains the value of α approaching 90° in Fig. 7.8. Hubbert further shows that a relation exactly equal to Eq. 7.15′ holds between the points of intersection of any equi-

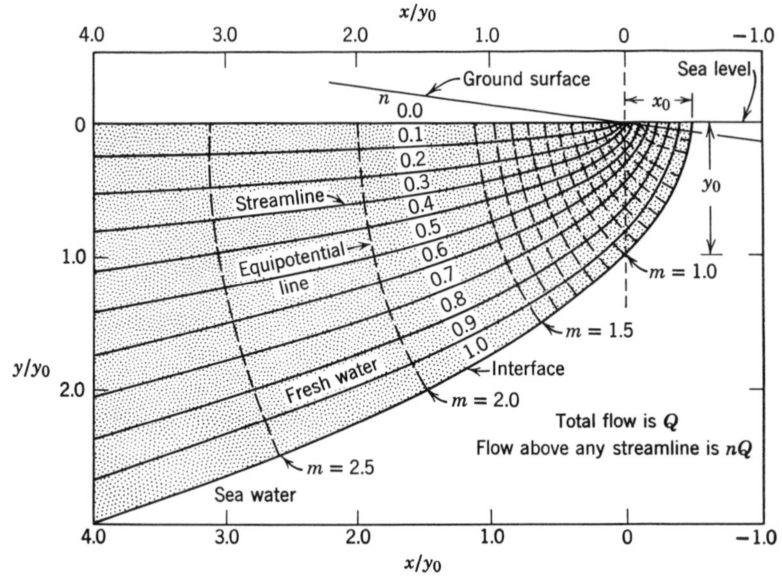

Fig. 7-9 Flow patterns near a beach. [After Glover.]

potential line of the flowing fresh water with the water table and with the fresh water–salt water interface, as indicated by points *A* and *B* of Fig. 7.8. These points do not lie in a vertical line and because the equipotential line is curved the vertical distance between the water table and the fresh water–salt water interface will be greater than that given by Eq. 7.15.

(C) GLOVER'S MODEL

Glover [7] suggested a close representation of the flow conditions near a beach (Fig. 7.9). He adopted the image of the flow through a semi-infinite underdrained earthdam resting on horizontal bedrock to the present problem. The analogy is not perfect, however, as there is no exact correspondence between all the boundary conditions for the moving fluid in both problems. It is possible to have exactly matching boundary conditions for the fresh water–salt water interface and the free surface in the dam, provided the water table in the present problem is perfectly horizontal

(Fig. 7.9). In this case it is reasonable to write down the equation of the interface as

$$y^2 = \frac{2Qx}{K\left(\dfrac{\gamma_s - \gamma_f}{\gamma_f}\right)} + \frac{Q^2}{K^2\left(\dfrac{\gamma_s - \gamma_f}{\gamma_f}\right)^2} \tag{7.18}$$

which may be compared with Eq. 20 of reference 8. Equation 7.18 could be derived by the use of complex variables and a velocity potential

$$\Phi_1 = \frac{\gamma_s - \gamma_f}{\gamma_f} Kh$$

instead of $\Phi = Kh$ of Eq. 4.26. The width of the gap through which the fresh water escapes to the ocean is

$$x_0 = -\frac{Q}{2K\left(\dfrac{\gamma_s - \gamma_f}{\gamma_f}\right)} \tag{7.19}$$

which corresponds to the length of the effective drain (p. 1050, reference 8). If the flow Q of fresh water to the sea increases because of increased replenishment of the water table through precipitation and percolation, the width of the gap increases. On the other hand, in times of drought the loss to the sea decreases so that there is a stabilizing mechanism to conserve the fresh water body.

7.3 *Dispersion in Salt Water Encroachment*

In section 7.2 immiscibility of fresh and salt water and static conditions for salt water have been assumed. Cooper [9] has advanced the hypothesis that where miscibility of the fluids involved is assumed salt water is not static but circulates back and forth between the floor of the sea and a zone of dispersion of mixed salt and fresh water. (Figs. 7.10 and 7.11). The principles of dispersion-diffusion phenomena are briefly reviewed [10]. It has been shown by actual measurements that these phenomena may be responsible for significant fluctuations of the water table [11].

Hydrodynamic dispersion in porous media describes the mechanical or convective process by which one fluid displaces another fluid while mixing with it in a zone of transition called the dispersion zone. This phenomenon is studied by itself as opposed to that of immiscible fluid displacement where a sharp interface separates regions occupied by fluids of different nature. Dispersion results in a variation of concentration of the displacing fluid in the dispersion zone, essentially because individual fluid particles travel at variable velocities through the irregularly shaped pore channels of

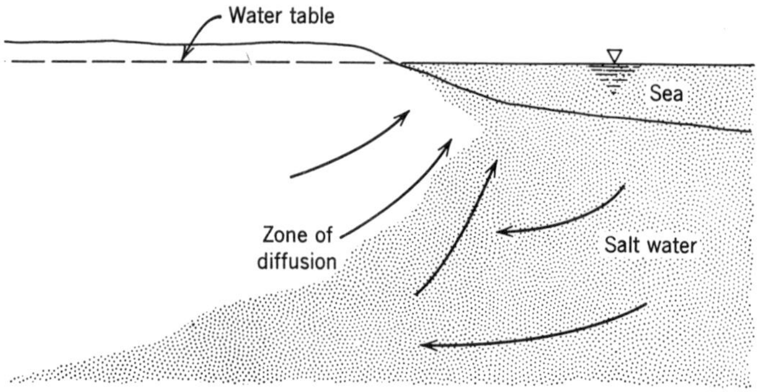

Fig. 7-10 Circulation of salt water from the sea to the zone of dispersion and back to the sea. [After Cooper.]

the medium. The random distribution and orientation of these channels impose a tortuous pattern upon the microscopic streamlines although these streamlines do not intersect each other.

Dispersion may be visualized by the introduction of a slug of dye (of initial volume V_0 and concentration c_0) in a fluid flowing through a porous medium. The center of the slug will travel along a conventional streamline with the average velocity of the entire fluid. As time goes on

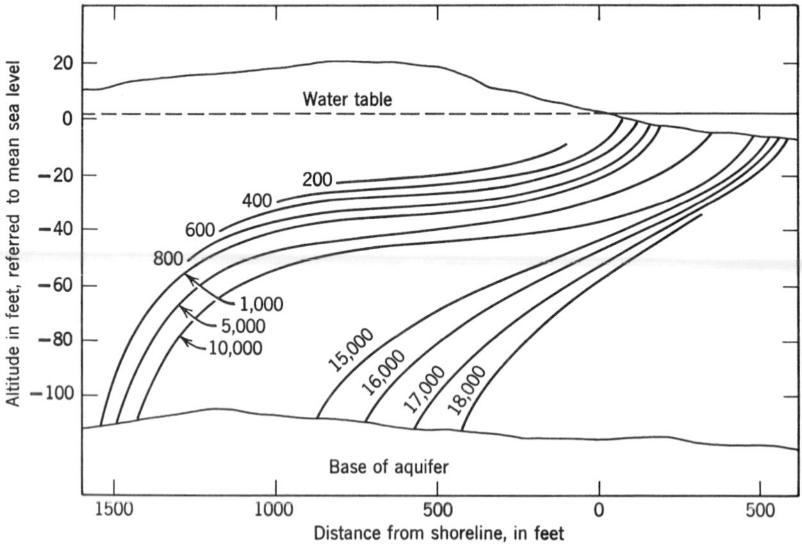

Fig. 7-11 Section through Cutler area, near Miami, Florida, showing the zone of dispersion, 1958; lines represent the chloride content in parts per million. [After Cooper.]

[12], the slug will increase in volume, displace the surrounding fluid, and mix with the fluid in such manner that its concentration c per unit of porous medium may be expressed as

$$\frac{c}{c_0} = (4\pi Dt)^{-3/2} e^{-r^2/4Dt} \tag{7.20}$$

where r indicates any ray emanating from the center of the slug, t represents the time, c_0 is the initial concentration, and D is the coefficient of dispersion introduced by Scheidegger [13]. D has the dimensions L^2T^{-1}. Since concentration may be thought of as the relative density of particles of a given substance in its surroundings, the right side of Eq. 7.20 may also be interpreted as the probability that a given particle after a time t will be found in a unit volume located a distance r away from the expected position of the center of the slug. This spatial distribution [12] of the particles is Gaussian with standard deviation σ

$$\sigma = \sqrt{2Dt} \tag{7.21}$$

If instead of the dispersion of a slug in three dimensions, unidirectional flow is considered and a study is made of the dispersion of a thin layer (say of unit thickness) of infinite extent moving with average velocity u in the direction x perpendicular to the plane of the layer, Eq. 7.20 is replaced by

$$\frac{c}{c_0} = (4\pi Dt)^{-1/2} e^{-(x-ut)^2/4Dt} \tag{7.22}$$

This equation shows that at least in this simplified framework of average unidirectional flow, lateral components of dispersion cancel out or neutralize one another, and the individual particles are normally distributed in the x-direction around the plane of maximum concentration that moves with average velocity $u = x/t$. Equation 7.22 is a solution to the partial differential equation (so-called diffusion equation, by analogy with heat conduction):

$$\frac{\partial c}{\partial t} = D \frac{\partial^2 c}{\partial X^2} \tag{7.23}$$

in which $X = x - ut$, as may be certified by direct insertion of Eq. 7.22 in Eq. 7.23. Here the origin of the coordinate axis $X = 0$ moves with the average velocity u; in other words it is attached to the dispersing layer.

Unidirectional flow with longitudinal dispersion has been tested in the laboratory by means of columns (lucite tubes filled with a water saturated porous medium) and at the entrance of which a tracer solution of given concentration is introduced. The dispersion equation in a coordinate system fixed to the column becomes

$$\frac{\partial c}{\partial t} = D\frac{\partial^2 c}{\partial x^2} - u\frac{\partial c}{\partial x} \tag{7.24}$$

in which u is again the average velocity of flow and in which the other symbols are as defined before.

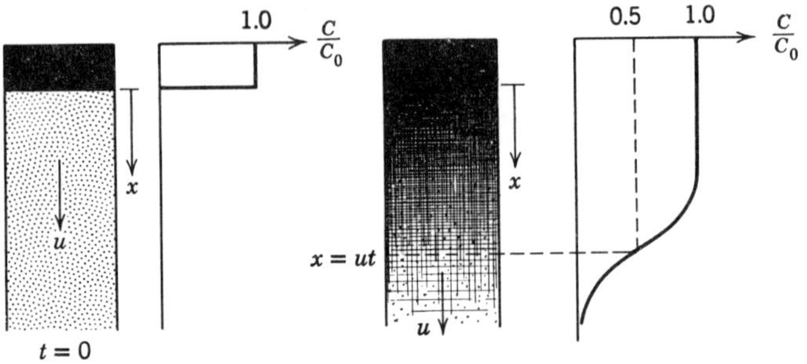

Fig. 7-12 Graphical representation of longitudinal dispersion. [After Harleman and Rumer.]

As shown by Rifai et al. [14], Harleman and Rumer [15], the solution of Eq. 7.24 for the

initial condition $\quad c(x, 0) = 0 \quad x > 0$

and boundary conditions $\quad c(0, t) = c_0 \quad t \geqslant 0$

$\quad c(\infty, t) = 0 \quad t \geqslant 0$

is given, as a first approximation, by (see Fig. 7.12)

$$\frac{c}{c_0} = \tfrac{1}{2}\,\mathrm{erfc}\left(\frac{x - ut}{2\sqrt{Dt}}\right) \tag{7.25}$$

where erfc (x) is the complementary error function =

$$1 - \mathrm{erf}\,(x) = 1 - \frac{2}{\sqrt{\pi}}\int_0^x e^{-\alpha^2}\,d\alpha$$

in which α is a dummy variable of integration.

Actually the solution to Eq. 7.24 for the given conditions contains a second term which may be neglected only for $D < ux$. Although the above variation of the concentration of the tracer in the displacing front is derived for a tube of infinite length, its typical sigmoid shape has been obtained in a similar way as a so-called breakthrough curve for a column of finite length. Here the ratio of the concentration c at the column outlet over the constant concentration c_0 at the column inlet is plotted against the ratio of effluent volume V_e over the water volume V_w in the voids of the column. (The latter ratio is commonly called pore-volume.)

Figure 7.13 represents an experimental breakthrough curve and a curve for piston flow, a fictitious flow in which all the velocities would be the same along parallel streamlines. If the effect of molecular diffusion were neglected, the inflection point of the sigmoid curve would have the coordinates $c/c_0 = 0.5$ and $V_e/V_w = 1.0$. This would essentially mean that

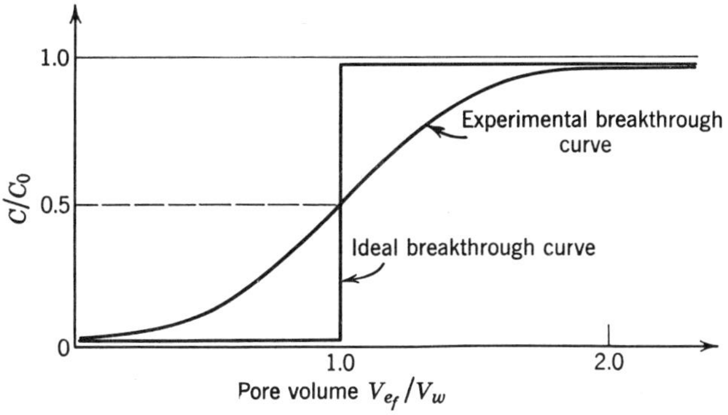

Fig. 7-13 Breakthrough curves. [After Rifai.]

the breakthrough curve has a mean concentration that moves with the average flow velocity. Biggar and Nielsen [16] in particular investigated the effects of molecular or ionic diffusion as part of the general dispersion phenomenon. They proved that molecular diffusion becomes important at small flow velocities and also when the medium consists of a natural soil skeleton instead of an assembly of washed sands or glass beads of uniform diameter size, and when the medium is not completely saturated. Owing to the presence of dead end pores, a significant fraction of the total pore volume of a soil does not contribute to the volume of effluent measured and this results in holdback, or a translation of the breakthrough curve to the left. This translation is a relative measure of the volume of water not displaced but remaining within the sample, and its magnitude depends on the amount of nonmoving water (a characteristic of the soil), so that clays would have breakthrough curves more displaced than sands. Ionic diffusion also becomes important at low flow velocities when the medium is not completely saturated, and this is manifested by a translation of the breakthrough curve to the left, with an early arrival of tracer particles in the effluent, and a slower asymptotic approach of the curve to the value $c/c_0 = 1.0$. Biggar and Nielsen's work clearly showed that intrinsic differences among porous materials such as clays and sands became evident through the shapes and positions of breakthrough curves obtained from saturated and unsaturated media.

Saffman [17] studied a statistical model of porous medium of a more general nature than that proposed by Scheidegger [13] and also suitable for the investigation of lateral dispersion. He objected to Scheidegger's assumption that a particle carries out a random walk consisting of statistically independent straight steps in equal small intervals of time because it would be expected that a particle stays longer in a region where the velocity is small than in one where it is large. Instead he proposed that the path followed by each particle would be a random walk in which length, direction, and duration of each step are random variables. The probability distribution function of the displacement of a fluid particle after a given time may then be calculated and a value for the dispersion obtained. Again the assumption was made that molecular diffusion would be small as compared to mechanical dispersion.

As to the expression for the coefficient D of longitudinal dispersion, there is general agreement among researchers [15] that it is of the form

$$D = D_m + \alpha u^n \tag{7.26}$$

where D_m is the effective molecular diffusion coefficient, u is the average flow velocity, and the value of the exponent n is close to one. For $n = 1.0$, α has the dimension of a length characteristic of the medium only. In a similar way an expression for the lateral dispersion coefficient may be derived. Both dispersion coefficients have been experimentally determined for various porous materials.

The study of miscible displacement is of particular interest to the civil engineer in charge of water supply from coastal aquifers where saline water from the sea intrudes the aquifer and mixes with fresh water. The study of the static and dynamic equilibrium between fresh water and salt water was first treated as the position of an interface between two immiscible fluids. Later [15] effects of dispersion were incorporated in the study using the result available from the above analysis in which the dispersion equation was solved for certain idealized boundary conditions.

7.4 *Flow of a Nonhomogeneous Fluid in a Porous Medium*

Another approach to the study of miscible flow is to consider the zone of transition between salt water and fresh water as occupied by a fluid of variable density and viscosity. Indeed it seems difficult to say whether the flow in this region should be conceived of as a miscible displacement process or as the flow of a nonhomogeneous fluid in a porous medium. So far the method of attack has been predominantly along the lines of a miscible displacement process, but an interesting paper using the approach of a nonhomogeneous fluid flowing in a porous medium was made available in 1960. In this paper Yih [18] tends to exclude completely the phenomenon of dispersion under conditions where the seepage flow has a large macroscopic scale compared with the dimensions of the pores and when the Peclet number[3] based on a macroscopic scale is large in comparison with one. Yih notices that variation in density causes two effects: an inertia effect due to the fact that the inertia of the fluid changes in direct proportion to the density of the fluid, and a gravity effect, also directly proportional to the change in density. Because gravity is not the only force acting on the fluid, inertia and gravity effects do not cancel each other. Furthermore, if viscosity and density of a fluid are variable the equations of the flow in porous medium are nonlinear and in general difficult to solve. However, in the case of steady flow Yih was able to derive an equation for a stream function Ψ', in two-dimensional flow,

$$\nabla^2\Psi' = \frac{kg}{\mu_0}\frac{d\rho}{d\Psi'}\frac{\partial\Psi'}{\partial x} \tag{7.27}$$

in which Ψ' determines a fictitious flow associated with the real flow and is a function of the density ρ alone (i.e., streamlines are isopicnic lines), k is the intrinsic permeability, μ_0 is a reference viscosity, and g is the gravitational constant. Because the velocities of the associated flow u', v' are given as $\frac{\mu}{\mu_0}u$, $\frac{\mu}{\mu_0}v$, streamlines in the real flow remain isopicnic lines and therefore this result closely agrees with findings by Baer and Todd [19], according to which velocities in each point of the transition zone are parallel to the lines of equal density.

A very significant contribution to the theory of flow of chemically inhomogeneous fluids has been made by Lusczynski [20]. Contingent upon the reading of water levels in some observation wells, it allows for the computation of the velocity vector field in a medium where the salt content of the water gradually varies. The value of this theory for engineering purposes cannot be overemphasized because the velocity vector field may be constructed regardless of the often complicated boundary conditions created by the geological nature of the water-bearing strata which preclude a complete analytical solution to the problem.

Lusczynski introduced the concepts of point-water head, fresh-water head, and environmental water head to compute the velocity distribution

[3] $Pe = ud_{50}/D$, in which D is the dispersion coefficient, u is the average flow velocity, and d_{50} is equal to the equivalent diameter of the aperture of a sieve through which passes 50% by weight of the sample.

in the zone of dispersion and to determine the front of the dispersed water in the fresh water. In the sequel we will use his concepts of point-water head and fresh-water head but will replace his definition of environmental water head by one of true environmental water head. This leads to a

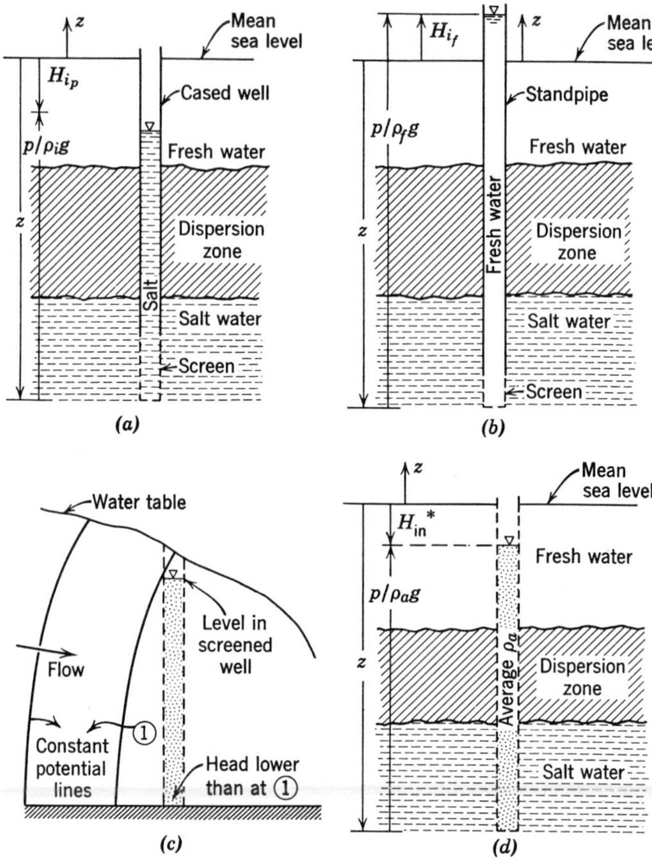

Fig. 7-14 (*a*) Point-water head. (*b*) Fresh-water head. [After Lusczynski.] (*c*) and (*d*) True environmental-water heads. [After De Wiest.]

simpler derivation of the velocity field but it no longer allows for the determination of the fresh water–dispersed water interface. The computation of the latter is given in section 7.5.

(A) POINT-WATER HEAD

Figure 7.14*a* shows a real well cased over its entire depth and screened at the point where the head as a function of the density of the water at the point is measured. Here the well is screened in salt water and filled only with salt water of uniform density as found at the bottom of the well. The overlying fresh or dispersed water that may have entered the well during its construction has been removed. Similarly, point-water head may be determined in the zones of fresh water and dispersed water. Point-water head H_{ip} at a point i in ground water of variable density ρ_i is defined as the water level, measured here from mean sea level as datum plane, in a well filled with an amount of water of the type found at the point and sufficient to balance the existing pressure p at the point. By definition

$$H_{ip} = z + \frac{p}{\rho_i g} \tag{7.28}$$

where the first subscript in H_{ip} refers to the point at which the head is measured and the second subscript indicates that the water in the well is that at the point in question.

(B) FRESH-WATER HEAD

If one end of a standpipe were introduced at the point i of Fig. 7.14*b* and if the pipe were filled with fresh water of density ρ_f only to such an extent that it would balance the existing pressure at point i, the level of the fresh water in the standpipe would measure the fresh-water head H_{if} defined as

$$H_{if} = z + \frac{p}{\rho_f g} \tag{7.29}$$

in which mean sea level again is chosen as datum plane for elevation. Fresh-water heads could only be measured in fresh water with such a device but not in salt nor in dispersed water, whereas point-water heads can always be measured. Therefore it is useful to express fresh-water heads in terms of measurable point-water heads. This is done by stating that the pressure p is unique and therefore the same in Eqs. 7.28 and 7.29. Elimination of p/g from these equations leads to

$$\rho_f H_{if} = \rho_i H_{ip} - z(\rho_i - \rho_f) \tag{7.30}$$

As is indicated in Figs. 7.14*a*,*b* the point-water head for a point in the salt water zone may be negative whereas the fresh-water head of the same point may be positive.

(C) TRUE ENVIRONMENTAL-WATER HEAD

This type of head may be visualized by the water level in a real well screened over its entire depth. Water of variable density would penetrate in such a well and occupy essentially the same position in depth as under natural conditions.

If, for example, a borehole is drilled in an unconfined aquifer of water of constant density and screened over its entire depth, it will fill with water up to a level slightly below the water table (Fig. 7.14*c*). Indeed this level is indicative of the average head over the vertical through a point of the water table because the equipotential lines are curved under dynamic flow conditions. If the water had a variable density with the denser water below, as is the case of Figs. 7.14*a,b* the head at the bottom of the well would even be smaller and hence the true environmental-water head would always be indicated in the well by a water level below the top of the zone of saturation. Define now

$$H_{in}^* = z + \frac{p}{\rho_a g} \tag{7.31}$$

in which H_{in}^* is the environmental-water head and ρ_a is the average density of water between point i and the free surface as indicated in Fig. 7.14*d*.

$$\rho_a = \frac{1}{H_{in}^* - z} \int_z^{H_{in}^*} \rho(x, y, l)\, dl \tag{7.32}$$

in which x, y, and z are the coordinates of the point i and l is a dummy variable of integration. Elimination of the pressure p from Eqs. 7.29 and 7.31 leads to

$$\rho_a H_{in}^* = \rho_f H_{if} - z(\rho_f - \rho_a) \tag{7.33}$$

If it is assumed that under conditions of complete miscibility the mixture in the zone of dispersion behaves as a single phase fluid, the generalized form of Darcy's law for flow of a chemically inhomogeneous fluid may be expressed [21] as

$$\begin{aligned} \mathbf{v} &= \frac{k_i}{\mu_i}(\rho_i \mathbf{g} - \operatorname{grad} p) \\ &= -\frac{k_i g}{\mu_i}\left(\rho_i \operatorname{grad} z + \frac{1}{g}\operatorname{grad} p\right) \end{aligned} \tag{7.34}$$

in which $\mathbf{v}$ is the specific discharge or bulk velocity in point i, k is the intrinsic permeability of the medium, and μ, ρ, p are respectively the dynamic viscosity, density, and pressure of the fluid. The validity of Eq. 7.34 has been questioned by Scheidegger [22] but so far no other expression has been made available to describe macroscopic flow through porous media.

In order to derive an equation for $\mathbf{v}$ in terms of H_{if}, H_{in}^*, ρ_f, and ρ_a, equivalent expressions of the right side of Eq. 7.34 are developed. First, the gradient of Eq. 7.33 is taken, and the result is

$$\rho_a \operatorname{grad} H_{in}{}^* = \rho_f \operatorname{grad} H_{if} + (\rho_a - \rho_f) \operatorname{grad} z + (z - H_{in}{}^*) \operatorname{grad} \rho_a$$

or $\rho_a \operatorname{grad} H_{in}{}^* - (z - H_{in}{}^*)\left[\frac{\partial \rho_a}{\partial x}\mathbf{u}_x + \frac{\partial \rho_a}{\partial y}\mathbf{u}_y\right]$

$$= \rho_f \operatorname{grad} H_{if} + (z - H_{in}{}^*)\frac{\partial \rho_a}{\partial z}\mathbf{u}_z + (\rho_a - \rho_f)\mathbf{u}_z \quad (7.35)$$

where $\mathbf{u}_x$, $\mathbf{u}_y$, $\mathbf{u}_z$ are unit vectors along the coordinate axes. Furthermore, differentiation of ρ_a with respect to z gives

$$\frac{\partial \rho_a}{\partial z} = \frac{\rho_a - \rho_i}{H_{in}{}^* - z} + \frac{\partial H_{in}{}^*}{\partial z}\frac{\rho_f - \rho_a}{H_{in}{}^* - z} \quad (7.36)$$

and Eq. 7.35 becomes, after insertion of Eq. 7.36:

$$\rho_a\left(\frac{\partial H_{in}{}^*}{\partial x}\mathbf{u}_x + \frac{\partial H_{in}{}^*}{\partial y}\mathbf{u}_y + \frac{\partial H_{in}{}^*}{\partial z}\mathbf{u}_z\right) + (\rho_f - \rho_a)\frac{\partial H_{in}{}^*}{\partial z}\mathbf{u}_z$$

$$- (z - H_{in}{}^*)\left[\frac{\partial \rho_a}{\partial x}\mathbf{u}_x + \frac{\partial \rho_a}{\partial y}\mathbf{u}_y\right] = \rho_f \operatorname{grad} H_{if} + (\rho_i - \rho_f)\mathbf{u}_z$$

and finally

$$\rho_a\left[\frac{\partial H_{in}{}^*}{\partial x}\mathbf{u}_x + \frac{\partial H_{in}{}^*}{\partial y}\mathbf{u}_y\right] + \rho_f\frac{\partial H_{in}{}^*}{\partial z}\mathbf{u}_z - (z - H_{in}{}^*)$$

$$\times\left[\frac{\partial \rho_a}{\partial x}\mathbf{u}_x + \frac{\partial \rho_a}{\partial y}\mathbf{u}_y\right] = \rho_f \operatorname{grad} H_{if} + (\rho_i - \rho_f)\mathbf{u}_z \quad (7.37)$$

Also, the gradient of Eq. 7.29 renders, after addition of $\rho_i \operatorname{grad} z$ to both members of the equation

$$\rho_i \operatorname{grad} z + \frac{1}{g}\operatorname{grad} p = \rho_f \operatorname{grad} H_{if} + (\rho_i - \rho_f)\mathbf{u}_z \quad (7.38)$$

In the case of an anisotropic medium, if k_x, k_y, and k_z are the intrinsic permeabilities in the principal directions, coinciding with the coordinate axes, the components of Eq. 7.34 may be written as

$$v_x = -\frac{k_x g}{\mu_i}\left[\rho_i \operatorname{grad} z + \frac{1}{g}\operatorname{grad} p\right]_x$$

$$v_y = -\frac{k_y g}{\mu_i}\left[\rho_i \operatorname{grad} z + \frac{1}{g}\operatorname{grad} p\right]_y$$

$$v_z = -\frac{k_z g}{\mu_i}\left[\rho_i \operatorname{grad} z + \frac{1}{g}\operatorname{grad} p\right]_z$$

Because the right sides of Eqs. 7.37 and 7.38 are the same, both sides of Eq. 7.37 may be used as equivalents of the left side of Eq. 7.38 and be inserted in Eq. 7.34 to render the velocity components of the specific discharge. The interesting feature of Eq. 7.37 is that its left side, when projected on a vertical, reduces to $\rho_f(\partial H_{in}^*/\partial z)$, which is simply a product of fresh-water density and a gradient defined by environmental-water heads. Also, along any horizontal l, the right side of Eq. 7.37 reduces to $\rho_f(\partial H_{if}/\partial l)$, which is simply a product of fresh-water density and a directional derivative defined by fresh-water heads. Therefore,

$$\begin{aligned} v_x &= -\frac{k_x g}{\mu_i}\rho_f\frac{\partial H_{if}}{\partial x} \\ v_y &= -\frac{k_y g}{\mu_i}\rho_f\frac{\partial H_{if}}{\partial y} \\ v_z &= -\frac{k_z g}{\mu_i}\rho_f\frac{\partial H_{in}^*}{\partial z} \end{aligned} \tag{7.39}$$

When the velocity components are computed by means of Eqs. 7.39 and when the resultant velocity is constructed, it will follow that the direction of flow is usually not parallel to the resultant hydraulic gradient.

To conclude, in this modification of Lusczynski's theory the advantage resides in the simple measurement of H_{in}^* and of the water level in the screened well as it penetrates deeper in the soil and as the density of the water changes.

7.5 *Correlation between Lusczynski's and Hubbert's Work*

To make this correlation easily it is necessary to illustrate Lusczynski's artificial but ingenious concept of environmental head. It is based on a standpipe device (Fig. 7.15) similar to the one used in the definition of fresh-water head. If in the standpipe of Fig. 7.14*b* the fresh water were replaced by dispersed water in the zone of dispersion and by salt water in the zone of salt water, the water level in the standpipe would drop and be measured by H_{in} above mean sea level, numerically smaller than H_{if} of Fig. 7.14*b*. H_{in} is Lusczynski's environmental-water head. Let z_r be the elevation of any reference point in the zone of fresh water and define the average density $\bar{\rho}_a$ of water between point i and point z_r as

$$\bar{\rho}_a = \frac{1}{z_r - z}\int_z^{z_r} \rho(x, y, l)\, dl \tag{7.40}$$

in which again x, y, and z are the coordinates of point i and l is a dummy variable of integration. Because the water above z_r is fresh, the pressure p in i may be expressed as

$$p = (z_r - z)\bar{\rho}_a g + (H_{in} - z_r)\rho_f g \qquad (7.41)$$

Elimination of p/g between Eqs. 7.29 and 7.41 gives

$$\rho_f H_{in} = \rho_f H_{if} - (z - z_r)(\rho_f - \bar{\rho}_a) \qquad (7.42)$$

It is evident that H_{in} cannot be measured in the field and that it should be

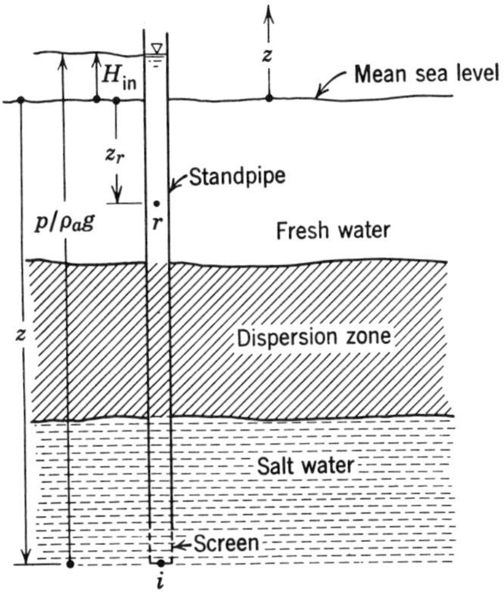

Fig. 7-15 Environmental-water head. [After Lusczynski.]

expressed in terms of H_{ip}. This is done by the insertion of the value of $\rho_f H_{if}$ from Eq. 7.30 in Eq. 7.42. The result is

$$\rho_f H_{in} = \rho_i H_{ip} - z(\rho_i - \bar{\rho}_a) - z_r(\bar{\rho}_a - \rho_f) \qquad (7.43)$$

Equation 7.43 may be used to determine the contact between zones of fresh water and dispersed water. Therefore it is combined with a computation (after measurement of point-water heads) of environmental-water heads at two points, one (1) arbitrarily taken in the fresh water zone and the other (2) taken along the same vertical as the first, somewhere in the salt water zone.

In these computations z_r assumes a special value in Eq. 7.43, namely z_d, the coordinate of the contact between fresh water and dispersed water along the vertical passing through points 1 and 2. For point 2 and the special choice of z_d, Eq. 7.43 becomes

$$\rho_f H_{2n} = \rho_2 H_{2p} - z_2(\rho_2 - \bar{\rho}_a) - z_d(\bar{\rho}_a - \rho_f) \tag{7.44}$$

in which $\rho_f = \rho_1$.

If the term $\rho_1(H_{1n} - H_{2n})$ is added to both sides of this equation, the result is

$$\rho_1 H_{1n} = \rho_1(H_{1n} - H_{2n}) + \rho_2 H_{2p} - z_2(\rho_2 - \bar{\rho}_a) - z_d(\bar{\rho}_a - \rho_1) \tag{7.45}$$

Although $\bar{\rho}_a$ in Eq. 7.45 depends on z_d it is possible to solve this equation for z_d by iteration. Indeed it is possible to measure or compute all quantities in this equation except $\bar{\rho}_a$ and z_d. A first approximation $z_d{}^*$ may be obtained from Eq. 7.45 by assuming $\bar{\rho}_a = \rho_2$. With this value of $z_d{}^*$, $\bar{\rho}_a{}^*$ is computed from Eq. 7.40, namely

$$\frac{1}{z_d{}^* - z_2}\int_{z_2}^{z_d{}^*} \rho\, dz = \bar{\rho}_a{}^*$$

This value of $\bar{\rho}_a{}^*$ is now inserted in Eq. 7.45 and a second value z_d^{**} is found. With the help of Eq. 7.40, a second value $\bar{\rho}_a^{**}$ is obtained, which again is inserted in Eq. 7.45 to render a third value z_d^{***}. The process is carried on until the values of z_d converge to a limit.

Equation 7.45 is the only simple equation available to compute the depth of the zone of dispersed water. Lucszynski compared the equation with Eq. 7.15 of the Ghyben-Herzberg theory which may be written in the form

$$\rho_1 H_{1n} = -z_d{}'(\rho_2 - \rho_1) \tag{7.46}$$

by a mere change of symbols. The prime of $z_d{}'$ is used to remind the reader that there is no zone of dispersion in this elementary theory. This would show that the Ghyben-Herzberg formula is but a special case of Eq. 7.45, in which $\bar{\rho}_a = \rho_2$, $H_{1n} = H_{2n}$ and $H_{2p} = 0$. These three conditions respectively require that there be no zone of dispersion, that there be no vertical velocity components due to differences in environmental-water head between points along the vertical in salt-water and fresh-water zones, and that the point-water head in the salt-water zone be negligible.

Perlmutter et al. [23] tried to determine the contact between fresh water and salt water using a formula adapted from Hubbert's Eq. 7.11. Indeed, if it were possible to establish two standpipes tied to each other so that their ends at depth z were to terminate upon the sharp interface between immiscible fresh and salt water, and if one pipe were to fill only with fresh water to a height h_f, the other to fill only with salt water to a height h_s, both measured from sea level, then (Fig. 7.16)

$$z = \frac{\rho_s}{\rho_s - \rho_f}\, h_s - \frac{\rho_f}{\rho_s - \rho_f}\, h_f \tag{7.47}$$

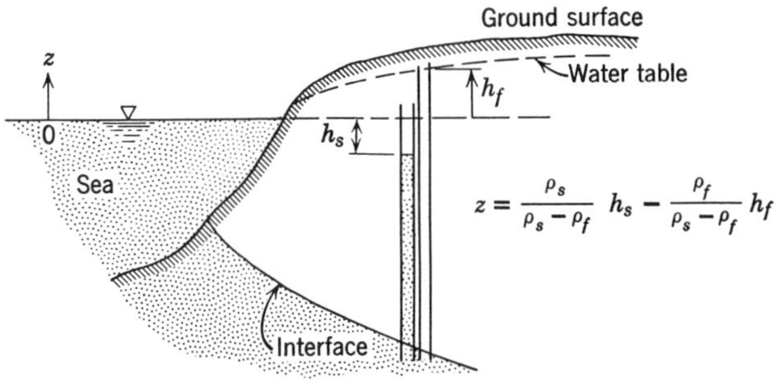

Fig. 7-16 Measuring of depth to interface with help of double standpipe.

Perlmutter et al. applied this formula to observations made in one well screened at points 350 ft apart above and below the zone of dispersion. They were aware of the fact that this procedure did not comply with the conditions under which Eq. 7.47 was derived and that the error thus introduced would be minimum for a narrow zone of dispersion and for screens as near as possible to each other but still separated by a distance equal to the thickness of the dispersion zone. Lusczynski noticed that Eq. 7.47 may be written as

$$\rho_f h_f = \rho_s h_s - z(\rho_s - \rho_f)$$

or in his notations as

$$\rho_1 H_{1n} = \rho_2 H_{2p} - z_d''(\rho_2 - \rho_1) \qquad (7.48)$$

which is again a special case of Eq. 7.45 in which $\bar{\rho}_a = \rho_2$, and $H_{1n} = H_{2n}$, i.e., one less condition than was required for the Ghyben-Herzberg formula. It means that Hubbert's equation does not neglect the point-water head in the salt-water zone. The double prime of z_d again is used to remind the reader that in Hubbert's approach the zone of dispersion is not considered. Finally, Lusczynski's theory was applied to data communicated by the Government Institute of Water Supply in the Netherlands (Fig. 7.17). This figure shows the various heads in a coastal ground-water system near The Hague, The Netherlands, and a picture of the wedge of dispersed water.

7.6 *Recent Developments*

Considerable progress has recently been made in the experimental and analytical investigation of transfers of water in porous media in multiple-phase conditions [24]. The treatment or even a review of this work is beyond the scope of this elementary textbook. It suffices to quote, beside some papers of researchers already mentioned before (references 25, 26,

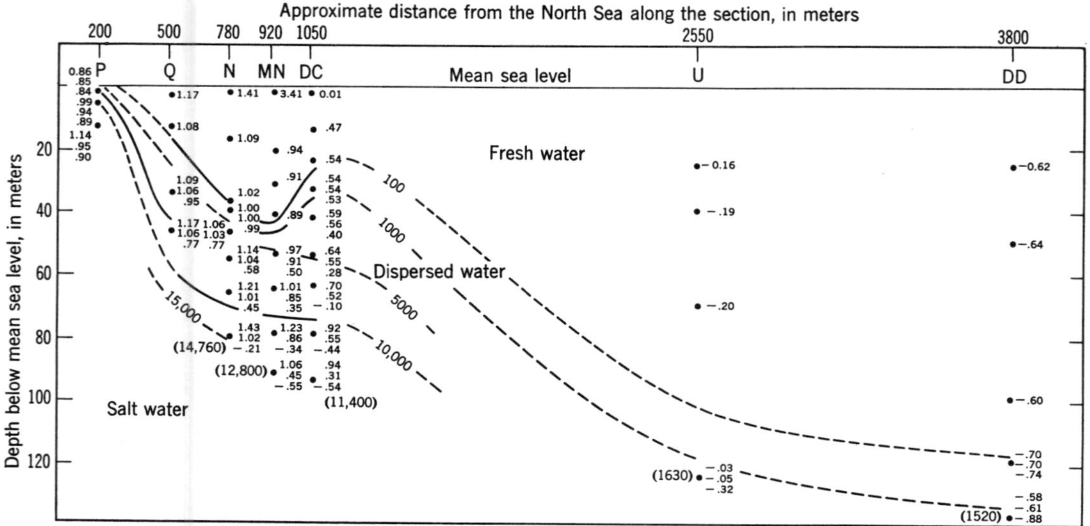

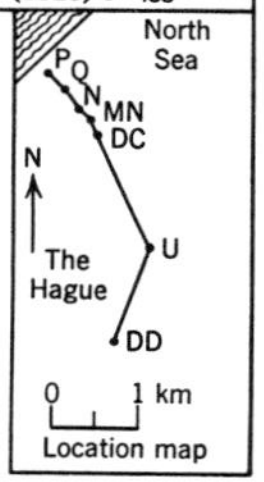

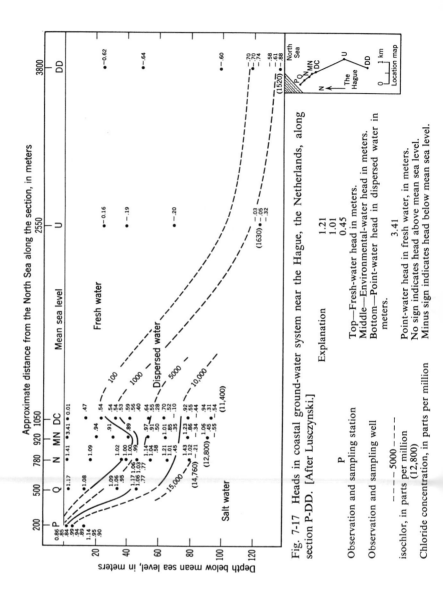

Fig. 7-17 Heads in coastal ground-water system near the Hague, the Netherlands, along section P-DD. [After Lusczynski.]

Explanation

P
Observation and sampling station

Observation and sampling well

– – – – 5000 – – – –
isochlor, in parts per million

(12,800)
Chloride concentration, in parts per million

1.21
1.01
0.45
Top—Fresh-water head in meters.
Middle—Environmental-water head in meters.
Bottom—Point-water head in dispersed water in meters.

3.41
Point-water head in fresh water, in meters.
No sign indicates head above mean sea level.
Minus sign indicates head below mean sea level.

27, 28, 29, 30, 40, 41, 42 and 43), the work by Josselin de Jong [31, 38, 39], Ogata and Banks [32, 33], and Simpson [34]. Stallman [35] reviewed those theories of multiphase fluids in porous media which are pertinent to hydrologic studies. His study emphasizes unsaturated flow conditions in which the thermodynamics of porous media play a vital role. The topic of fluid transfer in the vapor phase was intensively treated at the RILEM [36] Symposium on transfer of water in porous media.

REFERENCES

1. Hubbert, M. K., "Entrapment of Petroleum under Hydrodynamic Conditions," *Bulletin of the American Association of Petroleum Geologists*, Vol. 37, pp. 1954–2026 (1953).
2. Hubbert, M. K., "The Theory of Ground-Water Motion," *The Journal of Geology*, Vol. 48, No. 8, pp. 785–944 (1940).
3. Badon Ghyben, W., "Nota in Verband met de Voorgenomen Putboring nabij Amsterdam," *Tijdschrift van het koninklyk Instituut van Ingenieurs*, p. 21, The Hague, 1888–1889.
4. Herzberg, A., "Die Wasserversorgung einiger Nordseebader," *Journal Gasbeleuchtung und Wasserversorgung*, Vol. 44, pp. 815–819, 842–844, Munich, 1901.
5. Wentworth, C. K., "Storage Consequences of the Ghyben-Herzberg Theory," *Transactions American Geophysical Union*, pp. 683–693 (1942).
6. Hubbert, M. K., *The Theory of Ground-Water Motion*, p. 95.
7. Glover, R. E., "The Pattern of Fresh-Water Flow in a Coastal Aquifer," *Journal of Geophysical Research*, Vol. 64, No. 4, pp. 457–459 (April 1959).
8. De Wiest, R. J. M., "Free Surface Flow in Homogeneous Porous Medium," *ASCE Transactions*, Vol. 127, Chapter 1, pp. 1045–1089, Eq. 20, p. 1050 (1962).
9. Cooper, H. H., "A Hypothesis Concerning the Dynamic Balance of Fresh Water and Salt Water in a Coastal Aquifer," *Journal of Geophysical Research*, Vol. 64, No. 4, pp. 461–467 (April 1959).
10. De Wiest, R. J. M., *Sur les phénomènes de Dispersion et de Diffusion dans les Milieux Poreux*, Paper presented at the RILEM (Réunion Internationale des Laboratoires d'Essais et de Recherches sur les Matériaux et les Constructions) Symposium on Transfer of Water in Porous Media, Paris, April 8–10, 1964. For a translation of this paper, see *Ground-Water Journal of the NWWA*, Vol. 2, No. 3, 1964.
11. Kohout, F. A., "Fluctuations of Ground-Water Levels Caused by Dispersion of Salts," *Journal of Geophysical Research*, Vol. 66, No. 8, pp. 2429–2434 (August 1961).
12. Day, P. R., "Dispersion of a Moving Salt-Water Boundary Advancing through Saturated Sand," *Transactions American Geophysical Union*, Vol. 39, No. 1, pp. 67–74 (February 1958).
13. Scheidegger, A. E., "Statistical Hydrodynamics in Porous Media," *Journal of Applied Physics*, Vol. 25, No. 8, pp. 997–1001 (August 1954).
14. Rifai, M. N. E. et al., "Dispersion Phenomena in Laminar Flow through Porous Media," *Sanitary Engineering Research Laboratory Report No.* 3, University of California, Berkeley, 1956 (157 pp.).
15. Harleman, D. R. F. and R. R. Rumer, "The Dynamics of Salt-Water Intrusion in

Porous Media," *Civil Engineering Department Report No.* 55, Massachusetts Institute of Technology, August, 1962 (125 pp.).

16. Biggar, J. W. and D. R. Nielsen, "Diffusion Effects in Porous Materials," *Journal of Geophysical Research*, Vol. 65, pp. 2887–2895 (September 1960).
17. Saffman, P. G., "A Theory of Dispersion in a Porous Medium," *Journal of Fluid Mechanics*, Vol. 6, pp. 321–349 (1959).
18. Yih, C. S., "Flow of a Non-Homogeneous Fluid in a Porous Medium," *Journal of Fluid Mechanics*, Vol. 10, pp. 133–140 (1960).
19. Bear, J. and D. K. Todd, "The Transition Zone between Fresh and Salt Waters in Coastal Aquifers," *Water Resources Center Contribution No.* 29, University of California, Berkeley, September 1960.
20. Lusczynski, N. J., "Head and Flow of Ground Water of Variable Density," *Journal of Geophysical Research*, Vol. 66, No. 12, pp. 4247–4256 (December 1960).
21. Offeringa, T. and C. Van der Poel, *Transactions American Institute of Mining, Metallurgical Petroleum Engineers*, p. 310 (1954).
22. Scheidegger, A. E., *The Physics of Flow Through Porous Media*, pp. 256–258, Macmillan, 1960.
23. Perlmutter, N. M., J. J. Geraghty, and J. E. Upson, "The Relation between Fresh and Salty Ground-Water in Southern Nassau and Southeastern Queens Counties, Long Island, New York," *Economic Geology*, Vol. 54, pp. 416–435.
24. Knudsen, W. C., "Equations of Fluid Flow through Porous Media—Incompressible Fluid of Varying Density," *Journal of Geophysical Research*, Vol. 67, pp. 733–737 (February 1962).
25. Bear, J., "On the Tensor Form of Dispersion in Porous Media, *Journal of Geophysical Research*, Vol. 66, pp. 1185–1197 (April 1961).
26. Bear, J., "Some Experiments in Dispersion," *Journal of Geophysical Research*, Vol. 66, pp. 2455–2467 (August 1961).
27. Scheidegger, A. E., "General Theory of Dispersion in Porous Media," *Journal of Geophysical Research*, Vol. 66, pp. 3273–78 (October 1961).
28. Fara, H. D. and A. E. Scheidegger, "Statistical Geometry of Porous Media," *Journal of Geophysical Research*, Vol. 66, pp. 3279–3284 (October 1961).
29. Harleman, D. R. F. and R. R. Rumer, "Longitudinal and Lateral Dispersion in an Isotropic Porous Medium," *Journal of Fluid Mechanics*, Vol. 16, Part 3, pp. 385–394 (1963).
30. Rumer, R. R. and D. R. F. Harleman, "Intruded Salt-Water Wedge in Porous Media," *ASCE Journal of the Hydraulics Division*, pp. 193–220 (November 1963).
31. Josselin de Jong, G., "Longitudinal and Transverse Diffusion in Granular Deposits," *Transactions American Geophysical Union*, Vol. 39, pp. 67–74 (1958).
32. Ogata Akio and R. B. Banks, "A Solution of the Differential Equations of Longitudinal Dispersion in Porous Media," *USGS, Professional Paper* 411-A, Washington, 1961.
33. Ogata Akio, "Transverse Diffusion in Saturated Isotropic Granular Media," *USGS Professional Paper* 411-B, Washington, 1961.
34. Simpson, E. S., "Transverse Dispersion in Liquid Flow through Porous Media," *USGS Professional Paper*, 411-C, Washington, 1962.
35. Stallman, R. W., "Multiphase Fluids in Porous Media, A Review of Theories Pertinent to Hydrologic Studies," *USGS, Professional Paper* 411-E, Washington, 1964.
36. RILEM (Réunion Internationale des Laboratoires d'Essais et de Recherches sur les Matériaux et les Constructions), International Symposium on Transfer of Water in Porous Media, Paris, April 8–10, 1964.

37. Todd, D. K. and L. Huisman, "Ground-Water Flow in the Netherlands Coastal Dunes," *ASCE Journal of the Hydraulics Division*, pp. 63–81 (July 1959).
38. De Josselin de Jong, G., "L'entrainement de particules par le courant intersticiel," *International Association of Scientific Hydrology, Publication* No. 41, pp. 139–147 (1956).
39. De Josselin de Jong, G., "Singularity Distributions for the Analysis of Multiple-Fluid Flow Through Porous Media" *Journal of Geophysical Research*, Vol. 65, pp. 3739–3758 (November 1960).
40. Bear, J. and Zaslavsky, D. *Underground Storage and Mixing of Water*, *Progress Report No.* 1, Technion, Haifa, Israel, 1962 (102 pp.).
41. Jacobs, M. and S. Schmorak. *Sea Water Intrusion and Interface Determination Along the Coastal Plain of Israel*, Hydrological Service of Israel, Jerusalem, Israel, 1960 (12 pp.).
42. Bear, J. and G. Dagan. "Intercepting Fresh Water Above the Interface in a Coastal Aquifer," *International Association of Scientific Hydrology, Publication No.* 64, pp. 154–181 (1963).
43. Bear, J. and G. Dagan, "Some Exact Solutions of Interface Problems by Means of the Hodograph Method," *Journal of Geophysical Research*, Vol. 69, pp. 1563–1572 (April 1964).

CHAPTER EIGHT

Numerical and Experimental Methods in Ground-Water Flow

8.1 *Introduction*

This chapter deals almost exclusively with model techniques, the word model being used in the hydraulic sense [23] meaning the equivalent of an analog. Often analog and model are linked together or used as synonyms, although, as pointed out by Irmay [24], it is worthwhile to have consistency in the definitions of model, analog, and analogy. According to Irmay "a model of a porous medium is a simplified porous medium or system bearing some similarity to the prototype. Some models are called analogies, as they are very simplified porous media, e.g., bundles of parallel tubes, flow around a single sphere, etc. Analogs are physical systems or mathematical models obeying partial differential equations with boundary conditions similar to those in the prototype, e.g., the Hele-Shaw viscous flow analog; the electrolytic tray analog, etc." Irmay's definition has been observed by the author of this book in his studies of mathematical models of aquifers [17] and it is his intention to conform to Irmay's definition of the analog as a physical system obeying the aforementioned criteria.

Emphasis in this chapter is placed on analogs which are at present most frequently used in studies of ground-water flow. For a complete description of analogs which are not treated here, the reader is referred to the specialized text of Karplus [18].

8.2 *Grapho-Numerical Analysis—Finite Differences* [1]

The graphical method for two-dimensional flow which was treated extensively in Chapter 5 may be completed and refined with the help of a property of the harmonic function Φ which also follows when Laplace's equation is replaced by its corresponding finite-difference equation. In potential theory it is proved that the value of the function Φ in some point 0 is equal to the average of its values along a circle centered around the point 0

$$\Phi_o = \frac{1}{2\pi} \int_o^{2\pi} \Phi_c \, d\theta \tag{8.1}$$

in which Φ_0 is the value of the harmonic function in point 0 and Φ_c is its value in any point C of the contour. In first approximation, the value of Φ_o is

$$\Phi_o = \frac{\Phi_1 + \Phi_2 + \cdots + \Phi_n}{n} \tag{8.2}$$

in which $\Phi_1, \Phi_2, \cdots, \Phi_n$ are values of Φ in equidistant points along the perimeter. In particular, if only four points are taken,

$$\Phi_o = \frac{\Phi_1 + \Phi_2 + \Phi_3 + \Phi_4}{4} \tag{8.3}$$

where the points are in the vertices of a square, Φ_o is the value of the function in the center of the square.

The same result is obtained when a finite-difference approximation of $\nabla^2\Phi = 0$ is obtained. Therefore (Fig. 8.1), an equilateral coordinate grid is superimposed upon the flow field and the value of the potential Φ is only sought at the node points of the grid. Approximations for the first

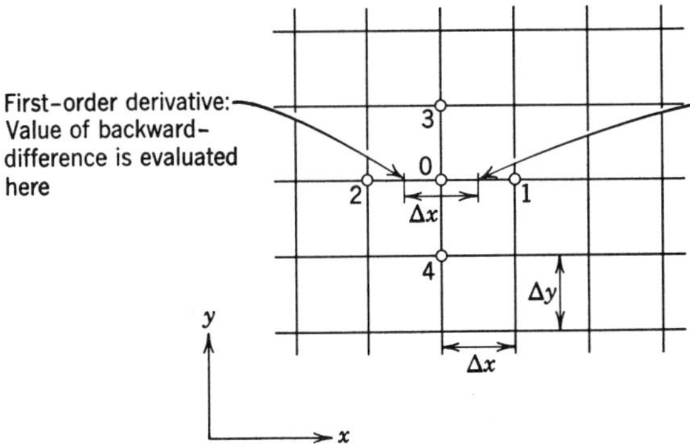

Fig. 8-1 Finite difference approximation of $V^2\Phi = 0$. Equilateral grid. The second order derivative in the x-direction is

$$\left(\frac{\partial^2\Phi}{\partial x^2}\right)_0 \cong \frac{\left(\dfrac{\partial\Phi}{\partial x}\right)_{0-1} - \left(\dfrac{\partial\Phi}{\partial x}\right)_{2-0}}{\Delta x}.$$

and second order derivatives are found in terms of the node potentials. If potentials Φ_0, Φ_1, Φ_2, Φ_3, Φ_4 are assumed to exist at nodes 0, 1, 2, 3, 4 respectively, the average values for the first order derivatives at points midway between node 0 and the four adjacent nodes are approximated as

$$\left(\frac{\partial\Phi}{\partial x}\right)_{0-1} \cong \frac{\Phi_1 - \Phi_0}{\Delta x} \qquad \left(\frac{\partial\Phi}{\partial y}\right)_{0-3} \cong \frac{\Phi_3 - \Phi_0}{\Delta y} \tag{8.4}$$

$$\left(\frac{\partial\Phi}{\partial x}\right)_{2-0} \cong \frac{\Phi_0 - \Phi_2}{\Delta x} \qquad \left(\frac{\partial\Phi}{\partial y}\right)_{4-0} \cong \frac{\Phi_0 - \Phi_4}{\Delta y} \tag{8.5}$$

Equations 8.4 are called forward-difference approximations and Eqs. 8.5 are backward-difference approximations. The second order derivatives follow as the rate of change of the first order derivatives (see Fig. 8.1), or as the difference between forward- and backward-differences divided by the distance, i.e.,

$$\begin{aligned}\left(\frac{\partial^2\Phi}{\partial x^2}\right)_0 &\cong \frac{1}{(\Delta x)^2}(\Phi_1 + \Phi_2 - 2\Phi_0)\\ \left(\frac{\partial^2\Phi}{\partial y^2}\right)_2 &\cong \frac{1}{(\Delta y)^2}(\Phi_3 + \Phi_4 - 2\Phi_0)\end{aligned} \tag{8.6}$$

The Laplacian in two dimensions then becomes

$$\nabla^2\Phi = \frac{\partial^2\Phi}{\partial x^2} + \frac{\partial^2\Phi}{\partial y^2} \cong \frac{1}{(\Delta x)^2}(\Phi_1 + \Phi_2 - 2\Phi_0) + \frac{1}{(\Delta y)^2}(\Phi_3 + \Phi_4 - 2\Phi_0) \tag{8.7}$$

For an equilateral grid, $\Delta x = \Delta y$, and Laplace's equation may be replaced for node 0 and its adjacent nodes by

$$\Phi_1 + \Phi_2 + \Phi_3 + \Phi_4 - 4\Phi_0 = 0 \tag{8.8}$$

which may be generalized for the potential $\Phi_{k,i}$ at the intersection of the kth horizontal and ith vertical line of the grid to

$$\Phi_{k+1,i} + \Phi_{k;i+1} + \Phi_{k-1,i} + \Phi_{k,i-1} - 4\Phi_{k,i} = 0 \tag{8.9}$$

Equations 8.8 and 8.3 are obviously equivalent. They are used to correct the flow net of Fig. 8.2*a*, which is a first trial, in the following way. Let the value of the potential Φ along *AB* be 4.5, along *CD* zero (the normal derivatives of Φ vanish along *AFC* and *BE*). By interpolation between the equipotential lines approximate values of Φ in the nodes of the grid are found. The potential at any node is now found in terms of that of the

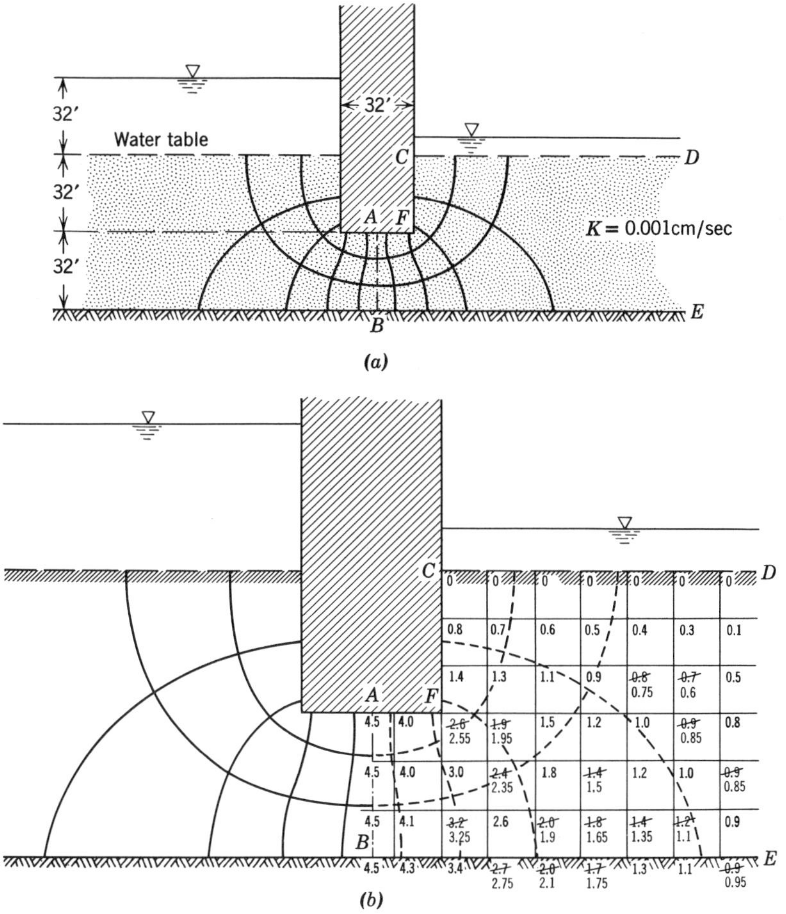

Fig. 8-2 Grapho-numerical analysis of seepage below a cut-off wall. (*a*) above: first trial net; (*b*) below: net corrected by relaxation method.

four adjacent nodes by successive relaxation of each node, whereby the new values of Φ after a series of relaxations converge to the true value of Φ at the node. The more accurate the first trial, the fewer steps will be needed in the relaxation process. Figure 8.2*b* shows the results of the successive steps. Finally the equipotential lines Φ = constant are redrawn, based on the adjusted values of Φ.

The streamlines Ψ = constant may be drawn in the same way, independently, by assigning, for example, a value of 3 to Ψ along *AFC* and a value of zero along *BE*. Orthogonality of the two nets thus drawn constitutes a check of the method.

8.3 *The Parallel-Plate Model or Hele-Shaw Apparatus*

(A) INTRODUCTION

In 1899 Sir G. G. Stokes [2] presented a mathematical analysis of the steady flow of a viscous liquid between closely-spaced parallel plates and thus proved that such flow can be derived from a potential as had been assumed by Hele-Shaw [3, 4], who designed the first model of this kind in 1897. Since then the Hele-Shaw model has been used extensively to simulate two-dimensional laminar flow of water through porous soil which may be expressed by the same differential equation as the flow of the liquid in the model. The proof presented by Stokes may be found in

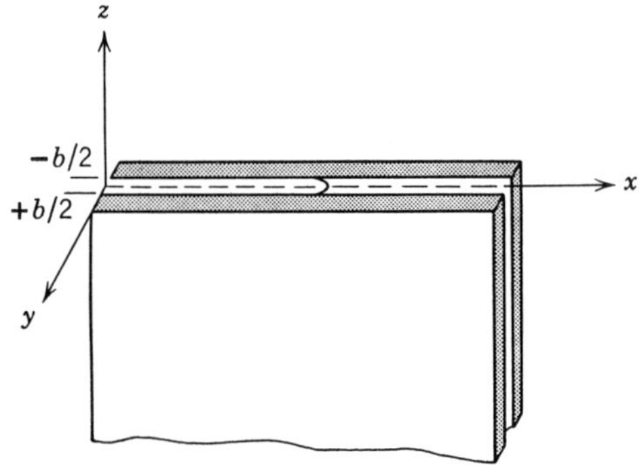

Fig. 8-3 Sketch of flow between two vertically erected parallel plates.

many textbooks [5, 6, 7], and valuable information about the construction and scaling of the model has become available through numerous publications [8, 9, 10, 11, 12, 13, 14, 48, 49]. Stokes' analysis for the case of a model with vertical plates (Fig. 8.3) may be summarized as follows.

The Navier-Stokes equations for the flow between the plates follow by extension to three dimensions from Eqs. 4.27 and 4.28, in which $n = 1$ and in which u, v, and w are the velocity components respectively in the x,y,z-direction. The velocity component v perpendicular to the plates is zero and therefore the Navier-Stokes equations become

$$\begin{aligned}
\frac{\partial u}{\partial t} + u\frac{\partial u}{\partial x} + w\frac{\partial u}{\partial z} &= -\frac{1}{\rho}\frac{\partial p}{\partial x} + \frac{\mu}{\rho}\left(\frac{\partial^2 u}{\partial x^2} + \frac{\partial^2 u}{\partial y^2} + \frac{\partial^2 u}{\partial z^2}\right) \\
0 &= -\frac{1}{\rho}\frac{\partial p}{\partial y} \\
\frac{\partial w}{\partial t} + u\frac{\partial w}{\partial x} + w\frac{\partial w}{\partial z} &= -\frac{1}{\rho}\frac{\partial p}{\partial z} + \frac{\mu}{\rho}\left(\frac{\partial^2 w}{\partial x^2} + \frac{\partial^2 w}{\partial y^2} + \frac{\partial^2 w}{\partial z^2}\right) - g
\end{aligned} \tag{8.10}$$

The viscous forces are so large for this kind of flow that the inertia terms (which constitute the left sides of the equation) may be neglected. Furthermore, the changes of u and w and of their derivatives in the y-direction are much larger than those of u and w in the x-z plane. Therefore the second derivatives of the velocity components in the x- and z-directions are neglected in comparison with $\partial^2 u/\partial y^2$ and $\partial^2 w/\partial y^2$, and Eqs. 8.10 become

$$0 = -\frac{\partial p}{\partial x} + \mu \frac{\partial^2 u}{\partial y^2}$$

$$0 = -\frac{\partial p}{\partial y} \tag{8.11}$$

$$0 = -\frac{\partial p}{\partial z} + \mu \frac{\partial^2 w}{\partial y^2} - \rho g$$

From the second of Eqs. 8.11, it follows that the pressure p is not a function of y. The remaining Eqs. 8.11, which are in fact ordinary differential equations of u and w in function of y, may be integrated without difficulty. It is noticed that u and w vanish for $y = \pm b/2$, in which b is the spacing of the plates. For example, the equation for w renders, after twofold integration

$$\mu w = \frac{y^2}{2} \frac{\partial}{\partial z}(p + \rho g z) + c_1 y + c_2$$

The boundary conditions require $c_1 = 0$ and $c_2 = -\dfrac{b^2}{8} \dfrac{\partial}{\partial z}(p + \rho g z)$ so that

$$w = \frac{1}{2\mu}\left(y^2 - \frac{b^2}{4}\right)\frac{\partial}{\partial z}(p + \rho g z) \tag{8.12}$$

In a similar way, the expression for u is derived as

$$u = \frac{1}{2\mu}\left(y^2 - \frac{b^2}{4}\right)\frac{\partial}{\partial x}(p + \rho g z) \tag{8.13}$$

The average velocities over the distance b between the two plates are found by simple integration

$$u_{\text{ave}} = \frac{1}{b}\int_{-b/2}^{b/2} u\, dy = \frac{1}{2\mu b}\left(\frac{y^3}{3} - \frac{b^2}{4} y\right)_{-\frac{b}{2}}^{\frac{b}{2}} \frac{\partial}{\partial x}(p + \rho g z)$$

or

$$u_{\text{ave}} = -\frac{b^2 \gamma}{12\mu}\frac{\partial}{\partial x}\left(z + \frac{p}{\gamma}\right)$$

$$w_{\text{ave}} = -\frac{b^2 \gamma}{12\mu}\frac{\partial}{\partial z}\left(z + \frac{p}{\gamma}\right) \tag{8.14}$$

It is possible to rewrite Eqs. 8.14 as follows

$$u_{\text{ave}} = -K_m \frac{\partial h_m}{\partial x}$$
$$w_{\text{ave}} = -K_m \frac{\partial h_m}{\partial z} \tag{8.15}$$

in which
$$K_m = \frac{b^2}{12\mu}\gamma \tag{8.16}$$

and
$$h_m = z + \frac{p}{\gamma} \tag{8.17}$$

in perfect analogy with Eq. 4.24.

For constant ρ and μ, the insertion of Eqs. 8.15 into the continuity equation

$$\frac{\partial u_{\text{ave}}}{\partial x} + \frac{\partial w_{\text{ave}}}{\partial z} = 0$$

shows that the head $h_m = z + p/\gamma$ in the model satisfies Laplace's equation. In this case, the existence of a velocity potential

$$\Phi_m = K_m h_m = \frac{b^2\gamma}{12\mu}\left(z + \frac{p}{\gamma}\right) \tag{8.18}$$

in analogy with Eq. 4.26 is established. When the plates are horizontal Eq. 8.18 reduces to

$$\Phi_m = \frac{b^2}{12\mu}p = \frac{b^2\gamma}{12\mu}\frac{p}{\gamma} \tag{8.19}$$

which shows that K_m has the same form for both vertical and horizontal Hele-Shaw models. The requirement of constant ρ and μ in order to have potential flow in the model is hard to satisfy because it is difficult in many cases to keep the temperature of the fluid in the model constant and because the viscosity of most oils and glycerine is very much temperature dependent. Hele-Shaw models utilizing such fluids should be set up in temperature controlled rooms and, if the fluid is pumped, on some occasions it may be necessary to cool the fluid after it leaves the pump.

The analogy between Eqs. 4.26 and 8.18 shows that the average flow in the Hele-Shaw model, far enough from obstacles placed between the plates, may simulate the flow of ground water. In the immediate vicinity of obstacles, such as cut-off walls below a concrete weir represented in the model, the hypothesis that the variations of the velocity components in the x and z-directions are negligible as compared to variations of those components in the y-direction, is no longer valid. Both Santing [11] and Aravin [15] have proposed a limit of 1,000 for the Reynolds number

$$N_R = \frac{V_{\text{ave}} b}{\nu}$$

above which the flow would cease to be laminar. This requires a very narrow spacing around one millimeter or fractions of a millimeter when liquids of low viscosity such as light oils and water are used, and a spacing of a few millimeters when heavy oils or glycerine is used.

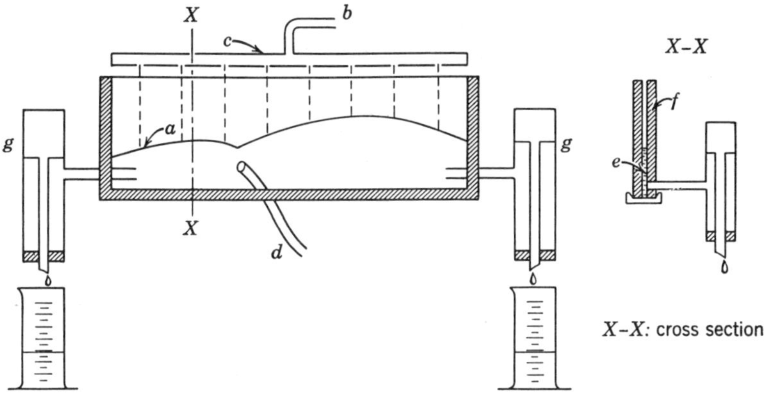

Fig. 8-4 Vertical type of Hele-Shaw analog. [After Santing.] *a*, ground-water table; *b*, supply of rain; *c*, sprinkler; *d*, discharge tube; *e*, viscous liquid; *f*, transparent plates; *g*, vertically adjustable overflow.

The versatility of the model as a tool for investigation of both steady and unsteady ground-water flow lies in the easy simulation of different values of the hydraulic conductivity of the soil through variation of the components of K_m for the model, i.e., through variation of the spacing b and of the fluid properties γ and μ. Regional variations of K may be accounted for by the insertion of thin strips to reduce the spacing in the desired region. Figure 8.4 [11] shows a sketch of a vertical model in which rainfall is imitated by sprinkling over the length of the channel and ground water withdrawal or artificial replenishment is reproduced by withdrawing or adding liquid at the respective places in the channel. The simple boundary condition of constant (in place) and variable (in time) head may be materialized by means of vessels which are vertically adjustable and connected to the interspace; impervious boundaries in the field are represented by impervious boundaries in the model. Two fluids of slightly

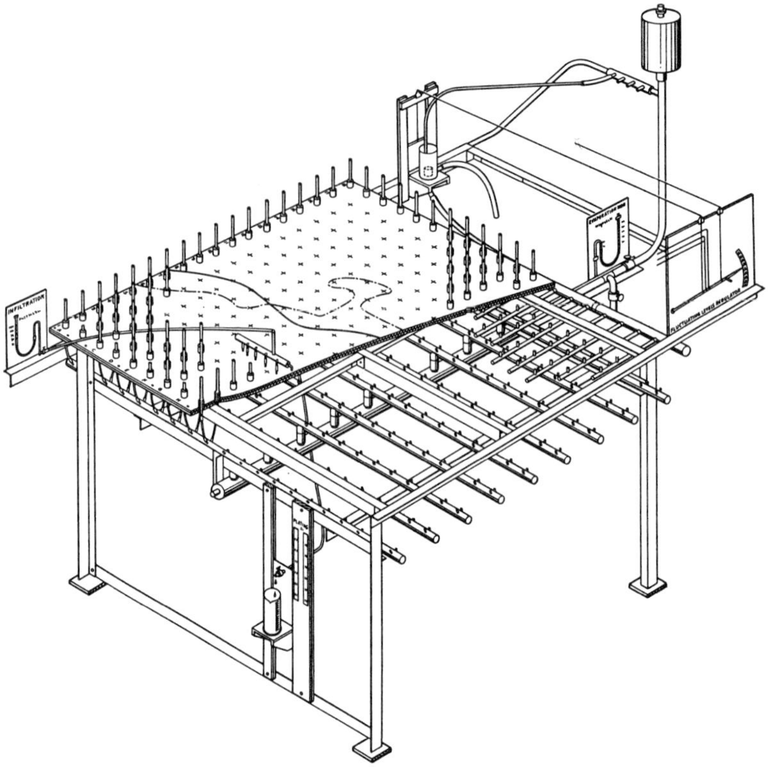

Fig. 8-5 Horizontal type of Hele-Shaw analog. [After Santing.]

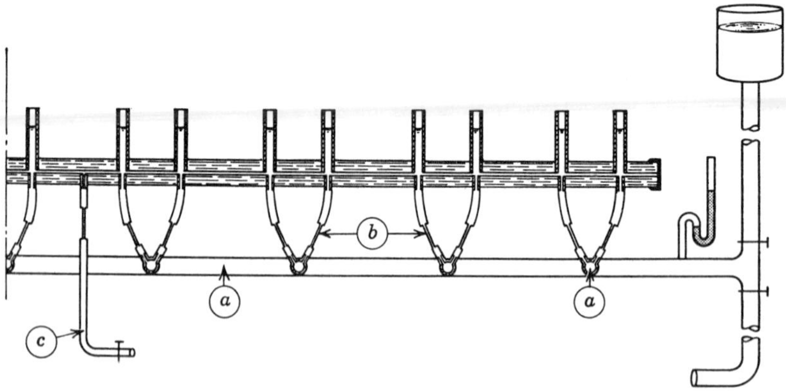

Fig. 8-6 Close-up of horizontal model, with simulation of storage capacity, replenishment and evaporation. [After Santing.] *a*, replenishment and drainage system to simulate precipitation and evaporation; *b*, capillaries to increase the resistance to flow; *c*, drainage tube to simulate water withdrawal by pumpage.

different density may be used for imitating fresh and salt ground water; fluids of different colors but of equal density and viscosity may be used to study multiple (connected or parallel) aquifers.

(B) HORIZONTAL MODELS—SCALES

Santing [11, 16] used the horizontal model to investigate the effects of artificial replenishment of the ground-water table in the well fields of the waterworks company of Zealand Flanders. To make the model suitable for the study of unconfined aquifers the assumption has to be made that the fluctuations in the ground-water level are negligibly small as compared to the thickness of the aquifer. This is equivalent to assuming a transmissivity $T = K\tilde{h}$ for the water-table aquifer in which $\tilde{h}$ is the average depth of flow [17].

Figure 8.5 represents a sketch of the horizontal model used by Santing. A grid of vertical vessels open at the top is established on top of the upper plate, each vessel being connected to the interspace of the model as indicated in Fig. 8.6. These vessels introduce storage capacity: a rise of the level of the liquid in the vessels means that liquid is stored, a drop means a release from storage. In most cases the storage vessels will be all of the same size and equally spaced on the upper plate of the model, representing an area with uniform storage. In the case of two parallel overlying aquifers, however, storage of the lower aquifer may be introduced by connecting the storage vessels to the lower plate of the bottom aquifer. Variations in storage capacity may be simulated by varying either the spacing or the diameter of the storage vessels. The storage coefficient S_m of the model may be defined in complete analogy with the coefficient S defined in section 4.11. It is the amount of liquid in storage released from a column of interspace with unit cross section under a unit decline of head. Because there is one storage vessel per area A_m of the model, it follows that

$$S_m A_m = \pi r_m^2 \cdot 1 \tag{8.20}$$

in which the subscript m refers to model characteristics. In nature the storage capacity of the aquifer is more or less uniformly distributed; in the model it is concentrated at the intersections of a grid, as in a finite difference approach. Theoretically, the closer the spacing the better the approximation of the true conditions becomes. There is a practical limit however, to the spacing as well as to the diameter of the storage vessels.

Replenishment of the water table through percolating rain water and evaporation from the water table may be reproduced by means of a distribution system (Fig. 8.6). Withdrawal by pumping is accomplished by draining the interspace as indicated in Fig. 8.6.

Scales:

α. *Confined aquifer* (*S*, *T*), *unsteady flow without radial symmetry.* The differential equation is given by Eq. 4.51, or in polar coordinates

$$\frac{\partial^2 h}{\partial r^2} + \frac{1}{r}\frac{\partial h}{\partial r} + \frac{1}{r^2}\frac{\partial^2 h}{\partial \theta^2} = \frac{S}{T}\frac{\partial h}{\partial t} \tag{8.21}$$

The corresponding equation for the model is:

$$\frac{\partial^2 h_m}{\partial {r_m}^2} + \frac{1}{r_m}\frac{\partial h_m}{\partial r_m} + \frac{1}{{r_m}^2}\frac{\partial^2 h_m}{\partial {\theta_m}^2} = \frac{S_m}{T_m}\frac{\partial h_m}{\partial t_m} \tag{8.22}$$

in which the subscript m refers to model characteristics. The scales are taken as the ratios of the characteristics of the model over those of the prototype, therefore

$$\bar{u}_r = \frac{r_m}{r}, \quad \bar{u}_z = \frac{h_m}{h}, \quad \bar{u}_t = \frac{t_m}{t}, \quad \bar{u}_S = \frac{S_m}{S}, \quad \bar{u}_T = \frac{T_m}{T} \tag{8.23}$$

in which $\bar{u}$ indicates scale and the subscript refers to the quantity under consideration. The scales are determined by transforming Eq. 8.21 by means of Eq. 8.23 and by expressing compatibility between the transformed equation and Eq. 8.22. This is done as follows:

$$\frac{\partial h}{\partial r} = \frac{\bar{u}_r}{\bar{u}_z}\frac{\partial h_m}{\partial r_m}$$

$$\frac{\partial^2 h}{\partial r^2} = \frac{\partial}{\partial r}\frac{\partial h}{\partial r} = \bar{u}_r \frac{\partial}{\partial r_m}\frac{\bar{u}_r}{\bar{u}_z}\frac{\partial h_m}{\partial r_m} = \frac{{\bar{u}_r}^2}{\bar{u}_z}\frac{\partial^2 h_m}{\partial {r_m}^2}$$

In the same way it is found that

$$\frac{\partial h}{\partial t} = \frac{\bar{u}_t}{\bar{u}_z}\frac{\partial h_m}{\partial t_m}$$

$$\frac{\partial h}{\partial \theta} = \frac{1}{\bar{u}_z}\frac{\partial h_m}{\partial \theta_m} \quad \text{and} \quad \frac{\partial^2 h}{\partial \theta^2} = \frac{1}{\bar{u}_z}\frac{\partial^2 h_m}{\partial {\theta_m}^2}, \quad \text{since } \bar{u}_\theta = 1$$

The transformed Eq. 8.21 becomes:

$$\frac{{\bar{u}_r}^2}{\bar{u}_z}\frac{\partial^2 h_m}{\partial {r_m}^2} + \frac{{\bar{u}_r}^2}{\bar{u}_z}\frac{1}{r_m}\frac{\partial h_m}{\partial r_m} + \frac{{\bar{u}_r}^2}{\bar{u}_z}\frac{1}{{r_m}^2}\frac{\partial^2 h_m}{\partial {\theta_m}^2} = \frac{S_m}{T_m}\frac{\bar{u}_T}{\bar{u}_S}\frac{\bar{u}_t}{\bar{u}_z}\frac{\partial h_m}{\partial t_m} \tag{8.24}$$

Equation 8.24 may be multiplied by $\bar{u}_z/\bar{u}_r^2$ and then compared with Eq. 8.22. Both equations are compatible if and only if

$$\frac{\bar{u}_T}{\bar{u}_S}\frac{\bar{u}_t}{{\bar{u}_r}^2} = 1 \tag{8.25}$$

It should be noted that the compatibility condition does not contain $\bar{u}_z$ so that $\bar{u}_z$ may be chosen freely in any case of confined flow, as was to be expected. The scale $\bar{u}_S$ is chosen to have a reasonable spacing of the storage vessels and a vessel diameter which permits easy readings during the experiment. Also, $\bar{u}_T$ is chosen after considering practical values of K_m in regard to the interspace and the fluid properties. Finally because $\bar{u}_t$ and $\bar{u}_r$ are related through Eq. 8.25, only one of them may be freely chosen. Equation 8.25 determines the alternate choice of these scales.

β. Unconfined aquifer (S_y, T), unsteady flow without radial symmetry, and replenishment R.

Under the assumption that the fluctuations in the ground-water level are negligibly small as compared to the thickness of the aquifer, an average transmissivity is used

$$T = K\tilde{h} \tag{8.26}$$

in which $\tilde{h}$ is the average depth of flow. For a replenishment R (flowrate per unit area) to the water table and a specific yield S_y of the aquifer, the differential equation of the ground-water flow becomes (see Eq. 96 reference 53):

$$\nabla^2 h + \frac{R}{T} = \frac{S_y}{T}\frac{\partial h}{\partial t} \tag{8.27}$$

The replenishment introduces a new scale $\bar{u}_R = R_m/R$ and a new compatibility relation. It is left as an exercise for the student to show that this compatibility relation may be expressed as

$$\frac{\bar{u}_T}{\bar{u}_R}\frac{\bar{u}_z}{\bar{u}_r^{\,2}} = 1 \tag{8.28}$$

If it is convenient to choose $\bar{u}_R$ freely, then $\bar{u}_z$ is no longer arbitrary but imposed by Eq. 8.28 after $\bar{u}_T$, $\bar{u}_R$, and $\bar{u}_r$ have been determined.

γ. Leaky aquifer (B, S, T), unsteady flow without radial symmetry.

Leakage in the horizontal model may be simulated by a perforated plate between the two plates of the model or by perforating the upper plate, in which case the fluid of the model communicates with the free surface [16] and simulates a water-table aquifer on top of an aquitard overlying a confined stratum. The equation governing the flow is given by Eq. 6.66

$$\nabla^2 s - \frac{s}{B^2} = \frac{S}{T}\frac{\partial s}{\partial t} \tag{8.29}$$

in which s is the drawdown and B is the leakage factor as defined in

Chapter 6. A new scale $\bar{u}_B$ is introduced and a new compatibility condition arises besides Eq. 8.25 which must be satisfied in the first place. It is left as an exercise for the student to show that this compatibility relation may be expressed as

$$\bar{u}_r = \bar{u}_B \tag{8.30}$$

If the aquifer overlying the aquitard is unconfined and replenished at a flowrate R per unit area it is also necessary to satisfy Eq. 8.28 for that part of the model.

(c) *Vertical models—Scales*

Vertical models are described in particular in references 8, 10, 12, and 13 and have been used more extensively than horizontal models. Figure 8.7 shows the model used by the author [13] to study the unsteady behavior

Fig. 8-7 Vertical type of Hele-Shaw analog simulating flow through underdrained earthdam. [After De Wiest.]

of the free surface of the flow through an underdrained earth embankment. The originality of the model resides in the fact that it has been scaled not after the differential equation governing the flow but according to the free surface boundary conditions. This peculiar feature prohibited distortion of the model as will be shown hereafter.

With the symbols defined in section 5.10 (although Dupuit's simplifying assumptions are not made here as they would lead to distortion of the model), the basic equations at the free surface are [13]

$$h(x, z, t) = z_f(x, t) + \frac{1}{\gamma} p(x, z, t) \tag{8.31}$$

$$\frac{\varepsilon}{K} \frac{\partial z_f}{\partial t} = \frac{\partial h}{\partial x} \frac{\partial z_f}{\partial x} - \frac{\partial h}{\partial z} \tag{8.32}$$

in which z_f is the ordinate of the free surface, x, z, t are the independent variables, and h is the head.

The corresponding equations in the model are

$$h_m = z_{f,m} + \frac{p_m}{\gamma} \tag{8.33}$$

and

$$\frac{\varepsilon_m}{K_m} \frac{\partial z_{f,m}}{\partial t_m} = \frac{\partial h_m}{\partial x_m} \frac{\partial z_{f,m}}{\partial x_m} - \frac{\partial h_m}{\partial z_m} \tag{8.34}$$

It is noted that $\varepsilon_m = 1$ and Eq. 8.32 is transformed by means of the scales $\bar{u}_z$ (for h and z_f), $\bar{u}_t$, $\bar{u}_x$, $\bar{u}_K$ as was demonstrated for the horizontal models. It is left as an exercise for the student to show that two compatibility conditions arise.

$$\frac{1}{\varepsilon} \frac{\bar{u}_x^{\,2}}{\bar{u}_z \bar{u}_t \bar{u}_K} = 1 \tag{8.35}$$

and

$$\frac{1}{\varepsilon} \frac{\bar{u}_z}{\bar{u}_t \bar{u}_K} = 1 \tag{8.36}$$

From Eqs. 8.35 and 8.36 it follows that $\bar{u}_x$ must be equal to $\bar{u}_z$ in order to have a unique time scale. This means that the model cannot be distorted, which, as far as the author knows, it was in most of the previously constructed models of this kind. Since ε and K are given for each case under investigation a suitable time scale $\bar{u}_t$ is found after a judicious choice of $\bar{u}_x = \bar{u}_z$ and of K_m (model spacing and viscous liquid characteristics).

In conclusion it may be said that Hele-Shaw models are extremely useful in ground-water flow investigations. At present the author has a horizontal model under construction to check analytical results for two parallel overlying confined aquifers. A drawback in using the model as a tool is that it is commonly built for one specific problem and that it is expensive to remodel for other applications.

8.4 *Resistance-Capacity Network Analogs*

The theory of the R-C network analog may be found in many papers [25], some of which have been condensed in books on analog simulation.

The most recent one by Karplus [18] gives a thorough and comprehensive treatment of the subject. Skibitzke [19], Stallman [20] and Walton [21] applied the R-C analog to problems in ground-water flow, and the author is using the same tool to corroborate his analytical studies on leaky aquifers involving Dirac's delta function [17 and 53].

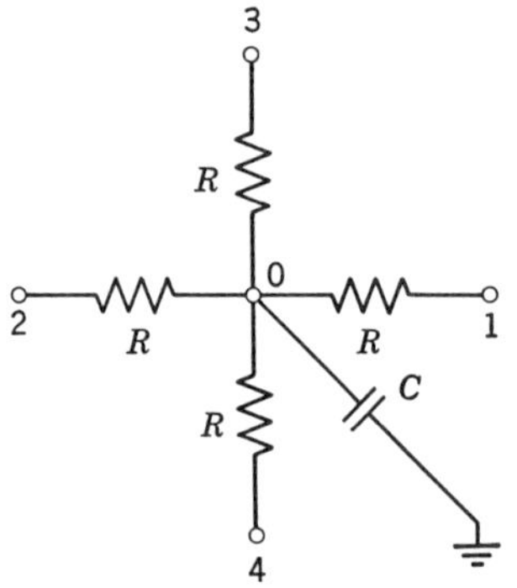

Fig. 8-8 Simulation of ground-water flow with help of R-C analog. Two-dimensional flow.

In the R-C analog network, solutions to the difference equation replacing the differential equation of ground-water flow may be visualized directly on an oscilloscope. Lumped electric-circuit elements are used to simulate the distributed properties of the subsurface strata; solutions to the problem are only found for these points of the medium which correspond to the nodes of the network that consists of an assembly of resistors and capacitors. Voltage and current sources simulating the excitations are applied to the boundaries of the network and at the interior nodes when required.

The resistor-capacitance network dissipates electrical energy in somewhat the same way a porous medium consumes ground-water energy to let water travel through its voids. Electrical conductance and hydraulic conductivity are of the same nature; electrical charges are stored in capacitors while the storage of water in an aquifer is related to its storage coefficient, so that a simple relationship between capacitance of R-C analog network and storage coefficient of an aquifer can be established. The head h in an aquifer is analogous to the electric potential Φ in the network.

The analogy becomes more apparent when the differential equations governing ground-water flow and electrical current respectively in the aquifer and in the network are compared. Therefore a finite difference approach is used in the x, y coordinates of Eq. 4.51 which may be written in two dimensional form as

$$\nabla^2 h = \frac{S}{T}\frac{\partial h}{\partial t} \tag{4.51}$$

The area of the aquifer is covered with an equilateral grid of mesh size $\Delta x = \Delta y = a$, whereby a^2 is small compared with the area of the aquifer. By analogy with Eq. 8.7, Eq. 4.51 may be rewritten as

$$\frac{h_1 + h_2 + h_3 + h_4 - 4h_0}{a^2} = \frac{S}{T}\frac{\partial h_0}{\partial t} \tag{8.37}$$

Consider now Fig. 8.8 in which four resistors of equal magnitude R and one capacitor C are connected to a common terminal 0; the capacitor is also connected to ground. Application of Kirchhoff's current law to node 0 gives [18]:

$$\frac{\Phi_1 - \Phi_0}{R} + \frac{\Phi_2 - \Phi_0}{R} + \frac{\Phi_3 - \Phi_0}{R} + \frac{\Phi_4 - \Phi_0}{R} = C\,\frac{\partial \Phi_0}{\partial t} \tag{8.38}$$

in which Φ_0, Φ_1, Φ_2, Φ_3, Φ_4 are the potentials at the nodes 0, 1, 2, 3, 4. Equations 8.37 and 8.38 are similar if R is made proportional to the reciprocal of T and if C is chosen to be proportional to the product of a^2 and S.

CONVERSION FACTORS

A number of conversion factors are needed to express the measured electrical quantities in equivalent ground-water terms [22]. The conversion factors for practical U.S. hydraulic units (gallons for quantity q; gallons per day for flowrate Q; feet for head loss Δh) and corresponding electrical units (coulombs for charge Q^*; coulombs per second, or amperes, for current I; volts for potential loss $\Delta\Phi$) may be defined as follows:

$$q = C_1 Q^* \tag{8.39}$$

$$h = C_2 \Phi \tag{8.40}$$

$$Q = C_3 I \tag{8.41}$$

and

$$t_d = C_4 t_s \tag{8.42}$$

in which t_d expresses the time in days, t_s denotes the time in seconds, and C_4 is the fraction of a day equivalent to one second. Likewise C_1 refers to gallons per coulomb, C_2 stands for feet per volt, and C_3 is gallons per day per amperes. The relation between the conversion factors C_1, C_3, and C_4 may be found as follows. By definition

$$Q = q/t_d \tag{8.43}$$

If Q, q, and t_d in Eq. 8.43 are replaced by their respective values from Eqs. 8.41, 8.39, and 8.42, and if it is noticed that $Q^*/t_s = I$,

$$C_3 C_4 / C_1 = 1 \tag{8.44}$$

DESIGN OF ELECTRICAL CIRCUIT ELEMENTS

The value of the resistance R may be determined from Ohm's law and the aforementioned conversion formulas. Ohm's law states

$$R = \Phi/I \tag{8.45}$$

in which $\Phi = h/C_2$ and $I = Q/C_3$, so that

$$R = hC_3/QC_2 \tag{8.46}$$

However, by definition $T = Q/h = $ (gal/day)/ft, and Eq. 8.46 may be rewritten as

$$R = C_3/C_2T \tag{8.47}$$

in which R is expressed in ohms because Φ is expressed in volts and I in amperes.

In a similar way the capacitance C may be determined from Coulomb's law and the aforementioned conversion formulas. Coulomb's law for the electrical charge of a capacitor states

$$C = Q^*/\Phi \tag{8.48}$$

in which $Q^* = q/C_1$ and $\Phi = h/C_2$, so that

$$C = \frac{q}{h}\frac{C_2}{C_1} \tag{8.49}$$

However q/h has the dimensions of L^2, and therefore may be replaced by a^2S. The conversion is computed as follows:

$$\left[\frac{q}{h}\right]^{\frac{\text{gal}}{\text{ft}}} \times \frac{\text{gal}}{\text{ft}} = [a^2S]^{\text{ft}^2} \times \text{ft}^2 \quad \text{or} \quad \left[\frac{q}{h}\right]^{\frac{\text{gal}}{\text{ft}}} = [a^2S]^{\text{ft}^2} \times \frac{\text{ft}^3}{\text{gal}} = 7.48[a^2S]^{\text{ft}^2}$$

Therefore

$$C = 7.48a^2S\,\frac{C_2}{C_1} \tag{8.50}$$

in which a is expressed in feet, S is the dimensionless storage coefficient, and C is the capacitance in farads because Q^* is expressed in coulombs and because Φ is in volts.

EXTENSION TO THREE-DIMENSIONAL FLOW. ANISOTROPIC AND LEAKY ARTESIAN CONDITIONS

The analog may be extended without difficulty to three dimensional flow in a confined aquifer of thickness b as indicated in Fig. 8.9a. Anisotropy introduces no particular problem as it merely requires a different value for the resistors in the horizontal plane of the aquifer (x-y plane) and for the resistors in the z-direction, assuming that the principal values of the hydraulic conductivity are $K_x = K_y = K_z$. For grid dimensions Δx, Δy, and Δz, the resistors may be computed by

$$R_x = C_5\,\frac{(\Delta x)^2}{K_x} \quad ; \quad R_y = C_5\,\frac{(\Delta y)^2}{K_y} \tag{8.51}$$

in which C_5 stands for ohms per units of length and time. For $\Delta x = \Delta y$, $K_x = K_y$, the resistors R_x and R_y are equal to R_h. Also,

$$R_z = C_5 \frac{(\Delta z)^2}{K_z} \qquad (8.52)$$

These formulas follow from the difference equation derived from Eq. 4.47 in three dimensions and from the generalization of Eq. 8.38 for different resistors R_x, R_y and R_z. The capacitance C in this case becomes proportional to S_s of Eq. 4.47.

Leakage through semiconfining beds or aquitards into the main aquifer is vertical and proportional to the drawdown. It adds a term, proportional to the unknown potential, to Laplace's equation (see Eq. 6.64) and therefore may easily be simulated [18, 21] by the addition of resistors connected to ground and to each node of the network (see Fig. 8.9*b*, resistance R_L). By analogy with Eq. 8.47 the value of R_L may be expressed as

$$R_L = \frac{C_3}{C_2\left(\dfrac{K'}{b'}\right)a^2} \qquad (8.53)$$

in which K'/b' is the leakance as defined in section 6.9 and in which R_L is expressed in ohms.

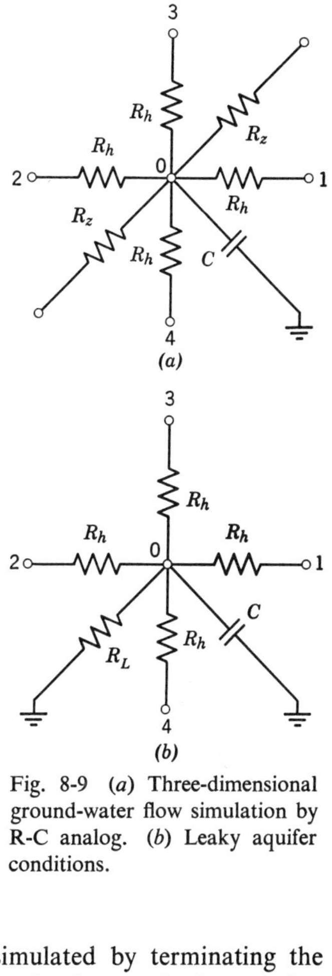

Fig. 8-9 (*a*) Three-dimensional ground-water flow simulation by R-C analog. (*b*) Leaky aquifer conditions.

BOUNDARY CONDITIONS [18, 21]

Boundaries of constant head may be simulated by terminating the corresponding part of the network in a short circuit. Barrier boundaries, across which no flow takes place, may be reproduced by an open circuit. A radiant boundary condition in which the head along the boundary is proportional to its normal derivative along that boundary may be simulated by connecting resistors between the nodes along the boundary and the ground. Irregular shapes of the boundary are duplicated with the help of the vector volume technique [18] whereby resistors and capacitors in the proximity of the boundary are modified to suit the correspondence between network and aquifer parameters. Network boundaries can be extended to infinity through the use of termination strips [18]. In the case of a water table aquifer the free boundary condition may be observed as explained by Stallman [20].

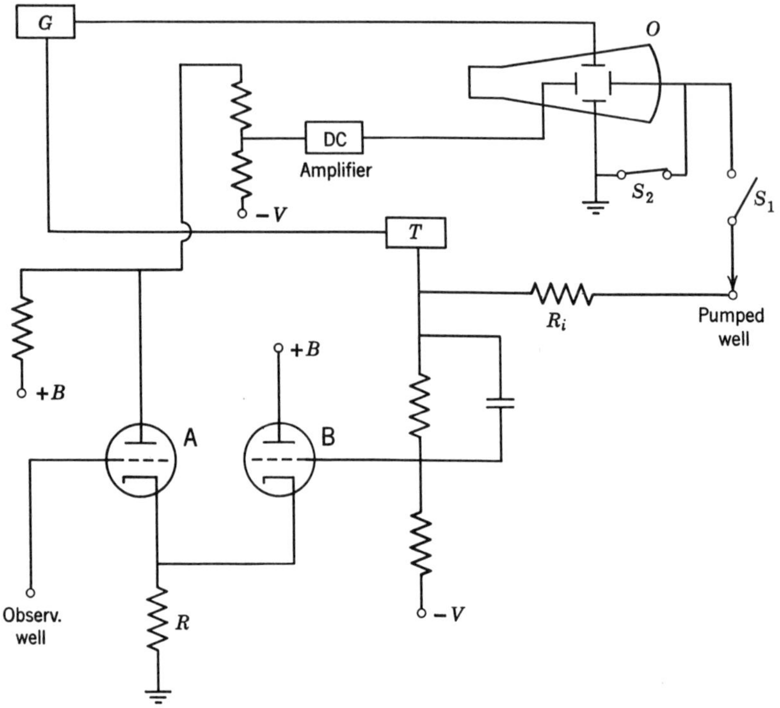

Fig. 8-10 Excitation-response apparatus for R-C analog. [After Karplus.]

EXCITATION-RESPONSE APPARATUS

The excitation-response apparatus that forces electrical energy in the proper time phase into the network is comprised of a waveform generator, a pulse generator, and an oscilloscope [21] which essentially operate as indicated in Fig. 8.10. The variation of drawdown in observation wells (nodes of the R-C network) is made visual on the screen of an oscilloscope when a step-function type change of pumpage is applied to a well or a well system as follows. A sawtooth generator G provides the horizontal sweep of oscilloscope O and also a single initiating pulse that is fed into a trigger T circuit of the Jordan and Eccles type. A large negative pulse of rectangular shape is released by the trigger and applied over a resistor R_i to the node representing the pumped well. The same pulse is also brought to the grid of tube B and cuts off the current in this tube. Before this happened the plate current of tube B caused such a large voltage drop across the cathode resistor of both tubes A and B that the voltage difference between the plate and cathode of tube A was too low to ignite tube A. The voltage of the grid of tube A which is connected to any node of the system (observation well) was not transferred to the oscilloscope. Once B is cut off, however, the voltage difference between the plate and cathode of A becomes sufficiently high for this tube to ignite and to transfer the voltage of the node connected to its grid via a d-c amplifier to the vertical plates of the oscilloscope. When S_2 is closed and S_1 is open this voltage represents the drawdown in an observation well resulting from a sudden change ΔQ in yield from the pumped well. ΔQ is measured for S_2 open and S_1 closed by the voltage drop $\Delta\Phi$ across R_i. By virtue of Eq. 8.41,

$$\Delta Q = C_3 \, \Delta\Phi / R_i \tag{8.54}$$

in which ΔQ is measured in gallons per day, $\Delta\Phi$ in volts, and R_i in ohms. To express ΔQ in gallons per minute, Eq. 8.54 is transformed into

$$\Delta Q = C_3 \, \Delta\Phi / 1440 R_i \tag{8.55}$$

Figure 8.11 shows the sequence of events in two computational cycles of the R-C network analog. The reason why this kind of R-C analog and energizing method was also used by the author lies in the direct analogy between Dirac's delta function applied in the theoretical analysis of ground-water problems and the rectangular shaped pulse applied to the node of the network that represents the pumped well. It was possible to check with this analog the complicated formulas arising from a mathematical analysis which even by means of high-speed digital computers do not lend themselves easily to numerical computations.

COST AND ACCURACY OF R-C NETWORK ANALOG

Cost of the analog and accuracy of the analog solutions depend on the quality of the resistors and capacitors. Electric circuit elements with tolerances of $\pm 10\%$ may be purchased in bulk quantities at low cost and may be entirely satisfactory for use in the network, especially because no network analog is better than the hydrogeologic data on which its construction is based. Walton [21] indicates that the overall-accuracy of the analog solutions can be increased by 1 or 2% through the use of precision resistors, capacitors, and refined excitation-response apparatus. Such precision is not warranted in view of the low incremental benefit-cost ratio. Finally, at present the total cost of a R-C network analog with excitation-response apparatus of the commercially available type may be estimated at $1,200 to $1,500.

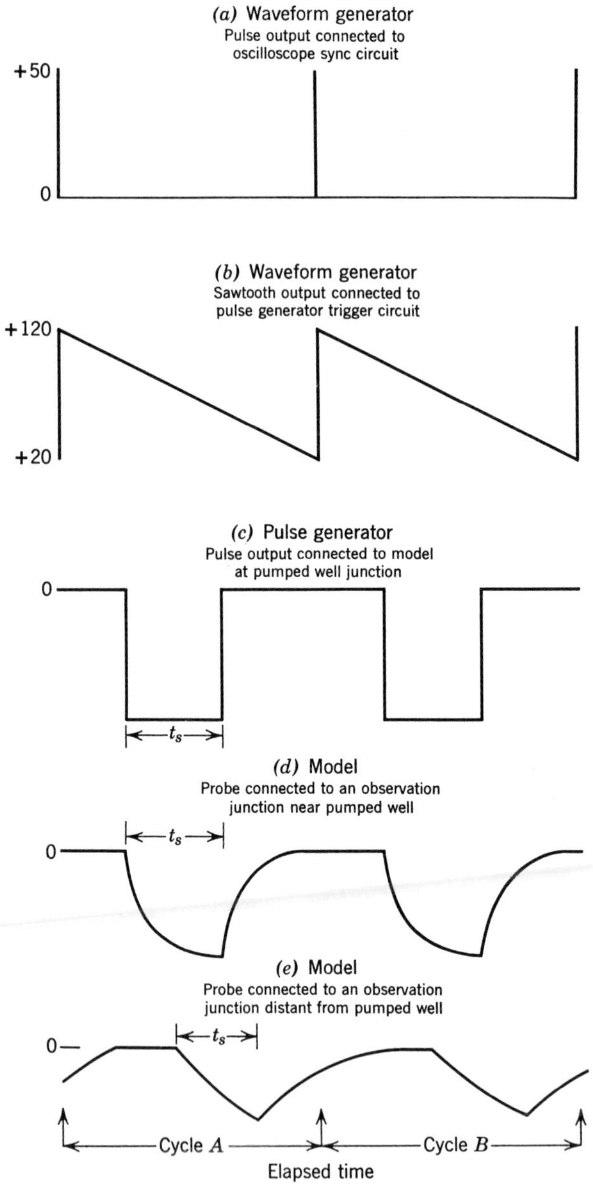

Fig. 8-11 Sequence of events in two computational cycles of the R-C network analog. [After Walton.] Voltages as seen by an oscilloscope at various points in the analog.

8.5 *Electrical Resistance Networks*

The electrical resistance network [32, 33, 50, 51] has been a favorite tool of agricultural engineers and soil physicists in studies of drainage and infiltration [26, 27, 28, 29, 30, 31]. In many cases the dimensions of the prototype are one order of magnitude smaller than those of interest to geological and civil engineers. Local conditions of soil nonuniformity and anisotropy become more important for smaller prototypes; for example, in some agricultural problems a distinction has to be made between fillable porosity (for rising water tables) and drainable porosity (for falling water tables) [31]. To simulate these variations in the properties of the medium use is made of a network of variable resistors or rheostats. The total voltage drop across the network is around 10 volts or higher but always below the danger level for electrocution, the source being ordinary 60-cycle, 120-volt alternating current rectified to give about 10 volts direct current or regular batteries of 6 or 12 volt. Potentials at the nodes of the network are measured with a vacuum tube voltmeter and the total current through the network is measured with an ammeter (milliammeter).

BOUNDARY CONDITIONS

Figure 8.12 [26] gives a block diagram of a resistance network and a sketch of the conditions along an impermeable boundary. If it is assumed that the current through AD is related to the diamond shaped square with horizontal hatching, by comparison the current through BC relates to the triangular shaped area only. Such a condition may be materialized by adjusting the rheostats on impermeable boundaries to twice the resistances of rheostats lying in the interior region.

Similar considerations lead to the determination of the resistance along a boundary between layers of different hydraulic conductivities. In Fig. 8.12*c* if R_1 and R_2 are the basic network resistances respectively for the media of hydraulic conductivity K_1 and K_2, then $2R_1$ and $2R_2$ are the resistances corresponding to the hachured triangular areas in media 1 and 2. The resistance R of AB along the boundary will be representative of the inverse of the average hydraulic conductivity $(K_1 + K_2)/2$ if R is taken as the resistance of branches $2R_1$ and $2R_2$ in parallel, say

$$R = \frac{2R_1R_2}{R_1 + R_2} \tag{8.56}$$

Streamlines may be obtained without changing the basic arrangement of

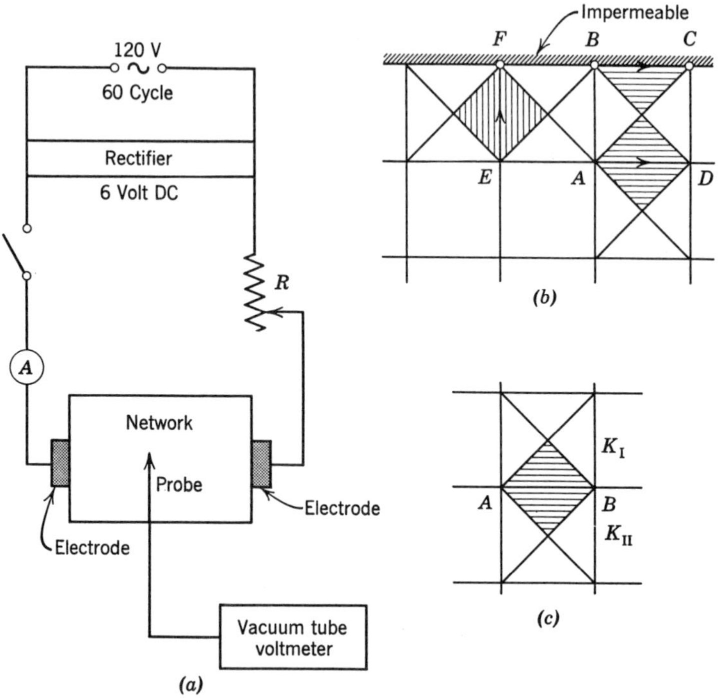

Fig. 8-12 (*a*) Block diagram of resistance network. [After Luthin.] (*b*) Points on impermeable boundary. (*c*) Points on boundary of different hydraulic conductivities.

the network because streamfunction Ψ and potential function Φ are connected through the Cauchy-Riemann equations (see Chapter 5, Eq. 5.8).

$$\frac{\partial \Phi}{\partial x} = \frac{\partial \Psi}{\partial z}; \qquad \frac{\partial \Phi}{\partial z} = -\frac{\partial \Psi}{\partial x} \tag{5.8}$$

Along a vertical impervious boundary, for example, $\partial\Phi/\partial x = 0$, or because of the above equations, $\partial\Psi/\partial z = 0$. This means that the streamfunction is constant along the vertical boundary and therefore this boundary is a streamline. The electrical analog for the streamfunction has an electrode (line of imposed equipotential) on this boundary since absence of any change in the streamfunction indicates presence of an equipotential surface in the analog.

On the other hand, an equipotential surface such as a water table parallel to the x-axis means $\partial\Phi/\partial x = 0$, hence $\partial\Psi/\partial z = 0$, so that Ψ does not change at the water table in the z-direction. Therefore the water table can be considered an impermeable to the Ψ function (i.e., in the absence of percolation to or evaporation from the water table).

These considerations may be summarized by the statement that an equipotential surface in the hydraulic head analog becomes an impermeable surface in the stream analog. An impermeable surface in the hydraulic head analog becomes an equipotential surface in the stream analog.

Seepage surfaces, in which the numerical value of the head in any point is equal to the elevation of that point above the datum plane for head, are simulated by resistance wires over which there is a linear drop in potential.

The free surface for the resistance network analog may be defined as a streamline along which the pressure is uniform and the derivative of the hydraulic head normal to the free surface is zero. In the resistance network the free surface is located by trial and error. First a guess is made as to the most probable shape of the free surface and the boundary condition for a streamline is satisfied along this first trial position. When the lines of equal pressure are drawn the free surface should fit in the family. Also, the isopiestic lines (lines of constant piezometric head) should be orthogonal to the free surface. The free surface is shifted until those conditions are satisfied. This free surface should not be confused with the moving water table [28, 34] treated as a succession of steady states.

Finally, resistance networks may be used to determine variations in transmissivity of nonhomogeneous aquifers [35, 36].

UNSTEADY STATE FLOW

The resistance network essentially yields solutions for steady flow conditions because it contains no electrical reactance (capacitance, inductance). Any transient flow must therefore be handled as a succession of steady states, e.g., moving ground-water mounds [31, 52] or water-table drawdown during tile drainage [28]. An example of computation of the falling water table by iteration is given in Fig. 8.13. The boundary condition at the water table is [34]:

$$\frac{\partial z_f}{\partial t} = \frac{K}{S_y}\left(\frac{\partial \Phi}{\partial z} - \frac{\partial \Phi}{\partial x}\tan\theta\right) \tag{8.57}$$

in which S_y is the specific yield, $\Phi = z + \dfrac{p}{\gamma}$, z_f is the height of the water table above the impervious datum plane, K is the hydraulic conductivity, θ the slope angle of the water table, and x, z, t the independent variables. The initial position of the water table must be assumed and from the measured potentials at the nodes near the water table $\tan\theta$, $\Delta\Phi/\Delta z$, and $\Delta\Phi/\Delta x$ are computed as functions of x. For a given Δt, $\Delta z_f(x)$ is found

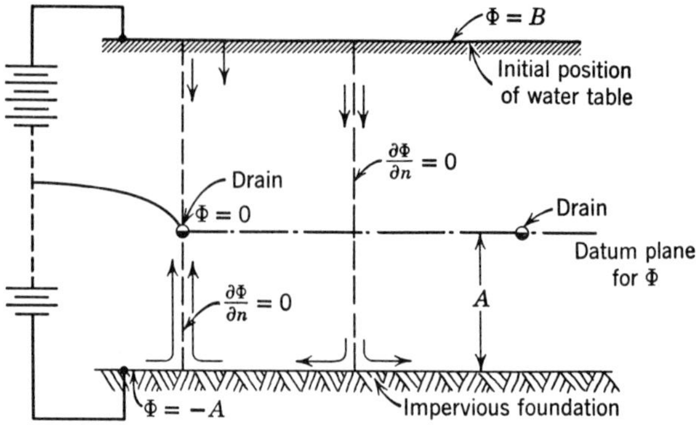

(a)

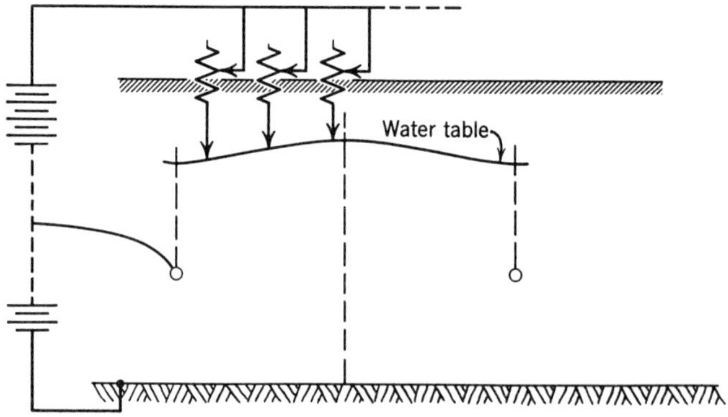

(b)

Fig. 8-13 (*a*) Drainage to parallel drains, watertable at ground level. (*b*) Falling watertable by iteration. [After Brutsaert.]

from Eq. 8.57. A second position of the water table is drawn with the help of the values $\Delta z_f(x)$. Along this new position of the water table the potential Φ must be proportional to the height above the drain. The variable potential along the water table is obtained by means of potentiometers. A convenient starting value is offered by an horizontal watertable, for which $\theta = 0$. (*Conversion factors:* see R-C analog network.)

8.6 *Conductive-Liquid and Conductive-Solid Analogs* [18]

The voltage distributions in a sheet or solid made of a conductive material as well as those in a tank filled with electrolyte satisfy Laplace's equation and may be used to simulate the potential fields of ground-water flow.

(A) CONDUCTIVE-LIQUID ANALOGS

The electrolytic tank normally consists of a watertight container of a nonconductive material, is of shallow depth and is filled with a few centimeters of electrolyte. The conducting liquid should have the following properties. It should have no electrical reactance; its resistivity must be uniform throughout the liquid and linear so that there is a linear relation between voltage and current; chemical reactions between liquid and electrodes must be impossible; the rate of evaporation of the liquid must be slow in order to prevent alteration of the resistivity in time.

Figure 8.14 [38] gives a sketch of a typical conductive liquid analog model showing the electrolytic tank, the electrical equipment, and the drawing board with pantograph. As in the case of the resistance network analog, the voltage is reduced to avoid electrocution and also to avoid heating of the electrolyte. A potential divider allows for accurate division of the voltage into a large number of subdivisions. The oscilloscope helps in the identification of the potential at the probe position. When the probe is at the desired potential no current flows through the oscilloscope and the pencil of the pantograph is depressed to leave a mark on the drawing paper. Equipotential lines may be sketched by connecting points of the same potential.

The theory underlying the use of the electrolyte tank has been known for a long time [37, 39]. Its application to ground-water flow is based on the similarity between the differential equations which describe the flow of ground water and those which govern the flow of electric current through conducting materials [40]. Let Ohm's law for the flow of electricity through a conducting material be written as

$$\mathbf{I} = -\kappa \operatorname{grad} V \tag{8.58}$$

in which $\mathbf{I}$ is the vector of electric current per unit area, κ is the electric conductivity of the material, and V is the electric potential. Darcy's law for the laminar flow of liquid through porous media on the other hand may be expressed by Eq. 4.24, or using the symbol $\mathbf{q}$ to avoid confusion,

$$\mathbf{q} = -K \operatorname{grad} h \tag{8.59}$$

in which $\mathbf{q}$ is the specific discharge, K is the hydraulic conductivity, and h is the head.

For steady flow in a region without sources or sinks the conservation equation for the electric charge requires

$$\operatorname{div} \mathbf{I} = 0 \tag{8.60}$$

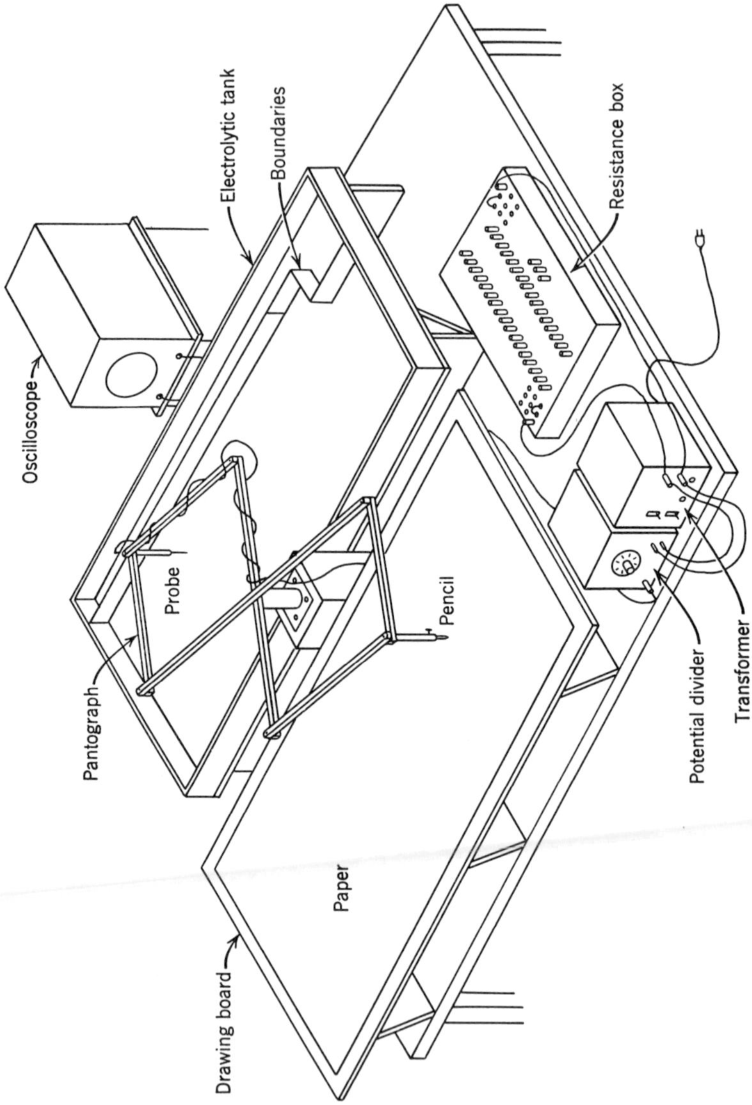

Fig. 8-14 Sketch of conductive liquid analog. [After Bear.]

and if κ is constant, i.e., for an isotropic conducting material, substitution of Eq. 8.58 into Eq. 8.60 gives

$$\nabla^2 V = 0 \tag{8.61}$$

This equation may be compared directly with Eq. 4.52, and proves the analogy between head and electric potential, hydraulic conductivity and electric conductivity, specific fluid discharge and electric current per unit area. In the case of a two-dimensional field it is possible to represent variable (i.e., in a plane) hydraulic conductivity by varying the depth of the electrolyte. The electric potential however, must not vary with the depth (normal to the plane in which variation is sought) and therefore it is better to use different electrolyte concentrations of the same depth in the analog.

A boundary of constant head in the prototype is reproduced by an electrode at constant potential in the model and impervious boundaries in the prototype require insulating boundaries in the model. When the head or its normal derivative vary along the boundary the corresponding boundary of the model is usually covered by a large number of small electrodes which are insulated from each other. In the case of variable head along the boundary each electrode is raised to a potential V corresponding to the average head of that part of the boundary reproduced by the electrode; for variable normal derivative of the head each electrode is set to allow a certain current I to flow.

Conditions of anisotropy are treated by a scale transformation similar to that obtained by Eq. 5.31. Maasland [41, 42] derived the equation

$$x_r/z_r = \sqrt{K_z/K_x} \tag{8.62}$$

in which $x_r = x_m/x_p$ is the length scale in the x-direction and $z_r = z_m/z_p$ is the length scale in the z-direction, m and p stand for model and prototype, and K_z and K_x are the hydraulic conductivities of the prototype. It suffices to assign the value of one to z_r in Eq. 8.62 to obtain the previously derived Eq. 5.31.

(B) CONDUCTIVE-SHEET ANALOGS [43, 44, 18]

The principle of the conductive-sheet analog does not differ from the underlying principle of the electrolyte tank. Sheets of electrically conducting material of sufficiently uniform, isotropic, and high resistance are suitable. They may consist of filter paper soaked in a suspension of colloidal graphite, of woven grids of metal wire and silk thread, of conductive rubber, of metallized paper, etc. The most commonly used sheet appears to be Teledeltos paper, manufactured by the Western Union Telegraph Company [18]. This paper is formed by adding carbon black, a conductive material, to paper pulp in the pulp-beating stage of the paper-manufacturing process. The conductive paper is then coated on one side with a lacquer which acts as an electrical insulator and on the other side with a layer of aluminum paint.

The advantages and disadvantages of the conductive-sheet analog are reviewed in the literature [18, 37].

8.7 *Other Analogs*

Among the other analogs which are still in use the sand box should be mentioned first. The effect of capillary rise, however, is greatly exaggerated in this kind of analog [1]; furthermore, the analog lacks versatility and has lost ground to the aforementioned analogs as far as application is concerned. It is still in use in soil mechanics and agricultural laboratories.

Stretched-membranes [47] for which the vertical deflection of the membrane is governed by Laplace's equation in two dimensions, have also been used successfully especially in the case of well fields [45, 46].

REFERENCES

1. Polubarinova-Kochina, P. Ya., *Theory of Ground-Water Movement*, pp. 433–436, Princeton University Press, Princeton, 1962.
2. Stokes, G. G., see article by Hele-Shaw on the "Streamline Motion of a Viscous Film," *Report of the British Association for the Advancement of Science*, 68th Meeting, p. 136 (1899).
3. Hele-Shaw, H. S., "Experiments on the Nature of the Surface Resistance in Pipes and on Ships," *Transactions Institute Naval Architects*, Vol. 39, pp. 145–156 (1897).
4. Hele-Shaw, H. S., "Investigation of the Nature of Surface Resistance of Water and of Streamline Motion under Certain Experimental Conditions," *Transactions Institute Naval Architects*, Vol. 40, pp. 21–46 (1898).
5. Dachler, R., *Grundwasserströmung*, Springer, Vienna, 1936 (141 pp.).
6. Polubarinova-Kochina, *Theory of Ground-Water Movement*, pp. 465–476.
7. Harr, M. E., *Ground Water and Seepage*, McGraw-Hill, New York, 1962 (315 pp.).
8. Dietz, D. N., "Een Modelproef ter Bestudering van Niet-Stationaire Bewegingen van Grondwater," *Water*, No. 23, 1941.
9. Gunther, E., "Lösung von Grundwasseraufgaben mit Hilfe der Strömung in dünnen Schichten," *Wasserkraft und Wasserwirtschaft*, Vol. 3, No. 3, pp. 49–55 (1940).
10. Santing, G., "Modèle pour l'étude des problèmes de l'écoulement simultané des eaux souterraines douces et salées," *IASH*, Vol. 2, pp. 184–193, Assembleé Générale de Bruxelles, 1951.
11. Santing, G., "A Horizontal Scale Model, Based on the Viscous Flow Analogy, for Studying Ground-Water Flow in an Aquifer having Storage," *IASH*, pp. 105–114, General Assembly, Toronto, 1957.
12. Todd, D. K., "Unsteady Flow in Porous Media by Means of a Hele-Shaw Viscous

Fluid Model," *Transactions American Geophysical Union*, Vol. 35, No. 6, p. 905 (1954).
13. De Wiest, R. J. M., "Free Surface Flow in Homogeneous Porous Medium," *ASCE Transactions*, Vol. 127, Chapter I, pp. 1045–1089 (1962).
14. Sternberg, Y. and V. Scott, "The Hele-Shaw Model as a Tool in Ground-Water Research," *NWWA Conference*, San Francisco, September 1963.
15. Aravin, V. I., "Basic Problems in the Experimental Investigation of the Flow of Groundwater by Means of a Parallel Plate Model," Izvestiya NIIG, Vol. 23, 1938.
16. Santing, G., "Recente Ontwikkelingen op bet Gebied van de Spleet-Modellen voor het Onderzoek van Grondwaterstromingen," *Water*, No. 15, July 1958.
17. De Wiest, R. J. M., "Replenishment of Aquifers Intersected by Streams," *ASCE Journal of the Hydraulics Division*, pp. 165–191 (November 1963).
18. Karplus, W. J., *Analog Simulation*, McGraw-Hill, New York, 1958 (427 pp.).
19. Skibitzke, H. E., "Electronic Computers as an Aid to the Analysis of Hydrologic Problems," *IASH Publication* 52, 1961.
20. Stallman, R. W., "Electric Analog of Three-Dimensional Flow to Wells and its Application to Unconfined Aquifers," *U.S. Geological Survey Water Supply Paper* 1536-H, pp. 205–242 (1963).
21. Walton, W. C. and T. A. Prickett, "Hydrogeologic Electric Analog Computers," *ASCE Journal of Hydraulic Division*, pp. 67–91 (November 1963).
22. Bermes, B. J., "An Electric Analog Method for Use in Quantitative Studies," *U.S. Geological Survey Unpublished Report*, 1960.
23. Todd, D. K., *Ground Water Hydrology*, Wiley, 1959 (336 pp.).
24. Irmay, S., *Theoretical Models of Flow through Porous Media*, Rilem Symposium on the Transfer of Water in Porous Media, Paris, April 7–10, 1964.
25. Johnson, A. I., "Selected References on Analog Models for Hydrologic Studies," *Appendix F, Proceedings of the Symposium on Transient Ground Water Hydraulics*, Colorado State University, July 25–27, 1963.
26. Luthin, J. N., "An Electrical Resistance Network Solving Drainage Problems," *Soil Science*, Vol. 75, pp. 259–74 (1953).
27. Worstell, R. V. and J. N. Luthin, "A Resistance Network Analog for Studying Seepage Problems," *Soil Science*, Vol. 88, pp. 267–269 (1959).
28. Brutsaert, W., G. S. Taylor, and J. N. Luthin, "Predicted and Experimental Water Table Drawdown during Tile Drainage," *Hilgardia*, Vol. 31, pp. 389–418 (November 1961).
29. Bouwer, H., "A study of Final Infiltration Rates from Cylinder Infiltrometers and Irrigation Furrows with an Electrical Resistance Network," *Transactions International Society Soil Science 7th Congress*, Vol. 1, Paper 6, Madison, Wisc., 1960.
30. Bouwer, H. and W. C. Little, "A Unifying Numerical Solution for Two-Dimensional Steady Flow Problems in Porous Media with an Electrical Resistance Network," *Proceedings Soil Science Society of America*, Vol. 23, p. 91 (1959).
31. Bouwer, H., "Analyzing Ground-Water Mounds by Resistance Network," *ASCE Journal of the Irrigation and Drainage Division*, pp. 15–36 (September 1962).
32. Liebmann, G., "Solution of Partial Differential Equations with a Resistance Network Analogue," *British Journal of Applied Physics*, Vol. 1, pp. 92–103 (1950).
33. Liebmann, G., "Resistance Network Analogues with Unequal Meshes or Subdivided Meshes," *British Journal of Applied Physics*, Vol. 5, pp. 362–366 (1954).
34. Kirkham, D. and R. E. Gaskell, "Falling Water Table in Tile and Ditch Drainage," *Proceedings Soil Science Society of America*, Vol. 15, pp. 37–42 (1950). (Published in 1951.)

35. Stallman, R. W., "Calculation of Resistance and Error in an Electric Analog of Steady Flow through Nonhomogeneous Aquifers," *U.S. Geological Survey Water Supply Paper* 1544-G, pp. 1–20 (1963).
36. Sammel, E. A., "Evaluation of Numerical-Analysis Methods for Determining Variations in Transmissivity," *International Association of Scientific Hydrology Publication No.* 64, *Subterranean Waters*, pp. 239–251.
37. Van Everdingen, R. O. and B. K. Bhattacharya, "Data for Ground-Water Model Studies," *Geological Survey of Canada Paper* 63–12, December 1963 (31 pp.).
38. Todd, D. K. and J. Bear, "River Seepage Investigation," *Water Resources Center Contribution No.* 20, University of California, Berkeley, 1959.
39. Muskat, M., *The Flow of Homogeneous Fluids through Porous Media*, p. 318.
40. Malavard, L. C., "The Use of Rheoelectrical Analogies in Aerodynamics," *Agardograph* 18, NATO Advisory Group for Aeronautical Research and Development, Paris, 1956.
41. Maasland, M., *Soil Anisotropy and Land Drainage;* see "Drainage of Agricultural Lands," J. N. Luthin (Ed.), pp. 216–285, American Society of Agronomy, Madison, Wisconsin, 1957.
42. Todd, D. K. and J. Bear, "Seepage through Layered Anisotropic Porous Media," *ASCE Journal of the Hydraulics Division*, pp. 31–57 (May 1961).
43. Childs, E. C., "The Water Table, Equipotentials, and Streamlines in Drained Land," *Soil Science*, Vol. 56, pp. 317–330 (1943); Vol. 59, pp. 313–327 (1945); Vol. 59, pp. 405–415 (1945); Vol. 62, pp. 183–192 (1946); Vol. 63, pp. 361–376 (1947); Vol. 71, pp. 233–237 (1951).
44. Childs, E. C., "The Equilibrium of Rain-Fed Groundwater Resting on Deeper Saline Water; The Ghyben-Herzberg Lens," *Journal of Soil Sciences*, Vol. 1, No. 2, pp. 173–181 (1950).
45. Hansen, V. E., "Complicated Well Problems Solved by the Membrane Analogy," *Transaction American Geophysical Union*, Vol. 33, No. 6, pp. 912–916 (1952).
46. Zee, Chong-Hung, D. F. Peterson, and R. O. Bock, "Flow into a Well by Electric and Membrane Analogy," *Transactions American Society Civil Engineering*, Vol. 122, pp. 1088–1112 (1957).
47. De Josselin de Jong, G., "Moiré Patterns of the Membrane Analogy for Groundwater Movement Applied to Multiple Fluid Flow," *Journal of Geophysical Research*, Vol. 66, pp. 3625–3628 (1961).
48. Bear, J., "Scales of Viscous Analogy Models for Ground Water Studies," *ASCE Journal of the Hydraulics Division*, pp. 11–23 (February 1960).
49. Naor, I. and J. Bear, *Model Investigations of Coastal Ground-Water Interception*, Tahal, Tel-Aviv, Israel, 1963 (54 pp.).
50. DeJong, J., "Electrische Analogie Modellen voor het Oplossen van Geo-hydrologische Problemen," *Water*, Vol. 46, pp. 43–45 (1962).
51. DeJong, J., "Een Eenvoudige Methode voor het Nabootsen van een Oneindig Potentiaalveld," *Water*, Vol. 46, pp. 185–186 (1962).
52. Bouwer, H. and J. Van Schilfgaarde, "Simplified Method of Predicting Fall of Water Table in Drained Land," *Transactions American Society Agricultural Engineers*, Vol. 6, pp. 288–296 (1963).
53. De Wiest, R. J. M., "Replenishment of Aquifers Intersected by Streams; Closure of Discussion," *ASCE Journal of the Hydraulics Division*, pp. 161–168 (September 1964).

APPENDIX A

Proof of Hantush's Method[1]

The theoretical graph of drawdown s versus time t, Eq. 6.67 on semi-logarithmic paper, as in Fig. 6.24, has the following properties.

(a) Its slope m at any point is given by

$$m = \frac{\Delta s}{\Delta \log t} = 2.3 \frac{Q}{4\pi T} e^{-u-r^2/4B^2u} \tag{A.1}$$

This follows from differentiation of Eq. 6.67 with respect to log t

$$\frac{\Delta s}{\Delta \log t} = 2.3 \frac{\Delta s/\Delta t}{\Delta \log_e t/\Delta t} = 2.3t(\Delta s/\Delta t)$$

and from the application of Leibnitz' rule for differentiation of an integral:

$$\frac{\partial s}{\partial t} = \left[-\frac{1}{u} e^{-u-r^2/4B^2u} \frac{r^2S}{4T}\left(-\frac{1}{t^2}\right)\right]\frac{Q}{4\pi T} = \frac{Q}{4\pi T}\frac{1}{t} e^{-u-r^2/4B^2u}$$

(b) The curve has an inflection point at which the following relations hold:

$$u_i = \frac{r^2S}{4Tt_i} = \frac{r}{2B} \tag{A.2}$$

This result is obtained by equating to zero the second derivative of s with respect to log t and solving for u.

$$\frac{\partial^2 s}{\partial(\log t)^2} = \frac{\partial}{\partial \log t}\frac{\partial s}{\partial \log t} = 2.3t\frac{\partial}{\partial t}\left(\frac{\partial s}{\partial \log t}\right) = 4.6\frac{tQ}{4\pi T}\frac{\partial}{\partial t} e^{-u-r^2/4B^2u}$$

$$= 4.6\frac{tQ}{4\pi T} e^{-u-r^2/4B^2u}\left(-\frac{\partial u}{\partial t} + \frac{r^2}{4B^2u^2}\frac{\partial u}{\partial t}\right)$$

$\frac{\partial^2 s}{\partial(\log t)^2} = 0$ requires $\frac{r^2}{4B^2u^2} - 1 = 0$ or $u_i = \frac{r}{2B}$ and $u_i = \frac{r^2S}{4Tt_i}$ by definition.

[1] See section 6.10 and reference 20 at the end of Chapter 6.

(c) With the value $u_i = \dfrac{r}{2B}$, the slope m_i at the inflection point becomes from (a):

$$m_i = 2.3 \frac{Q}{4\pi T} e^{-r/B} \tag{A.3}$$

(d) Assume that

$$s_i = \tfrac{1}{2} s_{\max} = \frac{Q}{4\pi T} K_0\left(\frac{r}{B}\right). \tag{A.4}$$

A direct proof is given in reference 30, Chapter 6.

(e) Division of the results of (d) and (c) leads to

$$\frac{2.3 s_i}{m_i} = e^{r/B} K_0\left(\frac{r}{B}\right).$$

APPENDIX B

Functions Occurring in the Theory of Leaky Aquifers

Table B.1 Values of the Functions of e^x, $K_0(x)$, $-Ei(-x)$, and $Ei(-x)e^x$

x	e^x	$K_0(x)$	$e^xK_0(x)$	$-Ei(-x)$	$-Ei(-x)e^x$	x	e^x	$K_0(x)$	$e^xK_0(x)$	$-Ei(-x)$	$-Ei(-x)e^x$	x	e^x	$K_0(x)$	$e^xK_0(x)$	$-Ei(-x)$	$-Ei(-x)e^x$
0.010	1.0101	4.7212	4.7687	4.0379	4.0787	0.10	1.1052	2.4271	2.6823	1.8229	2.0147	1.0	2.7183	0.4210	1.1445	.2194	.5964
11	1.0111	4.6260	4.6771	3.9436	3.9874	11	1.1163	2.3333	2.6046	1.7371	1.9391	1.1	3.0042	.3656	1.0983	.1860	.5588
12	1.0121	4.5390	4.5938	3.8576	3.9044	12	1.1275	2.2479	2.5345	1.6595	1.8711	1.2	2.3201	.3185	1.0575	.1584	.5259
13	1.0131	4.4590	4.5173	3.7785	3.8282	13	1.1388	2.1695	2.4707	1.5889	1.8094	1.3	3.6693	.2782	1.0210	.1355	.4972
14	1.0141	4.3849	4.4467	3.7054	3.7578	14	1.1503	2.0972	2.4123	1.5241	1.7532	1.4	4.0552	.2437	0.9881	.1162	.4712
15	1.0151	4.3159	4.3812	3.6374	3.6925	15	1.1618	2.0300	2.3585	1.4645	1.7015	1.5	4.4817	.2138	.9582	.1000	.4482
16	1.0161	4.2514	4.3200	3.5739	3.6317	16	1.1735	1.9674	2.3088	1.4092	1.6537	1.6	4.9530	.1880	.9309	.0863	.4275
17	1.0171	4.1908	4.2627	3.5143	3.5746	17	1.1853	1.9088	2.2625	1.3578	1.6094	1.7	5.4739	.1655	.9059	.0747	.4086
18	1.0182	4.1337	4.2088	3.4581	3.5209	18	1.1972	1.8537	2.2193	1.3098	1.5681	1.8	6.0496	.1459	.8828	.0647	.3915
19	1.0192	4.0797	4.1580	3.4050	3.4705	19	1.2093	1.8018	2.1788	1.2649	1.5295	1.9	6.6859	.1288	.8614	.0562	.3758
0.020	1.0202	4.0285	4.1098	3.3547	3.4225	0.20	1.2214	1.7527	2.1408	1.2227	1.4934	2.0	7.3891	.1139	.8416	.0489	.3613
21	1.0212	3.9797	4.0642	3.3069	3.3771	21	1.2337	1.7062	2.1049	1.1829	1.4593	2.1	8.1662	.1008	.8230	.0426	.3480
22	1.0222	3.9332	4.0207	3.2614	3.3340	22	1.2461	1.6620	2.0710	1.1454	1.4273	2.2	9.0250	.0893	.8057	.0372	.3356
23	1.0233	3.8888	3.9793	3.2179	3.2927	23	1.2586	1.6199	2.0389	1.1099	1.3969	2.3	9.9742	.0791	.7894	.0325	.3242
24	1.0243	3.8463	3.9398	3.1763	3.2535	24	1.2713	1.5798	2.0084	1.0762	1.3681	2.4	11.0232	.0702	.7740	.0284	.3135
25	1.0253	3.8056	3.9019	3.1365	3.2159	25	1.2840	1.5415	1.9793	1.0443	1.3409	2.5	12.1825	.0623	.7596	.0249	.3035
26	1.0263	3.7664	3.8656	3.0983	3.1799	26	1.2969	1.5048	1.9517	1.0139	1.3149	2.6	13.4637	.0554	.7459	.0219	.2942
27	1.0274	3.7287	3.8307	3.0615	3.1452	27	1.3100	1.4697	1.9253	.9849	1.2902	2.7	14.8797	.0493	.7329	.0192	.2854
28	1.0284	3.6924	3.7972	3.0261	3.1119	28	1.3231	1.4360	1.9000	.9573	1.2666	2.8	16.4446	.0438	.7206	.0169	.2773
29	1.0294	3.6574	3.7650	2.9920	3.0800	29	1.3364	1.4036	1.8758	.9309	1.2441	2.9	18.1742	.0390	.7089	.0148	.2693
0.030	1.0305	3.6235	3.7339	2.9591	3.0494	0.30	1.3499	1.3725	1.8526	.9057	1.2226	3.0	20.0855	.0347	.6978	.0131	.2621
31	1.0315	3.5908	3.7039	2.9273	3.0196	31	1.3634	1.3425	1.8304	.8815	1.2018	3.1	22.1980	.0310	.6871	.0115	.2551
32	1.0325	3.5591	3.6749	2.8965	2.9908	32	1.3771	1.3136	1.8089	.8583	1.1820	3.2	24.5325	.0276	.6770	.0101	.2485
33	1.0336	3.5284	3.6468	2.8668	2.9631	33	1.3910	1.2857	1.7883	.8361	1.1630	3.3	27.1126	.0246	.6673	.0089	.2424
34	1.0346	3.4986	3.6196	2.8379	2.9362	34	1.4050	1.2587	1.7685	.8147	1.1446	3.4	29.9641	.0220	.6580	.0079	.2365
35	1.0356	3.4697	3.5933	2.8099	2.9101	35	1.4191	1.2327	1.7493	.7942	1.1270	3.5	33.1155	.0196	.6490	.0070	.2308
36	1.0367	3.4416	3.5678	2.7827	2.8848	36	1.4333	1.2075	1.7308	.7745	1.1101	3.6	36.5982	.0175	.6405	.0062	.2254
37	1.0377	3.4143	3.5430	2.7563	2.8603	37	1.4477	1.1832	1.7129	.7554	1.0936	3.7	40.4473	.0156	.6322	.0055	.2204
38	1.0387	3.3877	3.5189	2.7306	2.8364	38	1.4623	1.1596	1.6956	.7371	1.0779	3.8	44.7012	.0140	.6243	.0048	.2155
39	1.0398	3.3618	3.4955	2.7056	2.8133	39	1.4770	1.1367	1.6789	.7194	1.0626	3.9	49.4025	.0125	.6166	.0043	.2108
0.040	1.0408	3.3365	3.4727	2.6813	2.7907	0.40	1.4918	1.1145	1.6627	.7024	1.0478	4.0	54.5982	.0112	.6093	.0038	.2063
41	1.0419	3.3119	3.4505	2.6576	2.7688	41	1.5068	1.0930	1.6470	.6859	1.0335	4.1	60.3403	.0100	.6022	.0033	.2021
42	1.0429	3.2879	3.4289	2.6344	2.7474	42	1.5220	1.0721	1.6317	.6700	1.0197	4.2	66.6863	.0089	.5953	.0030	.1980
43	1.0439	3.2645	3.4079	2.6119	2.7267	43	1.5373	1.0518	1.6169	.6546	1.0063	4.3	73.6998	.0080	.5887	.0026	.1941
44	1.0450	3.2415	3.3874	2.5899	2.7064	44	1.5527	1.0321	1.6025	.6397	.9933	4.4	81.4509	.0071	.5823	.0023	.1903
45	1.0460	3.2192	3.3673	2.5684	2.6866	45	1.5683	1.0129	1.5886	.6253	.9807	4.5	90.0171	.0064	.5761	.0021	.1866
46	1.0471	3.1973	3.3478	2.5474	2.6672	46	1.5841	0.9943	1.5750	.6114	.9685	4.6	99.4843	.0057	.5701	.0018	.1832
47	1.0481	3.1758	3.3287	2.5268	2.6483	47	1.6000	.9761	1.5671	.5979	.9566	4.7	109.9472	.0051	.5643	.0016	.1798
48	1.0492	3.1549	3.3100	2.5068	2.6300	48	1.6161	.9584	1.5489	.5848	.9451	4.8	121.5104	.0046	.5586	.0014	.1766
49	1.0502	3.1343	3.2918	2.4871	2.6120	49	1.6323	.9412	1.5363	.5721	.9338	4.9	134.2898	.0041	.5531	.0013	.1734
												5.0	148.4132	.0037	.5478	.0011	.1704

0.050	1.0513	3.1142	3.2739	2.4679	2.5945	0.50	1.6487	.9244	1.5241	.5598	.9229
51	1.0523	3.0945	3.2564	2.4491	2.5773	51	1.6653	.9081	1.5122	.5478	.9123
52	1.0534	3.0752	3.2393	2.4306	2.5604	52	1.6820	.8921	1.5006	.5362	.9019
53	1.0544	3.0562	3.2226	2.4126	2.5440	53	1.6989	.8766	1.4892	.5250	.8919
54	1.0555	3.0376	3.2062	2.3948	2.5278	54	1.7160	.8614	1.4781	.5140	.8820
55	1.0565	3.0194	3.1901	2.3775	2.5120	55	1.7333	.8466	1.4673	.5034	.8725
56	1.0576	3.0015	3.1744	2.3604	2.4964	56	1.7507	.8321	1.4567	.4930	.8631
57	1.0587	2.9839	3.1589	2.3437	2.4811	57	1.7683	.8180	1.4664	.4830	.8541
58	1.0597	2.9666	3.1437	2.3273	2.4663	58	1.7860	.8042	1.4363	.4732	.8451
59	1.0608	2.9496	3.1288	2.3111	2.4516	59	1.8040	.7907	1.4264	.4637	.8365
0.060	1.0618	2.9329	3.1142	2.2953	2.4371	0.60	1.8221	.7775	1.4167	.4544	.8280
61	1.0629	2.9165	3.0999	2.2797	2.4230	61	1.8404	.7646	1.4073	.4454	.8179
62	1.0640	2.9003	3.0858	2.2645	2.4092	62	1.8589	.7520	1.3980	.4366	.8116
63	1.0650	2.8844	3.0719	2.2494	2.3956	63	1.8776	.7397	1.3889	.4280	.8036
64	1.0661	2.8688	3.0584	2.2346	2.3822	64	1.8965	.7277	1.3800	.4197	.7960
65	1.0672	2.8534	3.0450	2.2201	2.3691	65	1.9155	.7159	1.3713	.4115	.7882
66	1.0682	2.8382	3.0319	2.2058	2.3562	66	1.9348	.7043	1.3627	.4036	.7809
67	1.0693	2.8233	3.0189	2.1917	2.3434	67	1.9542	.6930	1.3543	.3959	.7737
68	1.0704	2.8086	3.0062	2.1779	2.3310	68	1.9739	.6820	1.3461	.3883	.7665
69	1.0714	2.7941	2.9937	2.1643	2.3188	69	1.9937	.6711	1.3380	.3810	.7596
0.070	1.0725	2.7798	2.9814	2.1508	2.3067	0.70	2.0138	.6605	1.3301	.3738	.7528
71	1.0736	2.7657	2.9693	2.1376	2.2949	71	2.0340	.6501	1.3223	.3668	.7461
72	1.0747	2.7519	2.9573	2.1246	2.2832	72	2.0544	.6399	1.3147	.3599	.7394
73	1.0757	2.7382	2.9455	2.1118	2.2717	73	2.0751	.6300	1.3072	.3532	.7329
74	1.0768	2.7247	2.9340	2.0991	2.2603	74	2.0959	.6202	1.2998	.3467	.7266
75	1.0779	2.7114	2.9226	2.0867	2.2492	75	2.1170	.6106	1.2926	.3403	.7204
76	1.0790	2.6983	2.9113	2.0744	2.2381	76	2.1383	.6012	1.2855	.3341	.7144
77	1.0800	2.6853	2.9002	2.0623	2.2273	77	2.1598	.5920	1.2785	.3280	.7084
78	1.0811	2.6726	2.8894	2.0503	2.2165	78	2.1815	.5829	1.2716	.3221	.7027
79	1.0822	2.6599	2.8786	2.0386	2.2062	79	2.2034	.5740	1.2649	.3163	.6969
0.080	1.0833	2.6475	2.8680	2.0269	2.1957	0.80	2.2255	.5653	1.2582	.3106	.6912
81	1.0844	2.6352	2.8575	2.0155	2.1856	81	2.2479	.5568	1.2517	.3050	.6856
82	1.0855	2.6231	2.8472	2.0042	2.1754	82	2.2705	.5484	1.2452	.2996	.6802
83	1.0865	2.6111	2.8370	1.9930	2.1655	83	2.2933	.5402	1.2389	.2943	.6749
84	1.0876	2.5992	2.8270	1.9820	2.1557	84	2.3164	.5321	1.2326	.2891	.6697
85	1.0887	2.5875	2.8171	1.9711	2.1460	85	2.3397	.5242	1.2265	.2840	.6644
86	1.0898	2.5759	2.8073	1.9604	2.1364	86	2.3632	.5165	1.2205	.2790	.6593
87	1.0909	2.5645	2.7976	1.9498	2.1270	87	2.3869	.5088	1.2145	.2742	.6545
88	1.0920	2.5532	2.7881	1.9393	2.1176	88	2.4109	.5013	1.2086	.2694	.6495
89	1.0931	2.5421	2.7787	1.9290	2.1086	89	2.4351	.4940	1.2029	.2647	.6446
0.090	1.0942	2.5310	2.7694	1.9187	2.0994	0.90	2.4596	.4867	1.1972	.2602	.6400
91	1.0953	2.5201	2.7602	1.9087	2.0906	91	2.4843	.4796	1.1916	.2557	.6352
92	1.0964	2.5093	2.7511	1.8987	2.0818	92	2.5093	.4727	1.1860	.2513	.6306
93	1.0975	2.4986	2.7421	1.8888	2.0729	93	2.5345	.4658	1.1806	.2470	.6260
94	1.0986	2.4881	2.7333	1.8791	2.0643	94	2.5600	.4591	1.1752	.2429	.6218
95	1.0997	2.4776	2.7246	1.8695	2.0558	95	2.5857	.4524	1.1699	.2387	.6172
96	1.1008	2.4673	2.7159	1.8599	2.0473	96	2.6117	.4459	1.1647	.2347	.6130
97	1.1019	2.4571	2.7074	1.8505	2.0390	97	2.6379	.4396	1.1595	.2308	.6088
98	1.1030	2.4470	2.6989	1.8412	2.0307	98	2.6645	.4333	1.1544	.2269	.6046
99	1.1041	2.4370	2.6906	1.8320	2.0227	99	2.6912	.4271	1.1494	.2231	.6004
0.100	1.1052	2.4271	2.6823	1.8229	2.0147	1.00	2.7183	.4210	1.1445	.2194	.5964

Source. See reference 20 at the end of Chapter 6. (Courtesy of Mahdi S. Hantush.)

Table B.2 Values of the Function = $W(u, r/B) = \int_u^\infty (1/y) \exp(-y - r^2/4B^2y)\,dy$

u \ r/B	0	0.001	0.002	0.003	0.004	0.005	0.006	0.007	0.008	0.009	0.01
0	∞	14.0474	12.6611	11.8502	11.2748	10.8286	10.4640	10.1557	9.8887	9.6532	9.4425
.000001	13.2383	13.0031	12.4417	11.8153	11.2711	10.8283	10.4640	10.1557	9.8887		
.000002	12.5451	12.4240	12.1013	11.6716	11.2259	10.8174	10.4619	10.1554	9.8886	9.6532	
.000003	12.1397	12.0581	11.8322	11.5098	11.1462	10.7849	10.4509	10.1523	9.8879	9.6530	9.4425
.000004	11.8520	11.7905	11.6168	11.3597	11.0555	10.7374	10.4291	10.1436	9.8849	9.6521	9.4422
.000005	11.6289	11.5795	11.4384	11.2248	10.9642	10.6822	10.3993	10.1290	9.8786	9.6496	9.4413
.000006	11.4465	11.4503	11.2866	11.1040	10.8764	10.6240	10.3640	10.1094	9.8686	9.6450	9.4394
.000007	11.2924	11.2570	11.1545	10.9951	10.7933	10.5652	10.3255	10.0862	9.8555	9.6382	9.4361
.000008	11.1589	11.1279	11.0377	10.8962	10.7151	10.5072	10.2854	10.0602	9.8398	9.6292	9.4313
.000009	11.0411	11.0135	10.9330	10.8059	10.6416	10.4508	10.2446	10.0324	9.8219	9.6182	9.4251
.00001	10.9357	10.9109	10.8382	10.7228	10.5725	10.3963	10.2038	10.0034	9.8024	9.6059	9.4176
.00002	10.2426	10.2301	10.1932	10.1332	10.0522	9.9530	9.8386	9.7126	9.5781	9.4383	9.2961
.00003	9.8371	9.8288	9.8041	9.7635	9.7081	9.6392	9.5583	9.4671	9.3674	9.2611	9.1499
.00004	9.5495	9.5432	9.5246	9.4940	9.4520	9.3992	9.3366	9.2653	9.1863	9.1009	9.0102
.00005	9.3263	9.3213	9.3064	9.2818	9.2480	9.2052	9.1542	9.0957	9.0304	8.9591	8.8827
.00006	9.1440	9.1398	9.1274	9.1069	9.0785	9.0426	8.9996	8.9500	8.8943	8.8332	8.7673
.00007	8.9899	8.9863	8.9756	8.9580	8.9336	8.9027	8.8654	8.8224	8.7739	8.7204	8.6625
.00008	8.8563	8.8532	8.8439	8.8284	8.8070	8.7798	8.7470	8.7090	8.6661	8.6186	8.5669
.00009	8.7386	8.7358	8.7275	8.7138	8.6947	8.6703	8.6411	8.6071	8.5686	8.5258	8.4792
.0001	8.6332	8.6308	8.6233	8.6109	8.5938	8.5717	8.5453	8.5145	8.4796	8.4407	8.3983
.0002	7.9402	7.9390	7.9352	7.9290	7.9203	7.9092	7.8958	7.8800	7.8619	7.8416	7.8192
.0003	7.5348	7.5340	7.5315	7.5274	7.5216	7.5141	7.5051	7.4945	7.4823	7.4686	7.4534
.0004	7.2472	7.2466	7.2447	7.2416	7.2373	7.2317	7.2249	7.2169	7.2078	7.1974	7.1859
.0005	7.0242	7.0237	7.0222	7.0197	7.0163	7.0118	7.0063	6.9999	6.9926	6.9843	6.9750
.0006	6.8420	6.8416	6.8403	6.8383	6.8353	6.8316	6.8271	6.8218	6.8156	6.8086	6.8009
.0007	6.6879	6.6876	6.6865	6.6848	6.6823	6.6790	6.6752	6.6706	6.6653	6.6594	6.6527
.0008	6.5545	6.5542	6.5532	6.5517	6.5495	6.5467	6.5433	6.5393	6.5347	6.5295	6.5237
.0009	6.4368	6.4365	6.4357	6.4344	6.4324	6.4299	6.4269	6.4233	6.4192	6.4146	6.4094

.001	6.3315	6.3313	6.3305	6.3293	6.3276	6.3253	6.3226	6.3194	6.3157	6.3115	6.3069
.002	5.6394	5.6393	5.6389	5.6383	5.6374	5.6363	5.6350	5.6334	5.6315	5.6294	5.6271
.003	5.2349	5.2348	5.2346	5.2342	5.2336	5.2329	5.2320	5.2310	5.2297	5.2283	5.2267
.004	4.9482	4.9482	4.9480	4.9477	4.9472	4.9467	4.9460	4.9453	4.9443	4.9433	4.9421
.005	4.7261	4.7260	4.7259	4.7256	4.7253	4.7249	4.7244	4.7237	4.7230	4.7222	4.7212
.006	4.5448	4.5448	4.5447	4.5444	4.5441	4.5438	4.5433	4.5428	4.5422	4.5415	4.5407
.007	4.3916	4.3916	4.3915	4.3913	4.3910	4.3908	4.3904	4.3899	4.3894	4.3888	4.3882
.008	4.2591	4.2590	4.2590	4.2588	4.2586	4.2583	4.2580	4.2576	4.2572	4.2567	4.2561
.009	4.1423	4.1423	4.1422	4.1420	4.1418	4.1416	4.1413	4.1410	4.1406	4.1401	4.1396
.01	4.0379	4.0379	4.0378	4.0377	4.0375	4.0373	4.0371	4.0368	4.0364	4.0360	4.0356
.02	3.3547	3.3547	3.3547	3.3546	3.3545	3.3544	3.3543	3.3542	3.3540	3.3538	3.3536
.03	2.9591	2.9591	2·9591	2.9590	2.9590	2.9589	2.9589	2.9588	2.9587	2.9585	2.9584
.04	2.6813	2.6812	2.6812	2.6812	2.6812	2.6811	2.6810	2.6810	2.6809	2.6808	2.6807
.05	2.4679	2.4679	2.4679	2.4679	2.4678	2.4678	2.4678	2.4677	2.4676	2.4676	2.4675
.06	2.2953	2.2953	2.2953	2.2953	2.2952	2.2952	2.2952	2.2952	2.2951	2.2950	2.2950
.07	2.1508	2.1508	2.1508	2.1508	2.1508	2.1508	2.1507	2.1507	2.1507	2.1506	2.1506
.08	2.0269	2.0269	2.0269	2.0269	2.0269	2.0269	2.0269	2.0268	2.0268	2.0268	2.0267
.09	1.9187	1.9187	1.9187	1.9187	1.9187	1.9187	1.9187	1.9186	1.9186	1.9186	1.9185
.1	1.8229	1.8229	1.8229	1.8229	1.8229	1.8229	1.8229	1.8228	1.8228	1.8228	1.8227
.2	1.2227	1.2226	1.2226	1.2226	1.2226	1.2226	1.2226	1.2226	1.2226	1.2226	1.2226
.3	0.9057	0.9057	0.9057	0.9057	0.9057	0.9057	0.9057	0.9057	0.9056	0.9056	0.9056
.4	7024	7024	7024	7024	7024	7024	7024	7024	7024	7024	7024
.5	5598	5598	5598	5598	5598	5598	5598	5598	5598	5598	5598
.6	4544	4544	4544	4544	4544	4544	4544	4544	4544	4544	4544
.7	3738	3738	3738	3738	3738	3738	3738	3738	3738	3738	3738
.8	3106	3106	3106	3106	3106	3106	3106	3106	3106	3106	3106
.9	2602	2602	2602	2602	2602	2602	2602	2602	2602	2602	2602
1.0	0.2194	0.2194	0.2194	0.2194	0.2194	0.2194	0.2194	0.2194	0.2194	0.2194	0.2194
2.0	489	489	489	489	489	489	489	489	489	489	489
3.0	130	130	130	130	130	130	130	130	130	130	130
4.0	38	38	38	38	38	38	38	38	38	38	38
5.0	11	11	11	11	11	11	11	11	11	11	11
6.0	4	4	4	4	4	4	4	4	4	4	4
7.0	1	1	1	1	1	1	1	1	1	1	1
8.0	0	0	0	0	0	0	0	0	0	0	0

Table B.2 (continued)

u \ r/B	0.01	0.015	0.02	0.025	0.03	0.035	0.04	0.045	0.05	0.055	0.06	0.065	0.07	0.075	0.08	0.085	0.09	0.095	0.10
0	9.4425	8.6319	8.0569	7.6111	7.2471	6.9394	6.6731	6.4383	6.2285	6.0388	5.8658	5.7067	5.5596	5.4228	5.2950	5.1750	5.0620	4.9553	4.8541
.000001																			
.000002																			
.000003	9.4425																		
.000004	9.4422																		
.000005	9.4413																		
.000006	9.4394																		
.000007	9.4361	8.6319																	
.000008	9.4313	8.6318																	
.000009	9.4251	8.6316																	
.00001	9.4176	8.6313	8.0569																
.00002	9.2961	8.6152	8.0558	7.6111	7.2471														
.00003	9.1499	8.5737	8.0483	7.6101	7.2470														
.00004	9.0102	8.5168	8.0320	7.6069	7.2465	6.9394	6.6731												
.00005	8.8827	8.4533	8.0080	7.6000	7.2450	6.9391	6.6730												
.00006	8.7673	8.3880	7.9786	7.5894	7.2419	6.9384	6.6729	6.4383											
.00007	8.6625	8.3233	7.9456	7.5754	7.2371	6.9370	6.6726	6.4382	6.2285										
.00008	8.5669	8.2603	7.9105	7.5589	7.2305	6.9347	6.6719	6.4381	6.2284										
.00009	8.4792	8.1996	7.8743	7.5402	7.2222	6.9316	6.6709	6.4378	6.2283										
.0001	8.3983	8.1414	7.8375	7.5199	7.2122	6.9273	6.6693	6.4372	6.2282	6.0388	5.8658	5.7067	5.5596	5.4228	5.2950				
.0002	7.8192	7.6780	7.4972	7.2898	7.0685	6.8439	6.6242	6.4143	6.2173	6.0338	5.8637	5.7059	5.5593	5.4227	5.2949	5.1750	5.0620	4.9553	
.0003	7.4534	7.3562	7.2281	7.0759	6.9068	6.7276	6.5444	6.3623	6.1848	6.0145	5.8527	5.6999	5.5562	5.4212	5.2942	5.1747	5.0619	4.9552	4.8541
.0004	7.1859	7.1119	7.0128	6.8929	6.7567	6.6088	6.4538	6.2955	6.1373	5.9818	5.8309	5.6860	5.5476	5.4160	5.2912	5.1730	5.0610	4.9547	4.8539
.0005	6.9750	6.9152	6.8346	6.7357	6.6219	6.4964	6.3626	6.2236	6.0821	5.9406	5.8011	5.6648	5.5330	5.4062	5.2848	5.1689	5.0585	4.9532	4.8530
.0006	6.8009	6.7508	6.6828	6.5988	6.5011	6.3923	6.2748	6.1512	6.0239	5.8948	5.7658	5.6383	5.5134	5.3921	5.2749	5.1621	5.0539	4.9502	4.8510
.0007	6.6527	6.6096	6.5508	6.4777	6.3923	6.2962	6.1917	6.0807	5.9652	5.8468	5.7274	5.6081	5.4902	5.3745	5.2618	5.1526	5.0471	4.9454	4.8478
.0008	6.5237	6.4858	6.4340	6.3695	6.2935	6.2076	6.1136	6.0129	5.9073	5.7982	5.6873	5.5755	5.4642	5.3542	5.2461	5.1406	5.0381	4.9388	4.8430
.0009	6.4094	6.3757	6.3294	6.2716	6.2032	6.1256	6.0401	5.9481	5.8509	5.7500	5.6465	5.5416	5.4364	5.3317	5.2282	5.1266	5.0272	4.9306	4.8368

.001	6.3069	6.2765	6.2347	6.1823	6.1202	6.0494	5.9711	5.8864	5.7965	5.7026	5.6058	5.5071	5.4075	5.3078	5.2087	5.1109	5.0133	4.9208	4.8292
.002	5.6271	5.6118	5.5907	5.5636	5.5314	5.4939	5.4516	5.4047	5.3538	5.2991	5.2411	5.1803	5.1170	5.0517	4.9848	4.9166	4.8475	4.7778	4.7079
.003	5.2267	5.2166	5.2025	5.1845	5.1627	5.1373	5.1084	5.0762	5.0408	5.0025	4.9615	4.9180	4.8722	4.8243	4.7746	4.7234	4.6707	4.6169	4.5622
.004	4.9421	4.9345	4.9240	4.9105	4.8941	4.8749	4.8530	4.8286	4.8016	4.7722	6.7406	4.7068	4.6710	4.6335	4.5942	4.5533	4.5111	4.4676	4.4230
.005	4.7212	4.7152	4.7068	4.6960	4.6829	4.6675	4.6499	4.6302	4.6084	4.5846	4.5590	4.5314	4.5022	4.4713	4.4389	4.4050	4.3699	4.3335	4.2960
.006	4.5407	4.5357	4.5287	4.5197	4.5088	4.4960	4.4814	4.4649	4.4467	4.4267	4.4051	4.3819	4.3573	4.3311	4.3036	4.2747	4.2446	4.2134	4.1812
.007	4.3882	4.3839	4.3779	4.3702	4.3609	4.3500	4.3374	4.3233	4.3077	4.2905	4.2719	4.2518	4.2305	4.2078	4.1839	4.1588	4.1326	4.1053	4.0771
.008	4.2561	4.2524	4.2471	4.2404	4.2323	4.2228	4.2118	4.1994	4.1857	4.1707	4.1544	4.1368	4.1180	4.0980	4.0769	4.0547	4.0315	4.0073	3.9822
.009	4.1396	4.1363	4.1317	4.1258	4.1186	4.1101	4.1004	4.0894	4.0772	4.0638	4.0493	4.0336	4.0169	3.9991	3.9802	3.9603	3.9395	3.9178	3.8952
.01	4.0356	4.0326	4.0285	4.0231	4.0167	4.0091	4.0003	3.9905	3.9795	3.9675	3.9544	3.9403	3.9252	3.9091	3.8920	3.8741	3.8552	3.8356	3.8150
.02	3.3536	3.3521	3.3502	3.3476	3.3444	3.3408	3.3365	3.3317	3.3264	3.3205	3.3141	3.3071	3.2997	3.2917	3.2832	3.2742	3.2647	3.2547	3.2442
.03	2.9584	2.9575	2.9562	2.9545	2.9523	2.9501	2.9474	2.9444	2.9409	2.9370	2.9329	2.9284	2.9235	2.9183	2.9127	2.9069	2.9007	2.8941	2.8873
.04	2.6807	2.6800	2.6791	2.6779	2.6765	2.6747	2.6727	2.6705	2.6680	2.6652	2.6622	2.6589	2.6553	2.6515	2.6475	2.6432	2.6386	2.6338	2.6288
.05	2.4675	2.4670	2.4662	2.4653	2.4642	2.4628	2.4613	2.4595	2.4576	2.4554	2.4531	2.4505	2.4478	2.4448	2.4416	2.4383	2.4347	2.4310	2.4271
.06	2.2950	2.2945	2.2940	2.2932	2.2923	2.2912	2.2900	2.2885	2.2870	2.2852	2.2833	2.2812	2.2790	2.2766	2.2740	2.2713	2.2684	2.2654	2.2622
.07	2.1506	2.1502	2.1497	2.1491	2.1483	2.1474	2.1464	2.1452	2.1439	2.1424	2.1408	2.1391	2.1372	2.1352	2.1331	2.1308	2.1284	2.1258	2.1232
.08	2.0267	2.0264	2.0260	2.0255	2.0248	2.0240	2.0231	2.0221	2.0210	2.0198	2.0184	2.0169	2.0153	2.0136	2.0118	2.0099	2.0078	2.0056	2.0034
.09	1.9185	1.9183	1.9179	1.9174	1.9169	1.9162	1.9154	1.9146	1.9136	1.9125	1.9114	1.9101	1.9087	1.9072	1.9056	1.9040	1.9022	1.9003	1.8983
.1	1.8227	1.8225	1.8222	1.8218	1.8213	1.8207	1.8200	1.8193	1.8184	1.8175	1.8164	1.8153	1.8141	1.8128	1.8114	1.8099	1.8084	1.8067	1.8050
.2	1.2226	1.2225	1.2224	1.2222	1.2220	1.2218	1.2215	1.2212	1.2209	1.2205	1.2201	1.2196	1.2192	1.2186	1.2181	1.2175	1.2168	1.2162	1.2155
.3	0.9056	0.9056	0.9055	0.9054	0.9053	0.9052	0.9050	0.9049	0.9047	0.9045	0.9043	0.9040	0.9038	0.9035	0.9032	0.9029	0.9025	0.9022	0.9018
.4	7024	7023	7023	7022	7022	7021	7020	7019	7018	7016	7015	7014	7012	7010	7008	7006	7004	7002	7000
.5	5598	5597	5597	5597	5596	5596	5595	5594	5594	5593	5592	5591	5590	5588	5587	5586	5584	5583	5581
.6	4544	4544	4543	4543	4543	4542	4542	4542	4541	4540	4540	4539	4538	4537	4536	4535	4534	4533	4532
.7	3738	3738	3737	3737	3737	3737	3736	3736	3735	3735	3734	3734	3733	3733	3732	3732	3731	3730	3729
.8	3106	3106	3106	3106	3105	3105	3105	3105	3104	3104	3104	3103	3103	3102	3102	3101	3101	3100	3100
.9	2602	2602	2602	2602	2601	2601	2601	2601	2601	2600	2600	2600	2599	2599	2599	2598	2598	2597	2597
1.0	0.2194	0.2194	0.2194	0.2194	0.2193	0.2193	0.2193	0.2193	0.2193	0.2193	0.2192	0.2192	0.2192	0.2191	0.2191	0.2191	0.2191	0.2190	0.2190
2.0	489	489	489	489	489	489	489	489	489	489	489	489	489	489	489	489	489	488	488
3.0	130	130	130	130	130	130	130	130	130	130	130	130	130	130	130	130	130	130	130
4.0	38	38	38	38	38	38	38	38	38	38	38	38	38	38	38	38	38	38	38
5.0	11	11	11	11	11	11	11	11	11	11	11	11	11	11	11	11	11	11	11
6.0	4	4	4	4	4	4	4	4	4	4	4	4	4	4	4	4	4	4	4
7.0	1	1	1	1	1	1	1	1	1	1	1	1	1	1	1	1	1	1	1
8.0	0	0	0	0	0	0	0	0	1	0	0	0	0	0	0	0	0	0	0

Table B.2 (Continued)

u \ r/B	0.25	0.3	0.35	0.4	0.45	0.5		0.8	0.85	0.9	0.95	1.0
0	3.0830	2.7449	2.4654	2.2291	2.0258	1.8488	2	1.1307	1.0485	0.9735	0.9049	0.8420
.0001												
.0002												
.0003												
.0004												
.0005												
.0006												
.0007												
.0008												
.0009												
.001												
.002	3.0830	2.7449										
.003	3.0821	2.7448										
.004	3.0788	2.7444	2.4654	2.2291								
.005	3.0719	2.7428	2.4651	2.2290								
.006	3.0614	2.7398	2.4644	2.2289	2.0258							
.007	3.0476	2.7350	2.4630	2.2286	2.0257							
.008	3.0311	2.7284	2.4608	2.2279	2.0256	1.8488						
.009	3.0126	2.7202	2.4576	2.2269	2.0253	1.8487						
.01	2.9925	2.7104	2.4534	2.2253	2.0248	1.8486	1.2212	1.1307	1.0485			
.02	2.7658	2.5688	2.3713	2.1809	2.0023	1.8379	1.2210	1.1306	1.0484	0.9735	0.9049	
.03	2.5571	2.4110	2.2578	2.1031	1.9515	1.8062	1.2195	1.1299	1.0481	9733	9048	0.8420
.04	2.3802	2.2661	2.1431	2.0155	1.8869	1.7603	1.2146	1.1270	1.0465	9724	9044	8418
.05	2.2299	2.1371	2.0356	1.9283	1.8181	1.7075	1.2052	1.1210	1.0426	9700	9029	8409
.06	2.1002	2.0227	1.9369	1.8452	1.7497	1.6524	1.1919	1.1116	1.0362	9657	9001	8391
.07	1.9867	1.9206	1.8469	1.7673	1.6835	1.5973	1.1754	1.0993	1.0272	9593	8956	8360
.08	1.8861	1.8290	1.7646	1.6947	1.6206	1.5436	1.1564	1.0847	1.0161	9510	8895	8316
.09	1.7961	1.7460	1.6892	1.6272	1.5609	1.4918	1.1358	1.0682	1.0032	9411	8819	8259
.1	1.7149	1.6704	1.6198	1.5644	1.5048	1.4422	1.114	1.0505	0.9890	0.9297	0.8730	0.8190
.2	1.1789	1.1602	1.1387	1.1145	1.0879	1.0592	0.893	0.8575	8216	7857	7501	7148
.3	0.8817	0.8713	0.8593	0.8457	0.8306	0.8142	715	6932	6706	6476	6244	6010
.4	6874	6809	6733	6647	6551	6446	58	5653	5501	5345	5186	5024
.5	5496	5453	5402	5344	5278	5206	47	4658	4550	4440	4326	4210
.6	4472	4441	4405	4364	4317	4266	3	3871	3793	3712	3629	3543
.7	3685	3663	3636	3606	3572	3534	3	324	3183	3123	3060	2996
.8	3067	3050	3030	3008	2982	2953	2	273	2687	2641	2592	2543
.9	2572	2559	2544	2527	2507	2485	2	23	2280	2244	2207	2168
1.0	0.2171	0.2161	0.2149	0.2135	0.2120	0.2103	0.	0.1	943	0.1914	0.1885	0.1885
2.0	486	485	484	482	480	477			56	452	448	444
3.0	130	130	130	129	129	128			24	123	123	122
4.0	38	38	38	38	37	37			6	36	36	36
5.0	11	11	11	11	11	11			1	11	11	11
6.0	4	4	4	4	4	4				4	4	4
7.0	1	1	1	1	1	1				1	1	1
[illegible]										0		

Table B.2 (Continued)

u \ r/B	1.0	1.5	2.0	2.5	3.0	3.5	4.0	4.5	5.0	6.0	7.0	8.0	9.0
0	0.8420	0.4276	0.2278	0.1247	0.0695	0.0392	0.0223	0.0128	0.0074	0.0025	0.0008	0.0003	0.0001
.01													
.02													
.03	0.8420												
.04	8418												
.05	8409												
.06	8391												
.07	8360	0.4276											
.08	8316	4275											
.09	8259	4274											
.1	0.8190	0.4271	0.2278										
.2	7148	4135	2268	0.1247	0.0695								
.3	6010	3812	2211	1240	694								
.4	5024	3411	2096	1217	691	0.0392							
.5	4210	3007	1944	1174	681	390	0.0223						
.6	3543	2630	1774	1112	664	386	222	0.0128					
.7	2996	2292	1602	1040	639	379	221	127					
.8	2543	1994	1436	961	607	368	218	127	0.0074				
.9	2168	1734	1281	881	572	354	213	125	73				
1.0	0.1855	0.1509	0.1139	0.0803	0.0534	0.0338	0.0207	0.0123	0.0073	0.0025			
2.0	444	394	335	271	210	156	112	77	51	21	0.0008	0.0003	
3.0	122	112	100	86	71	57	45	34	25	12	6	2	
4.0	36	34	31	27	24	20	16	13	10	6	3	2	0.0001
5.0	11	10	10	9	8	7	6	5	4	2	1	1	0
6.0	4	3	3	3	3	2	2	2	2	1	1	0	
7.0	1	1	1	1	1	1	1	1	1	0	0		
8.0	0	0	0	0	0	0	0	0	0				

Source. See reference 20 at the end of Chapter 6. (Courtesy of Madhi S. Hantush.)

INDEX

Adams, F. D., 5, 12
Adnerson, L. J., 42, 126
Analogs, 318, 322, 331
Anderson, E. R., 42, 126
Anisotropic soil, 228
Antecedent-precipitation index, 75, 77
Aquiclude, 133
Aquifer, 133
  confined, 239, 260
  leaky, 271
  unconfined, 242
Aquitard, 122, 133
Aravin, V. I., 325
Aronovici, V. S., 172, 203
Artificial recharge, 147

Baker, D. M., 47, 126, 146, 155, 158, 159
Baker, M. N., 7, 13
Bakhmeteff, B. A., 174, 201
Banks, R. B., 316
Barksdale, H. C., 147, 159
Barnes, B. S., 71, 127
Barometric efficiency, 189
Batisse, M., 17, 126
Baumann, P., 153, 159
Bear, J., 202, 238, 305, 313, 315, 316, 317, 322, 344, 348
Bentall, R., 172, 203
Bermes, B. J., 333, 347
Bhattacharya, B. K., 348
Biggar, T. W., 303
Bird, R. B., 2, 12
Bittinger, M., 153, 286
Blanchard, R., 12
Bogomolov, G. V., 158
Bondurant, D. C., 117
Boreli, M., 286
Boundary effects, 132, 206, 270
Bouwer, H., 202, 347
Bowman, I., 12
Brantly, J. E., 12
Breakthrough curve, 303
Brown, R. H., 284
Brune, G. M., 118
Brutsaert, W., 342, 347

Calif. Dept. of Public Works, 127
Cap, P. A., 13
Capillary force, 199, 288
Capillary fringe, 140
Capillary rise, 199
Capillary water, 143, 197, 199
Casagrande, A., 211
Casagrande, L., 238
Childs, E. C., 348
Circle of influence, 241
Coaxial relations, storm runoff, 77
Collins, R. S., 126
Compressibility, granular skeleton, 181
  of fluid, 183
Conductive material analogs, 342
Confined flow, 239, 260
Conkling, H., 47, 126, 155, 159
Conservation of mass, 179
Consumptive use, 49
Cooper, H. H., 128, 285, 300

Dachler, R., 346
Dagan, G., 315
Dalton's law, 39
Dam, on infinitely thick stratum, 215
  with cut-off walls, 212
Dampier-Whetham, W. C. D., 13
darcy, 170
Darcy, H., 13, 162
Darcy's law, 168
  derivation of, 176
  range of validity, 178
Daubrée, A., 13

Day, P. R., 315
De Buchananne, G. D., 147
De Glee, G. J., 285
De Jong, J., 348
De Josselin de Jong, G., 316, 317, 348
Depression storage, 49
Depth-area-duration analysis, 29
Desorption curve, 141
Detention, surface, 48
Dew point, 40
De Wiest, R. J. M., 12, 13, 109, 122, 124, 177, 202, 238, 271, 286, 306, 315, 330, 347, 348
Dicker, D., 223
Dietz, D. N., 346
Diffusion, 303
Dispersion, 299
  coefficient of, 301
Donnan, W. W., 172, 203
Double-mass-curve technique, 27
Drainage basin, area-elevation curve, 20
  land slope, 21
  mean elevation, 20
  physiography of, 22
Drawdown, 241, 274
Dupuit, J., 13, 223, 233
Dupuit's assumptions, 223
Dupuit's parabola, 235

Eagleson, P. S., 128
Earth dam, 233
  rectangular, 235
Eisenstadt, R., 202
Electrical resistance network, 339
Electro-osmosis, 198
Emde, F., 284
Engelund, F., 179
Environmental water head, 307, 311
Equilibrium equation, 242
Equipotential lines, 195, 209
Equipotential surfaces, 195
Evaporation, factors affecting, 40
  influence on runoff, 42
  monthly, 43
Evaporation pans, 42
Evaporation, reservoir, 41
Evapo-transpiration, 49
Extensive confined aquifer, 260

Fair, G. M., 174
Fara, H. D., 316
Feodoroff, N. V., 174, 201
Ferris, J. G., 284
Finite differences, 319
Fleming, R., 126
Flood routing, 87
Flow-duration curves, 109
Flow nets, 206
  construction of, 210
  for wells, 236
  properties of, 208
  singular points, 219
  with free surface, 222
Force potential, 192
Forchheimer, Ph., 13, 205
Formation constants, 262, 275
Fourmarier, P., 12
Fraser, H. J., 136
Free surface, 222
Frequency analysis, 99
  Gumbel's method, 103
  N-year event, 101
  return period, 101
Fresh-water head, 307

Gaskell, R. E., 347
Geraghty, J. J., 126, 316
Ghyben, B. W., 295
Ghyben-Herzberg law, 295
Gilcrest, B. R., 127
Gilmann, C. S., 127
Glover, R. E., 298
Goldschmidt, M. J., 159, 160
Gradient, 192
Grapho-numerical analysis, 318
Graton, L. C., 136
Gravitational water, 143
Gray, D. M., 128
Groot, C. R., 149
Ground-water flow, boundaries of, 132, 206
  dimensional character, 129
  elementary theory of, 161
  time dependency, 131
Ground-water law, 154
Gumbel, E. J., 102, 104, 127
Gumbel's method, 103
Gunther, E., 346

Hackett, O. M., 13
Hadley, W. A., 202
Hagen, G., 12
Hall, H. P., 13
Hansen, V. E., 348
Hantush, M. S., 271, 275, 278, 280, 285, 286
Harden, R. W., 13, 126
Harleman, D. R. F., 302
Harr, M. E., 13
Hatch, L. P., 174
Hatschek, 201
Hazen, A., 127, 201
Hele-Shaw, H. S., 322, 346
Hele-Shaw analog, 322, 330
Herzberg, A., 295
Hoel, P. G., 127
Horizontal Hele-Shaw analog, 326
Horton, R. E., 7, 13, 126, 127
Hubbert, M. K., 13, 192, 238, 287, 297, 310
Huisman, L., 317
Humidity, relative, 40
Hydraulic conductivity, 169
  measurement of, 172
  soils with different hydraulic conductivity, 226
Hydraulic gradient, 165
Hydrocarbons, 287, 294
Hydrodynamic dispersion, 299
Hydrodynamical entrapment, 290
Hydrograph, 63
  composition, 63, 65
  duration of storm, 67
  separation, 65
  shape, 72
  time base, 67
  time of concentration, 71
  unit, 79
Hydrologic cycle, 14
Hygroscopic water, 143

Images, method of, 255, 270
Immiscible flow, 287
Ince, S., 201
Infiltration, 50
  capacity, 50
  indices, 51, 52
  rate, 51
Initial moisture conditions, 75
Interception, 49
Interface, 287
Interfacial energy, 288
Interflow, 50
Intergranular stress, 140, 183
Intrinsic permeability, 170
Irmay, S., 202, 318
Irrotational flow, 196
Iseri, K. T., 128
Isobaric surfaces, 195
Isohyets, 30
Isopleths, 18

Jacob, C. E., 179, 271, 275, 278, 280, 285, 286
Jacobs, M., 159, 317
Jahnke, E., 284
Jansa, O. V., 159
Johnson, A. I., 160, 347
Jones, K. R., 202
Jones, P. B., 13, 126
Judson, S., 23, 126

Karplus, W. J., 318
Kazmann, R. G., 286
Kirkham, D., 286, 347
Klinkenberg, L. J., 202
Klinkenberg effect, 175
Knowles, D. B., 284
Knudsen, C. N., 238
Knudsen, W. C., 316
Kohler, M. A., 126, 127
Kohout, F. A., 315
Kozeny, J., 201, 286
Krayenhoff van de Leur, D. A., 202, 203
Krick, I. P., 126
Krynine, P. D., 12

Lang, S. M., 285
Langbein, W. B., 126, 127
Laplace's equation, 187, 205
Leakage coefficient, 274
Leakage factor, 274
Leakance, 274
Leaky aquifers, 271
  formation constants of, 275

Lebedev, A. F., 143
Leet, L. D., 23, 126
Leopold, L. B., 126
Liebmann, G., 347
Lightfoot, E. N., 2, 12
Lindquist, E., 201
Linsley, R. K., 126, 127
Little, W. C., 347
Lobeck, A. K., 126
Lopez, C. J., 12
Lusczynski, N. J., 305, 306, 310, 316
Luthin, J. N., 347

Maasland, D. E., 286
Maasland, M., 348
Maddock, Th. Jr., 128
Malavard, L. C., 348
Mann, J. F., 158
Marciano, J. J., 42, 126
Mariotte, E., 13
Mass curve, 115
Massau, J., 127
McCarthy, G. T., 127
McDaniels, L., 13, 126
McEwen, G. F., 126
Mead, D., 12
Medium properties, 135
  degree of saturation, 135
  density, 137
  moisture content, 135
  porosity, 134
  specific weight, 137
  void ratio, 134
Meinzer, O., 12
Mesnier, G. N., 128
Meyer, A. F., 126
Migration of oil, 287
Miller, D. W., 126
Miscible displacement, 304
Mjatiev, A. N., 271
Model techniques, 318
  scales, 327, 333
Moisture content, 135
Morris, D. A., 160
Mortier, F., 158
Multilayered soil, 231
Multiple-phase flow, 287
Muskat, M., 13, 201, 238, 286, 348
Myers, L. E., 202

Naor, I., 348
Nelson, R. W., 238
New Jersey, State of, Div. of Water Policy and Supply, 128
Nielsen, D. R., 303
Nonequilibrium equation, 262
Nonhomogeneous flow, 304
Norton, W. H., 12
Numerical methods, 318

Offeringa, J., 316
Ogata, A., 316
Osmosis, 198

Parallel-plate model, 322
Partial penetration of well, 283
Paulhus, J. L., 126, 127
Pavlovsky, N. N., 13
Pellicular water, 143
Perlmutter, N. M., 316
Piezometric head, 138, 164
Pioger, R., 159
Pirson, S. J., 158
Playfair's law, 22
Plotnikov, N. A., 158
Point-water head, 306
Poiseuille flow, 162
Polubarinova-Kochina, P. Ya., 12, 126, 158, 201, 202, 238, 284, 285, 346
Pore pressure, 138, 183
Porosity, 134, 167
  effective, 142
Posey, C. J., 126
Potential, force, 192
  velocity, 175
Powell, R. W., 126
Precipitation, average areal, 27
  convective, 24
  cyclonic, 24
  data, 25
  mean annual, 32
  measurement, 25
  orographic, 25
  variations in, 37
Precipitation-runoff relationships, 74
Prickett, T. A., 347
Prill, R. C., 160
Primary migration of oil, 287
Prinz, E., 12

Probability distributions, 101
Psychrometric tables, 40

Radial flow, 239, 260
Rainfall data, 25
  sources of, 25
Rainfall intensity-duration-frequency curves, 111
Rainfall-runoff relationships, 74
Rain gages, 26
Rangeley, W. R., 158
Rating curves, 54
  constant fall, 57
  extension of, 59
  normal fall, 57
Recession curve, 67
Recharge, artificial, 147
Recharge well, 247
Recovery, Theis' method, 269
Refraction of streamlines, 226
Reisenauer, A. E., 238
Reservoirs, 112
  capacity loss, 116
  design of, 114
Reservoir rocks, 291
Resistance-capacitance network, 331
  conversion factors, 333
  design of, 333
Retention, surface, 50
Rice, R. C., 202
Rifai, M. N., 303, 315
Rights in ground water law, 154
  appropriative, 155
  prescriptive, 157
  riparian, 155
Rilem, 316
Rorabaugh, M. I., 128
Rose, H. E., 201
Rouse, H., 201, 238
Routing of streamflow, in river channels, 95
  through reservoir, 89
Rumer, R. R., 302
Runoff cycle, 49
Runoff, surface, 49

Safe yield, 145
Saffman, P. G., 316
Salt water encroachment, 295
Sammel, E. A., 348
Santing, G., 326, 346, 347
Saturation, degree of, 135
  zone of, 140
Saturation vapor pressure, 44
Scheidegger, A., 12, 316
Schiff, L., 159
Schmorak, S., 159, 317
Schneebeli, G., 178, 202
Schoeller, H., 12
Seepage, rate, 213
  surface, 207, 223
Semiconfining strata, 272
Sheet pile, 218
Sheppard, T., 13
Sherman, L. K., 127
Simpson, E. S., 316
Singular points, in flow nets, 219
Sink, flow to, 250
Skibitzke, H. E., 347
Slichter, C. S., 13, 201
Smith, W. O., 160
Snyder, F. F., 127
Source, flow from, 251
Specific storage, 185
Specific yield, 142
Spreading of ground water, 152
Stage-discharge relationship, 54
Stallman, R. W., 284, 316, 347, 348
Steady state flow, 204, 239, 279
Steggewentz, J. H., 285
Stern, W., 159
Sternberg, Y., 347
Stevens, J. C., 127
Stewart, W. E., 2, 12
Stokes, G. G., 322
Storage coefficient, 185, 186
Stream flow, records of, 60
  routing, 87
Streamlines, 195, 209
Streampattern, 23
  dendritic, 23
  rectangular, 23
  trellis, 23
Subsurface water, 129
  various kind of, 143
Surface tension, 199, 288

Sustained yield, 145
Sutcliffe, J. V., 158
Suter, M., 159
Swartzendruber, D., 202

Taylor, D. W., 201, 202, 238, 286
Taylor, G. S., 347
Theis, C. V., 260, 269
Thermo-osmosis, 197
Thevenin, J., 128
Thiem, G., 13, 284
Thiessen, A. H., 29
Thiessen's method, 28
Thomas, H. A., Jr., 108, 127
Tidal fluctuations, 191
Time base, 67
Todd, D. K., 159, 201, 238, 286, 305, 317, 347
Toth, J., 160
Tracer, 302
Transmissivity, 185
Transpiration, 47
Type curve, 263, 276, 279

Unconfined flow, 233, 242
Uniform flow, 252
Unit hydrographs, 79
  construction of, 79
  different duration, 81
  synthetic, 83
Unsteady flow, 260
Uplift pressure, 212, 216
Upson, J. E., 158, 316
U.S. Army Corps of Engineers, 126
U.S. Geological Survey, 60, 62, 171
U.S. Weather Bureau, 25, 26, 42

Van der Poel, C., 316
Van Everdingen, R. O., 348
Van Ness, B. A., 285
Van Schilfgaarde, J., 347
Velocity potential, 175

Walker, G. D., 13, 126
Walton, W. C., 159, 276, 286, 332, 347
Water Supply Papers, 62
Water year, 61
Watters, G., 109
Wayland, H., 238
Wegenstein, M., 159
Well flow, 239
  equilibrium equation, 242
  mechanics of, 239
  several wells, 244
  steady, 239
Well function, 262
  for confined aquifers, 263
  for leaky aquifers, 275
Wentworth, C. K., 296
Wenzel, L. K., 202, 284
Wilting coefficient, 47
Worstell, R. V., 347

Yih, C. S., 305

Zaslavsky, D., 202, 315
Zee, C. H., 348
Zunker, F., 201